Water Science & Technology
Waste Stabilisation Ponds VI

Waste Stabilisation Ponds VI

Selected Proceedings of the 6th IWA International Conference on Waste Stabilisation Ponds, held in Avignon, France, 28 September – 1 October 2004

Issue Editors: François Brissaud*, Bernadette Picot* and Andrew Shilton**
* Hydrosciences, University of Montpellier, France
** Centre for Environmental Technology and Engineering, Massey University, New Zealand

Scientific Committee

A. Shilton (*Chair*), Massey University, New Zealand
C. Boutin, CEMAGREF, France
F. Brissaud, Université de Montpellier, France
R. H. R da Costa, Universidade Federal de Santa Catarina, Brazil
R. Craggs, National Institute of Water and Atmospheric Research, New Zealand
T. Curtis, University of Newcastle-upon-Tyne, UK
B. El Hamouri, Institut Agronomique et Vétérinaire Hassan II, Morocco
H. J. Fallowfield, Flinders University, Australia
F. B. Green, Lawrence Berkeley National Laboratory, USA
B. J. Lloyd, University of Surrey, UK
D. D. Mara, University of Leeds, UK
K. L. Nelson, University of California, Berkeley, USA
G. Oron, Ben Gurion University, Israel
B. Picot, Université de Montpellier, France
C. Polprasert, Asian Institute of Technology, Thailand
Y. Racault, CEMAGREF, France
F. Torrella, University of Murcia, Spain
J. L. Vasel, Fondation Universitaire Luxembourgeoise, Belgium

Organizing Committee

J. H. M. Barbier, AGHTM
J. Bontoux, AGHTM
C. Boutin, CEMAGREF
F. Brissaud, Hydrosciences, Université de Montpellier
P. H. Dodane, CEMAGREF
C. Holyst, PACA Region
A. Iwema, Agence Rhone-Mediterranee-Corse
A. Lasalmonie, AGHTM
A. Lienard, CEMAGREF
S. Miquel, Herault County Council
B. Picot, Université de Montpellier, France
A. Shilton, Massey University, New Zealand
M. Tapiau, Hydrosciences, IRD, Montpellier
D. Villesot, AGHTM

Organized by

AGHTM
Agence de l'Eau Rhone-Mediterranee-Corse
CEMAGREF
Hydrosciences, Montpellier

British Library Cataloguing in Publication Data
A CIP catalogue record for this book is available from the British Library
ISBN 1 84339 499 5

Contents

ix **Preface**

Case studies and retrofitting

1 **Waste stabilisation ponds in France: state of the art and recent trends** Y. Racault and C. Boutin

11 **Changes in waste stabilisation pond performance resulting from the retrofit of activated sludge treatment upstream: part I – water quality issues** N. J. Cromar, D. G. Sweeney, M. J. O'Brien and H. J. Fallowfield

17 **Changes in waste stabilisation pond performance resulting from the retrofit of activated sludge treatment upstream: part II – management and operating issues** D. G. Sweeney, M. J. O'Brien, N. J. Cromar and H. J. Fallowfield

23 **Twenty years' monitoring of Mèze stabilisation ponds: part I – removal of organic matter and nutrients** B. Picot, T. Andrianarison, J. P. Gosselin and F. Brissaud

33 **Twenty years' monitoring of Mèze stabilisation ponds: part II – removal of faecal indicators** F. Brissaud, T. Andrianarison, J. L. Brouillet and B. Picot

43 **A performance review of small German WSPs identifying improvement options** M. Barjenbruch and C. Erler

51 **Performance of an intensive pond system treating municipal wastewater in a cold region** B. Wang, P. Qi, L. Wang, W. Lu, S. Liu and F. Zhao

61 **Primary facultative ponds in the UK: the effect of operational parameters on performance and algal populations** K. L. Abis and D. D. Mara

Design and microbial removal efficiency

69 **Implications for physical design: the effect of depth on the performance of waste stabilization ponds** H. W. Pearson, S. T. Silva Athayde, G. B. Athayde Jr and S. A. Silva

75 *Escherichia coli* **removal in waste stabilization ponds: a comparison of modern and classical designs** C. G. Banda, P. A. Sleigh and D. D. Mara

83 **Performance of very shallow ponds treating effluents from UASB reactors** M. von Sperling and L. C. A. M. Mascarenhas

91 **Removal of *E. coli* and helminth eggs in UASB: polishing pond systems in Brazil** M. von Sperling, R. K. X. Bastos and M. T. Kato

99 **Aerated rock filters for enhanced nitrogen and faecal coliform removal from facultative waste stabilization pond effluents** M. Johnson and D. D. Mara

103 **CFD (computational fluid dynamics) modelling of baffles for optimizing tropical waste stabilization pond systems** A. N. Shilton and D. D. Mara

107 **Virus removal in a pilot-scale 'advanced' pond system as indicated by somatic and F-RNA bacteriophages** R. J. Davies-Colley, R. J. Craggs, J. Park, J. P. S. Sukias, J. W. Nagels and R. Stott

Nutrient removal

111 **'Active' filters for upgrading phosphorus removal from pond systems** A. Shilton, S. Pratt, A. Drizo, B. Mahmood, S. Banker, L. Billings, S. Glenny and D. Luo

117 **Performance of a pilot-scale high rate algal pond system treating abattoir wastewater in rural South Australia: nitrification and denitrification** R. A. Evans, N. J. Cromar and H. J. Fallowfield

125 **Anaerobic reactor/high rate pond combined technology for sewage treatment in the Mediterranean area** F. El Hafiane and B. El Hamouri

133 **Improving nitrogen reduction in waste stabilisation ponds** H. E. Archer and B. M. O'Brien

139 **Modeling ammonia removal in aerated facultative lagoons** C. D. Houweling, L. Kharoune, A. Escalas and Y. Comeau

Hydraulics and physical processes

143 **Light attenuation parameters for waste stabilisation ponds** S. Heaven, C. J. Banks and E. A. Zotova

153 **Optical characteristics of waste stabilization ponds: recommendations for monitoring** R. J. Davies-Colley, R. J. Craggs, J. Park and J. W. Nagels

163 **Profiling and modelling of thermal changes in a large waste stabilisation pond** D. G. Sweeney, J. B. Nixon, N. J. Cromar and H. J. Fallowfield

173 **The development and calibration of a physical model to assist in optimising the hydraulic performance and design of maturation ponds** G. J. Aldana, B. J. Lloyd, K. Guganesharajah and N. Bracho

183 **Variations in BOD, algal biomass and organic matter biodegradation constants in a wind-mixed tropical facultative waste stabilization pond** C. G. R. Meneses, L. B. Saraiva, H. N. de S. Melo, J. L. S. de Melo and H. W. Pearson

Environmental impact

191 **Control of chironomid midge larvae in wastewater stabilisation ponds: comparison of five compounds** R. Craggs, L. Golding, S. Clearwater, L. Susarla and W. Donovan

201 **Mosquito development and biological control in a macrophyte-based wastewater treatment plant** I. M. Kengne Noumsi, A. Akoa, R. Atangana Eteme, J. Nya, P. Ngniado, T. Fonkou and F. Brissaud

205 **Reduction of odors from a facultative pond using two different operating practices** A. Truppel, J. L. M. Camargos, R. H. R. da Costa and P. Belli Filho

Sludge

213 **Determination of the sedimentation constants for total suspended solids and the algal component in a full-scale primary facultative pond operating at high wind velocities under tropical conditions** L. B. Saraiva, C. G. Ribeiro Meneses, H. N. de Souza Melo, A. L. Calado Araújo and H. Pearson

217 **Some observations on the effects of accumulated benthic sludge on the behaviour of waste stabilisation ponds** C. J. Banks, S. Heaven and E. A. Zotova

227 **Wastewater stabilisation ponds: sludge accumulation, technical and financial study on desludging and sludge disposal case studies in France** B. Picot, J. P. Sambuco, J. L. Brouillet and Y. Riviere

235 **Evaluation of sludge from pond system for treatment of piggery wastes** C. T. Zanotelli, R. H. R. Costa and C. C. Perdomo

239 **Investigating helminth eggs and *Salmonella sp.* in stabilization ponds treating septage** G. S. Sanguinetti, C. Tortul, M. C. Garcia, V. Ferrer, A. Montangero and M. Strauss

249 **Isolation of *Salmonella* sp. in sludge from septage treatment plant** G. S. Sanguinetti, V. Ferrer, M. C. García, C. Tortul, A. Montangero, D. Koné and M. Strauss

Anaerobic treatment and treatment of industrial effluents

253 **Potential biogas scrubbing using a high rate pond** G. Mandeno, R. Craggs, C. Tanner, J. Sukias and J. Webster-Brown

257 **Treatment of wastewater containing high phenol concentrations using stabilisation ponds enriched with activated sludge** M. S. Ramos, J. L. Dávila, F. Esparza, F. Thalasso, J. Alba, A. L. Guerrero and F. J. Avelar

261 **Photosynthetically oxygenated acetonitrile biodegradation by an algal-bacterial microcosm: a pilot-scale study**   R. Muñoz, C. Rolvering, B. Guieysse and B. Mattiasson

Floating macrophytes

267 **Microphyte and macrophyte-based lagooning in tropical regions**   I. M. K. Noumsi, J. Nya, A. Akoa, R. A. Eteme, A. Ndikefor, T. Fonkou and F. Brissaud

275 **The effect of water hyacinths for wastewater treatment under Cuban climatic conditions**   C. Rodriguez and P. D. Jenssen

283 **Contribution of floating macrophytes (*Lemna* sp.) to pond modelization**   H. Jupsin, H. Richard and J. L. Vasel

Waste stabilisation ponds and constructed wetlands

291 **Comparison of nutrient cycling in a surface-flow constructed wetland and in a facultative pond treating secondary effluent**   A. Sajn Slak, T. G. Bulc and D. Vrhovsek

299 **Integrated natural treatment systems for developing communities: low-tech N-removal through the fluctuating microbial pathways**   O. Shipin, T. Koottatep, N. T. T. Khanh and C. Polprasert

307 **Comparison of maturation ponds and constructed wetlands as the final stage of an advanced pond system**   C. C. Tanner, R. J. Craggs, J. P. S. Sukias and J. B. K. Park

315 **Performance of a combined eco-system of ponds and constructed wetlands for wastewater reclamation and reuse**   L. Wang, J. Peng, B. Wang and R. Cao

325 **Municipal wastewater treatment with pond–constructed wetland system: a case study**   X. Wang, X. Bai, J. Qiu and B. Wang

Preface

Waste stabilisation pond systems are recognised worldwide as a low-cost and effective technology for efficient wastewater treatment. Over several years they have undergone a renaissance in several countries in order to cope with new challenges: growing population, enhanced treatment requirements in terms of reliability and removal of nutrients and pathogens, treatment of agro-industrial effluents, footprint reduction, limitation of environmental impact, sludge management. These challenges have driven both technological and scientific developments.

- The increasing diversity of the technology offers the possibility to combine different types of ponds (anaerobic, facultative, maturation, deep, shallow, aerated, high-rate algal ponds... and wetlands). A new interest in facilities and devices such as baffles, recirculation systems and gravel filters is observed.
- The growing complexity of waste stabilisation pond technology and the demand for higher performance predictability require more in-depth knowledge of the mechanisms involved in the treatment process. Moreover, most of these mechanisms are highly dependent on the meteorology.

More than a hundred papers and posters presented by professionals and scientists from the five continents during the 6th International Conference on Waste Stabilisation Ponds held in Avignon, France provided the audience with an overview of the latest technological developments and the most significant advances in the knowledge of physical and biological mechanisms involved in the removal of nutrients and micro-organisms.

Owing to the growing interest in the possibilities offered by combinations of stabilisation ponds and constructed wetlands for further effluent polishing and improved treatment performances, a full day was devoted to joint sessions with the 9th International Conference on Wetland Systems for Water Pollution Control (the selected proceedings of which were published as *Water Science and Technology* **51**(9)).

The issue editors would like to thank the members of the scientific committee for their efforts in selecting the 45 papers presented in this issue.

François Brissaud,
Bernadette Picot and
Andy Shilton

Waste stabilisation ponds in France: state of the art and recent trends

Y. Racault* and C. Boutin**

* Cemagref, Research unit: Water quality, 50 avenue de Verdun, BP 3, 33612, Cestas cedex, France
(E-mail: yvan.racault@cemagref.fr)
** Cemagref, Research unit: Water quality and pollution prevention, 3 quai Chauveau, 69336 Lyon cedex 09, France (E-mail: catherine.boutin@cemagref.fr)

Abstract Waste stabilisation ponds represent 20% of the total number of wastewater treatment plants in France. Practical expertise acquired during these last 20 years has led to modification in the design of the first facultative basin of WSP systems. Its active surface area is now dimensioned at $6\,m^2(p.e.)^{-1}$ in order to limit the risk of malfunctioning. The cumulated surface of the 2nd and 3rd basin is maintained at $5\,m^2(p.e.)^{-1}$. Another practical point is also that WSPs must receive mainly diluted influents. Globally, the plants are on average far from their nominal loadings, which explains why the first sludge removals took place on average 13 years after being put in operation. Based on a representative sample of plants, i.e. 15% of the French WSPs, it has been possible to estimate the time, material means and cost needed for sludge removal as well as the amount of sludge accumulated. The sludge removed at the 1st yields on average $110\,L\,(p.e.)^{-1}$ which represents $12\,kg\,DM\,(p.e.)^{-1}$. The current trend of increasing the quality levels necessary for discharge into sensitive receiving bodies has led to adaptive solutions of polishing treatments by intermittent sand filter systems with or without the plantation of reeds.

Keywords Design; domestic wastewater; effluent quality; pond malfunction; sludge production; waste stabilisation pond

Introduction

The use of Waste Stabilisation Ponds (WSP) to treat wastewater has increased greatly in France since the end of the 1970s and is frequently chosen by small rural communities. There are now some 2,500–3,000 WSP installations each for on average 600 person equivalent (p.e.) representing 20% of sewage treatment plants. Some coastal plants are much bigger because of increased loading from tourists in summer. These 25 years of experience in the use of WSPs under French conditions has allowed us to study their functioning and better define the optimal conditions for their use. This article presents a summary of the results gathered either by Cemagref or local surveillance organisations (SATESE) together with the consequences resulting from design or operation of the plants.

Materials and methods

Different data sources were used:
- results gathered by SATESE whilst studying the quality of treated wastewater (Racault *et al.*, 1995), improvements in quality (Boutin *et al.*, 2003) and more recently sludge treatment;
- a summary of the results from monitoring carried out on 102 WSP over 3 years in the Ille-et-Vilaine area (department) of France (Delouvée, 2002);
- results from studies on several annual cycles of operation initially carried out to analyse the causes of malfunction of WSPs (Racault, 1993).

Results and discussion

Characteristics of WSP plants and feeding

During the last 20 years WSPs have assumed an important role in the treatment of domestic wastewater in France, especially in rural areas. Even though in total they handle only 1–2% of wastewater to be treated, they represent 20% of treatment sites in France, and up to or even more than 50% in certain rural areas.

A national survey (Boutin and Racault, 1986) was able to establish the exact number of WSPs in the different French departments 10 years after their introduction. Their rapid development was obvious. Their simplicity had answered the needs of many small communities where traditional methods had resulted in unreliable or unsatisfactory outcomes.

During the last 10 years the increase in installations has declined rapidly for several reasons: the very small size of communities still to be served; certain increased quality standards for treated wastewater which cannot be attained by WSPs (level D4 in Table 1); and finally the loss of interest by certain consultants.

Nevertheless, in certain departments where the technique has been used successfully; construction has continued steadily. The average size of WSP plants in France is still about 600 p.e. The average loading cannot be calculated accurately because measurements in such small plants where there is usually no electricity are limited.

A survey carried out by Cemagref in 1992 (Racault et al., 1995), found the average organic loading for all ponds to be 25 kg $BOD_5 ha^{-1} d^{-1}$ the majority of which did not exceed 50% of the expected loading. A recent study in the Ille-et-Vilaine department (Delouvée, 2002) showed a similar average BOD loading of: 27 kg $BOD_5 ha^{-1} d^{-1}$. The majority of WSPs in France are therefore under-loaded in comparison to their original design. The quantity of waste water treated is very variable. WSPs are often chosen for non-separated systems which accept water infiltration. Hydraulic loads can be high especially in winter (Racault et al., 1995).

Performance of existing plants

A survey of plants for which results were available in 1992 gave overall values for the quality of effluent for separated and combined networks as well as overall averages. These results showed large variations in all parameters which were indicative of variability of influent and seasonal differences with more or less dilution depending on the plant. Racault et al. (1995) showed that it can be difficult to obtain values in line with European standards (Table 1) if the networks are too separated.

Recent measurements (Delouvée, 2002) confirm the averages found in 1992. For these plants located in Brittany the BOD loading was between 16 and 48 kg $BOD_5 ha^{-1} d^{-1}$ and the average age was 11.6 years. Table 2 shows a series of 317 measurements of which 30% were taken in summer (June to September). The figures are close to those obtained previously from non-separated networks (Racault et al., 1995). This can be explained by the frequent influx of rain and drainage water into WSPs in Brittany.

Whatever the cause, the results must be seen in relation to the organic loading which is, on average, well below the nominal loading. Due to long retention times there are

Table 1 European and/or French standards

For load > 120 kg $BOD_5 d^{-1}$ European/French Standard*		COD_f** = 120 mg.L^{-1} SS = 150 mg.L^{-1}
For load < 120 kg $BOD_5 d^{-1}$ French Standard	D3 level* D4 level	Yields: 60% for COD and N-NK BOD_5 = 25 mg.L^{-1} COD = 125 mg.L^{-1}

*values for WSPs
**COD_f: filtered COD

Table 2 Average quality of outflow from Brittany WSPs in mg.L^{-1} (Delouvée, 2002)

Period		CODf*		TSS		N-NK		N-NH4$^+$		TP	
June–September	average std-dev	95	38	74	68	16.3	10.6	9.6	10	6.6	3.6
October – May	average std-dev	71	29	50	45	18.8	11.2	12.6	10	5.7	2.9
Total year	average std-dev	78	34	57	54	18.0	11.0	11.6	10.1	6.0	3.1

*CODf: filtered COD

disparities in flow between influent and effluent. And thus, the removal must be calculated on loads. This method of calculation is more favourable to WSPs, especially in summer, and more representative of the actual effect on the receiving body. It has been used in France since 1997 when the D3 level was made compulsory by legislation (Ministère de l'Equipement, 1997).

Performance is obviously very dependent on/linked to seasonal variations. Additionally, the summer season favours WSPs processes as regards certain parameters measured on filtered samples at the same time as the receiving bodies become more sensitive. The age of the plant, as long as there is no excess sludge, seems to have little effect on carbon or nitrogen levels.

Main causes of malfunction

A certain amount of malfunction has appeared in WSP plants over the years. The main signs are a disappearance of algae, absence of oxygenation and olfactory nuisance. The phenomenon of "red water" can result if the number of phototrophic sulphur bacteria increases. These effects are most noticeable in autumn when decrease in day length and temperature can lead to a more or less sudden disappearance of the algae biomass (Racault, 1993). There are several causes for this but the most common factors, either singly or together, are as follow.

- Concentrated domestic wastewater (annual average $BOD_5 > 300$ mg.L^{-1}) and presence of industrial wastewater are both risk factors.
- Organic overloading of the first pond even if only seasonal.
- Septic influents, which affect the stability of the pond's ecosystem.
- Too deep a first pond. The thermic stratification formed in spring and summer in small ponds protected from the wind is destroyed in autumn and the layers are mixed. The consequences are related to the amount of water in an anaerobic condition.
- The shape of the pond. An elongated form leads to plug flow and consequent organic overloading of the first part of the pond.
- Accumulation of sludge in the sedimentation cone or over the whole first pond can lead to local anaerobia and the beginning of malfunction.

These phenomena can be the result of operational faults/poor maintenance which can be rectified or design faults. The observation of these problems in existing WSPs led to the modification of the French design adopted between 1975 and 1995.

Design in France

As a result of experience several changes have been made to the design recommendations established in the 1980s (Figure 1).

To ensure adequate functioning 3 ponds in series are necessary. The third pond even if it improves the performance little ensures good quality treatment during sludge removal of the first pond. Some departments recommend 4 ponds to increase holding capacity and improve disinfection.

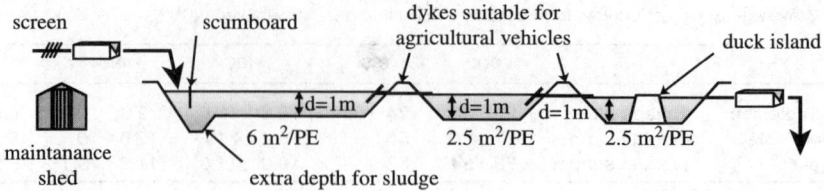

Figure 1 Typical French line design

The first pond. The main change in the first facultative pond is in regard to its dimensions. The behaviour of the first pond is essential to the performance of WSP systems. It was noticeable that there was a decrease in performance and increased risk of malfunction in the first pond when the loading was approaching the previous French recommendations. Consequently the recommended size for the first pond is now $6\,\text{m}^2(\text{p.e.})^{-1}$ (Table 3).

Taking the actual average load produced per person in rural areas to be 35–40 g $\text{BOD}_5 \text{d}^{-1}$ (Pujol and Liénard, 1990), the actual maximum surface load in the first pond rises to 58–67 kg $\text{BOD}_5 \text{ha}^{-1} \text{d}^{-1}$.

When there is a large volume of rain or drainage water and ensuring watertight conditions is not too expensive, the size can be increased to $7\,\text{m}^2(\text{p.e.})^{-1}$ to maintain good levels. For plants serving a variable population during hot weather the loads can temporarily exceed the nominal value for several days as long as it is not doubled.

The recommended depth for the first pond in the literature is 1.5–1.75 m (Mara and Pearson, 1998). However, to ensure reliability it is important to limit the effects of a sudden demand for oxygen. For example, following a thermal de-stratification at the end of summer, the mixing of the amount of anaerobic water may induce a dramatic increase of oxygen demand. In order to maintain a sufficient quantity of aerobic water the recommended depth is therefore 1 m with a maximum of 1.2 m.

Increasing the depth beneath the pond inlet is recommended to facilitate removal of accumulated sludge. An extra depth of 1 m is recommended. This area should be limited to a few square metres to allow for easy sludge removal with vacuum liquid manure tanker. In some departments the first pond is always preceded by a small settlement tank which is emptied several times a year.

The shape of the first pond must not lead to plug flow in order to prevent organic overloading near the inlet. An ovoid shape is therefore recommended with a ratio of length to width of ≤ 3. Baffles in the first pond generally serve no purpose and can even lead to dead zones (Racault and Boutin, 1984).

The third pond. During the 1980s, at the beginning of WSP development in France, putting shallow-rooted plants in the 3rd pond was advised. The presence of vegetation can

Table 3 Differences in nominal and maximal loadings in the first facultative pond

				Pond loading	
	Surface of 1st pond	Theoretic p.e.	Actual loading per person (p)	Nominal	Actual maximum
		(g $\text{BOD}_5.\text{d}^{-1}$)		(kg $\text{BOD}_5.\text{ha}^{-1}.\text{d}^{-1}$)	
Before 1997	$5\,\text{m}^2.(\text{p.e.})^{-1}$	50		100	
After 1997	$6\,\text{m}_2.(\text{p.e.})^{-1}$	50		83	
	$6\,\text{m}^2.\text{p}^{-1*}$		35–40		58 – 67

*p: person connected to the sewerage system

improve quality both by providing shade which minimises algal development and by taking nutrients out of the pond. Statistical comparison of planted and unplanted ponds could neither prove nor quantify this benefit (Cemagref, 1997).

To attain the desired effect the vegetation must be cut back and removed in autumn to prevent the build up of a layer of anaerobic decomposing organic matter. Such harvesting should be performed by someone standing in the water to use special cutting tools after which the vegetation must be removed. This operation is rarely carried out correctly, so in reality the third pond is no longer planted and is the same depth as the other two.

Putting an "island" in the third pond encourages the presence of ducks which, if introduced in the spring, stay and keep the pond clear of duckweed.

Sludge removal

It is essential to remove anaerobic sludge from the bottom of the ponds. A survey was carried out to define optimal conditions for this difficult operation in 5 French departments (A, B, C, D and E). This sample represents about 15% of the French WSP plants (Table 4). Very variable experiences were reported. The number of removal sites was between 6 and 26 representing 8–50% of the total number of plants per department. Analysis of the results for all parameters both globally and by department confirms the variability of local practice.

The ponds cleared were on average 590 p.e., in line with the average French size. They were generally cleared for the first time after about 13 years of operation whereas between 7 and 10 years was usually recommended. There were some late sludge removals (26 years) because of low organic loading. The 3 early clearances are harder to explain.

An estimation of the actual load is rarely available. The loading is calculated from the number of people actually using the system (easily available figures) in relation to the nominal capacity multiplied by a coefficient of 0.8 to take the difference between the actual number and theoretical p.e. into account (Table 3). The average organic loading at desludging was 90%, far above the usual figure of 50%, which shows that clearance is most necessary in plants operating close to their nominal loading.

The main reasons for desludging were the age of the installation plus additional signs, visual, such as sporadic surfacing of sludge or milky colour of the first pond or occasional olfactory nuisance. For small plants, physico-chemical measurements are not taken often enough to show deterioration as a result of excessive sludge.

Usually the first step in sludge management is to take bathymetric measurements of the sludge depth to identify areas of accumulation. This is carried out over all the ponds. These measurements are taken from a boat, usually guided by a kind of grid on the surface, every $25-100 \, m^2$ depending on the size of the pond. The successive steps are: estimation of the depth of sludge using an SS meter or ultrasonic probe; estimation of the depth of the pond using a rigid measure; and calculation of the difference between the two to obtain the actual depth of sludge.

This initial evaluation seems simple but requires a lot of organisation. For a medium-sized plant at least three people are needed for a whole day. Larger plants take longer but not in direct relation to the area to be measured. A technician is also needed for a day as his experience in overseeing and interpreting the findings is essential. Those who make these bathymetric measurements, usually SATESE operators, consider them to be the basis for good future operation. Contacts between municipal authorities and sludge clearance companies are now in fact based on them.

If the depth of sludge is more than 25 cm desludging is necessary. For water depths of between 1 and 1.2 m this amount of sludge represents 20–25% of usable volume. The

Table 4 Summary of results from sludge removal

Département	Number of ponds in department	Number of ponds cleared	Year put in service	Average values per department					Amount of sludge from the first pond				Cost of the operation (exclusive of VAT)		
				Size	Age	Organic loading at desludging								
				p.e.	year	%		cm	cm.y^{-1}	L(pe.)$^{-1}$ y^{-1}		€.m^{-3}	€ (pe)$^{-1}$ y^{-1}	
A	?	6	1985	850	19.0	46		31.5	1.7	115		59.1	5.12	
B	114	26	1987	605	11.9	?		21	1.6	73		25.5	2.08	
C	130	15 + 2*	1985	570	14.8	110		33.8	1.95	98		?	?	
D	30	16 + 5*	1983	500	11.5	81		25.9	2.75	165		19.8	4.30	
E	88	9	1986	575	14.1	97		25	1.80	90		22.4	3.54	
				Statistics from global data									
Mean			1986	590	13.1	91		25.5	2.05	111		23.85	3.19	
Max value			1998	3000	26	194		57	6.6	395		67	12	
Min value			1977	120	5	17		3	0.4	21		9.23	0.21	
standard deviation			4	460	4	37		13.7	1.36	80		15	2	
Number of values			71	71	67	45		27	27	27		25	26	

*number of ponds desludged twice

time between measurement and clearance is about a year due to financial constraints, it is not often provided for in the budget, and to technical aspects such as physico-chemical analysis and drawing up of land spreading plans. With the average loading conditions of the French WSPs, bathymetric measurements should therefore be taken after 10–12 years of operation.

During desludging wastewater is diverted to the second pond and various amounts of supernatant water remaining in the first pond are transferred to the following ponds by various means. The sludge is then removed by one of four techniques.

- Direct pumping out without emptying the pond. This method is discouraged as it does not permit the quality of the work to be seen.
- Supernatant water is partially removed leaving a depth of about 20 cm and a floating platform is used to guide the pump according to the texture of the evacuated sludge. A pipe takes the sludge directly to a liquid manure tanker which transports it to the chosen land-spreading area. This method is used in 35% of cases, even though it is difficult at the edges where sludge usually accumulates. It should always be used where a membrane could be damaged by use of machinery in the pond. There are no data related to this type of pond as they are too recent.
- Supernatant water is evacuated until the sludge is visible. A wetland bulldozer with large caterpillar tracks is then used inside the pond to push the sludge towards a pump. This is a well favoured method (38% of cases) as the results are visible.
- The above method is used but the sludge, instead of being pumped out, is left to dry for a considerable time in order to reduce its volume. This prevents the use of the pond for a long time so this technique is seldom used (5% of cases).

Usually the sludge is spread on the nearest available agricultural land, which has been clearly identified in the clearance plan, as it is removed. Several tankers are used in rotation. It is not always possible to organise this so in 10% of cases the sludge is temporarily left in a zone marked by small dykes to dry out for at least 6 months. The well dehydrated sludge is then spread on agricultural land.

To calculate the actual volume of sludge removed the technical services count the number of tankers filled and measure the surface area of the pond. The average depth of sludge measured was 25.5 cm. The large coefficient of variation (55%) shows the differences between departments. By taking the age of the plant into account the annual increase in sludge in the first pond was found to be 2 cm, in line with current estimations. If this annual volume of sludge is compared to the nominal capacity of the plant the average accumulated volume of sludge is 110 $L(p.e.)^{-1}y^{-1}$. Unfortunately these values do not take the increase in organic loading into account for which figures are not available. Whatever unit is used to express the volume, the large standard deviation clearly demonstrates the variability. These results are from 27 clearances of 1st ponds only.

There were only 12 whole WSP system desludging operations. Here the annual accumulation was large (2.3 cm y^{-1}) corresponding to a much higher volume, 200 $L(p.e.)^{-1}y^{-1}$. This surprising result can be explained by the following: in one case accumulated duckweed was not removed in the previous desludging 10 years before, and in the other case the desludging happened after 15.5 years of operation. It is therefore possible that very large hydraulic loadings had swept sludge out of the first pond into the others.

There were 19 mixed clearances where the sludge was removed from the whole first pond and a part of the others. Here the average sludge volume was 140 $L(p.e.)^{-1}y^{-1}$.

The sludge is systematically analysed to ensure it complies with regulations on land spreading. Such analyses should be viewed with caution, however, as samples could be taken: before evacuation of the water, during desludging when one cannot be sure that it is entirely representative or according to a protocol, after storage and before spreading. Data on dry weight in the first two cases is only based on 12 samples. The average DM of this small sample was 8.5%, ranging from 0.8% to 14.2%, with a standard deviation of 4.1. In the last case, before spreading, dry weight was more than $200\,g\,kg^{-1}$, i.e. 20% dry matter content (DM).

It is impossible to equate the DM to the method of desludging. Despite all these reservations it is possible to give an annual accumulation of sludge of about $12\,kg\,DM(p.e.)^{-1}$ approximately as the SS entering. This can be explained by the fact that the reduction due to anaerobic mineralization is balanced by the addition of the degrading algae and bacteria.

It is also difficult to calculate the average cost of desludging. Figures given usually include the cost of preliminary and spreading plans plus the whole clearance operation but not the chemical analyses. Sometimes the banks are repaired at the same time and this cost is included. The average cost of desludging and land spreading thus calculated was 24 € per m^3 before tax (Table 4). The annual budget should allow for 3.20 €.(p.e.)$^{-1}$ before tax. The factors already mentioned explain the large differences (coefficient of variance of 60%). In the year 2000 the investment cost, depending on the size of the plant, was 100–160 €.(p.e.)$^{-1}$ before tax and desludging represented an operating annual cost equivalent to 2–3% of this amount.

Future developments

The increased quality necessary for sensitive receiving bodies has led to the need for complementary treatments. At present research focuses on the use of intermittent sand filter (ISF) systems at the output of WSP plants. This system brings together the hydraulic capacities of WSP and the useful properties of ISF in the removal of residual organic matter and complementary nitrification. A quality of level D4 (Table 1) should be obtained. In the framework of a European Union LIFE programme, we set up a demonstration plant to try to solve the problems of small rural communities that are already equipped with networks, part of which are combined (Boutin and Liénard, 2004). The objective is to demonstrate a treatment system simple enough to be run by local people who have no specialist knowledge in the field of wastewater treatment but who are willing to invest some time. However, it must be noted that such a result can only be obtained by keeping strictly to alternate feeding of the filters which necessitates 2 visits a week, which a classical third pond does not. At the end of 2005, we hope to be able to define: the technical characteristics of new systems based on WSPs in combination with ISFs (one or two ponds, etc,) in terms of the quality of effluent. The hydraulic limitations of such plant. And the operation and maintenance costs of such plants.

Adding a vertical flow constructed wetland (VFCW) at the entry of an existing WSP should be considered when there is an organic overload of the first pond. Such a modification was studied by Liénard et al. (1993) and a quality of level D4 was obtained. Studies should be performed to confirm that algae discharge using these installations, $5\,m^2(p.e.)^{-1}$ pond after a $1\,m^2(p.e.)^{-1}$ VFCW, does not adversely affect COD and SS of effluent and that D4 quality levels can be maintained.

Conclusion

Waste stabilisation pond systems remain a widely used technique for the treatment of wastewater for rural areas in France. Recent data confirm that on average the plants receive an organic loading clearly lower than the one on which their design is based. Their performance in terms of removal calculated in flux achieves the D3 level of the French regulations. Nevertheless, there have been modifications of the design in the last few years to redress malfunctions observed. These changes are most notably the increase in size of the first pond to $6\,m^2(p.e.)^{-1}$ which should not receive a BOD loading of more than $70\,kg\ BOD_5 d^{-1}$. Also it seems that ponds function more reliably with a diluted influent which means that their use for sewerage networks which are totally separated, without any infiltration or inflow, is not recommended.

Sludge removal is still a major maintenance operation whose annual cost may represent 2–3% of the investment cost. Taking the low loadings received during the first few years of operation into account, the first sludge removal is usually after about 13 years of operation when the organic loading is reaching its nominal value. Statistical analysis shows a large variability of results despite which it is possible to estimate a figure of about $110\ L(p.e.)^{-1}$ which represents a dry matter value of $12\,kg\ SS(p.e.)^{-1}$. A second sludge removal seems to be necessary after about 10 years when sludge is removed from all ponds giving a possible total volume of $200\ L(p.e.)^{-1}$.

Acknowledgements

The authors wish to thank the SATESE (department 01, 22, 38, 57, 82) who provide and summarise data, especially on desludging. They also wish to thank Helen Burnett for her kind assistance.

References

Boutin, C. and Liénard, A. (2004). Waste stabilisation ponds and intermittent sand filters in series: a predicted evolution in France. *Water 21*, (June 2004) 56–58.

Boutin, C., Liénard, A., Billotte, N. and Naberac, J.P. (2003). Association de lagunes naturelles et d'infiltration-percolation: résultats des pilotes et perspectives. *Ingénieries*, **34**, (Juin 2003) 35–46.

Boutin, P. and Racault, Y. (1986). Le lagunage naturel, situation actuelle d'une technique d'épuration en France. *Techniques Sciences Méthodes*, **81**(6), 273–284.

Cemagref, SATESE, Ecole Nationale de la Santé Publique and Agences de l'Eau (1997). Le lagunage naturel: Les leçons tirées de 15 ans de pratique en France, *Ed.* Co-Editions Cemagref Editions, Agence de l'Eau Loire-Bretagne, Antony France.

Delouvée D (2002). Le lagunage naturel: Etude statistique de la qualité des rejets en Ille-et-Vilaine, traitements complémentaires au lagunage, mémoire de maîtrise, Université de Rennes 1, France.

Liénard, A., Boutin, C. and Bois, R. (1993). Coupling of reed bed filters and ponds: an example in France. *Wat. Sci. Tech.*, **28**(10), 159–167.

Mara D. and Pearson H (1998). *Design manual for waste stabilization ponds in Mediterranean Countries*. European Investment Bank (ed.), Lagoon Technology International, Leeds, UK, 112 p.

Ministère de l'Equipement, du Logement, du Tourisme et du Transport (1997). Circulaire d'application n° 97-31 du 17 février 1997 relative à l'assainissement collectif de communes-ouvrages de capacité inférieure à 120 kg DBO_5/jour. Parue au B.O. du 10 Mai 1997.

Pujol, R. and Liénard, A. (1990). Qualitative and quantitative characterization of waste water for small communities. *Wat. Sci. Tech.*, **22**(3/4), 253–260.

Racault, Y. (1993). Ponds malfunction: case study of three plants in the south west of France. *Wat. Sci. Tech.*, **28**(10), 183–192.

Racault, Y., Boutin, C. and Seguin, A. (1995). Waste stabilization ponds in France: a report on fifteen years experience. *Wat. Sci. Tech.*, **31**(12), 91–102.

Racault, Y. and Boutin, P. (1984). Etude par traçage du comportement hydraulique d'une lagune d'épuration; influence de la géométrie du bassin. *Rev. Fr. Sci. Eau*, **3**(2), 197–218.

Changes in waste stabilisation pond performance resulting from the retrofit of activated sludge treatment upstream: part I – water quality issues

N.J. Cromar*, D.G. Sweeney**, M.J. O'Brien** and H.J. Fallowfield*

*Department of Environmental Health, Flinders University, GPO Box 2100, Adelaide SA 5001, Australia
(E-mail: nancy.cromar@flinders.edu.au; howard.fallowfield@flinders.edu.au)

**United Water International, GPO Box 1875, Adelaide 5001, Australia
(E-mail: david.sweeney@uwi.com.au; mike.obrien@uwi.com.au)

Abstract This paper describes changes in effluent quality occurring before and after an upgrade to the Bolivar Wastewater Treatment Plant in South Australia. Trickling filters (TF) were replaced with an activated sludge (AS) plant, prior to tertiary treatment using waste stabilisation ponds (WSPs). The water quality in the WSPs following the upgrade was significantly improved. Reductions in total and soluble BOD, COD, TKN, suspended solids and organic nitrogen were recorded and the predominant form of inorganic nitrogen changed from NH_4-N to NO_2/NO_3-N. The reduction in ammonium and potentially toxic free ammonia removed a control upon the growth of zooplankton, which may have contributed to decreases in algal biomass in the final ponds and consequently lower dissolved oxygen. Additionally, changes in inorganic nitrogen speciation contributed to a slightly elevated pH which reduced numbers of faecal coliforms in WSPs. The AS pretreated influent recorded significantly lower inorganic molar N:P ratio (10–4:1) compared to those fed with TF effluent (17–13:1). Algae within the WSPs may now be nitrogen limited, a condition which may favour the growth of nitrogen-fixing cyanobacteria. The decrease in algal biomass and in dissolved oxygen levels may enhance sedimentary denitrification, further driving the system towards nitrogen limitation.

Keywords Waste stabilisation ponds; performance; activated sludge; trickling filters; water quality

Introduction

The Bolivar Wastewater Treatment Plant (WWTP) treats wastewater from a population of approximately 600,000, with an inflow of about 150 ML/d and a mean residence time > 30 d. Primary treatment includes screening, grit removal, pre-aeration and primary sedimentation. Secondary treatment from the mid-1960s was by trickling filtration followed by secondary clarification; the effluent then being further treated in six waste stabilisation ponds (WSPs) arranged in two parallel streams of three ponds each, with a total surface area of 346 ha. In February 2001 trickling filtration ceased at Bolivar and an activated sludge plant was commissioned. The effluent from this plant, following secondary clarification, provided influent to the WSPs. This paper compares the performance of the lagoons receiving secondary treated effluent initially from the trickling filters and subsequently from the activated sludge plant.

Comparison of waste stabilisation pond influent quality

Table 1 summarises lagoon influent composition over two year-long periods, covering the two different treatment processes. Comparison of the TF and AS secondary treatment regimes demonstrates that a 55% reduction in total biochemical oxygen demand (BOD_{tot}) loading occurred; along with a 94% reduction in soluble BOD (BOD_{sol}) and a 94% reduction in NH_4-N loading. There was, however, a greater than 1000-fold increase in oxides of nitrogen (NO_3-N and NO_2-N).

Table 1 Comparison of the influent composition from trickling filters and activated sludge plant to waste stabilisation ponds at Bolivar WWTP

Parameter (mg/L)	1997/98 (TF influent) (Median)	2001/02 (AS influent) (Median)
BOD_{tot}	97.0	43.5
BOD_{sol}	32.0	2.0
Suspended solids	84.0	59.0
TKN	51.4	13.5
NH_4-N	39.4	2.7
NO_2-N + NO_3-N	0.005	8.78

Comparison of waste stabilisation pond water quality

There was a significant reduction in the BOD_{tot} within the lagoons following the introduction of AS, whereas the BOD_{sol} remained largely unchanged (Figure 1). This decrease in BOD_{tot} in the lagoons is similarly reflected in the decline in suspended solids, COD (Figure 2) and organic nitrogen (Figure 3) following the change to AS influent. In addition, there was a small reduction in median total phosphorus concentration following the introduction of the AS plant (Table 2).

The introduction of the AS plant has resulted in a significant change in the nitrogen economy of the WSPs. The major change has been in the relative concentrations of NH_4-N and NO_2/NO_3-N before and after the introduction of the AS plant (Figure 4). When the lagoons were supplied with trickling filter-pretreated influent, the median NH_4-N concentration was 22 mg/L, whereas after the introduction of the AS plant it was 0.95 mg/L NH_4-N (Table 2). The WSPs clearly showed evidence of seasonal nitrification when the influent originated from the TF (Figure 4), indeed the median value for NO_2/NO_3-N within the ponds was higher at this time than after the introduction of the AS plant (Table 2).

These higher NO_2/NO_3-N values are somewhat unexpected given the highly nitrified effluent originating from the AS plant. The necessary metabolic reduction of nitrate to ammonia by algal cells for growth contributes to the increase in median pH following commissioning of the AS plant (Figure 5). While the difference between the median pH values pre- and post-AS is only 0.5, higher maximum pH values were recorded in WSPs

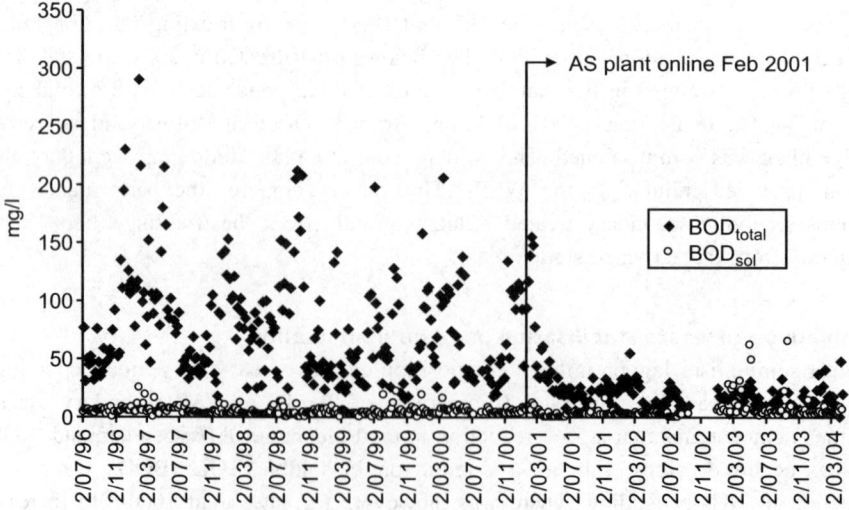

Figure 1 Biochemical oxygen demand (total and soluble; mg/L) in Bolivar WSP lagoons 1996–2004

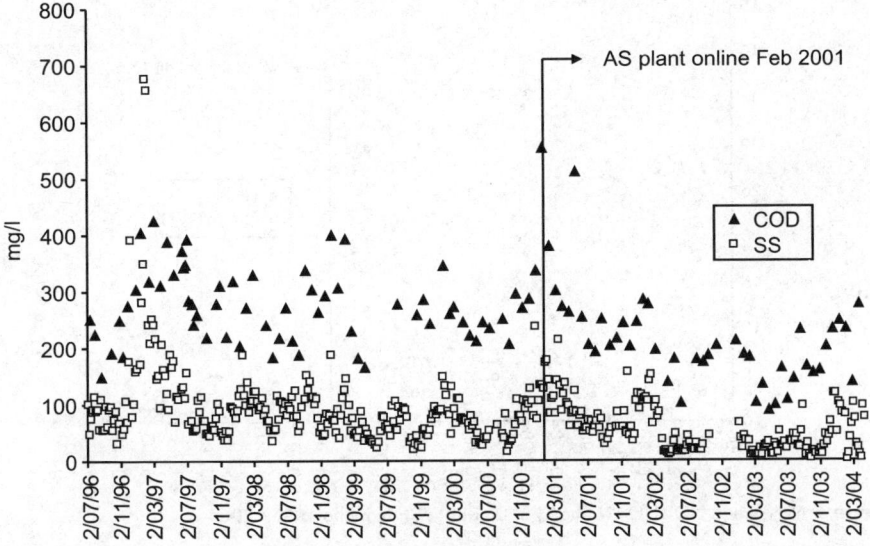

Figure 2 Chemical oxygen demand and suspended solids (mg/L) in Bolivar WSP lagoons 1996–2004

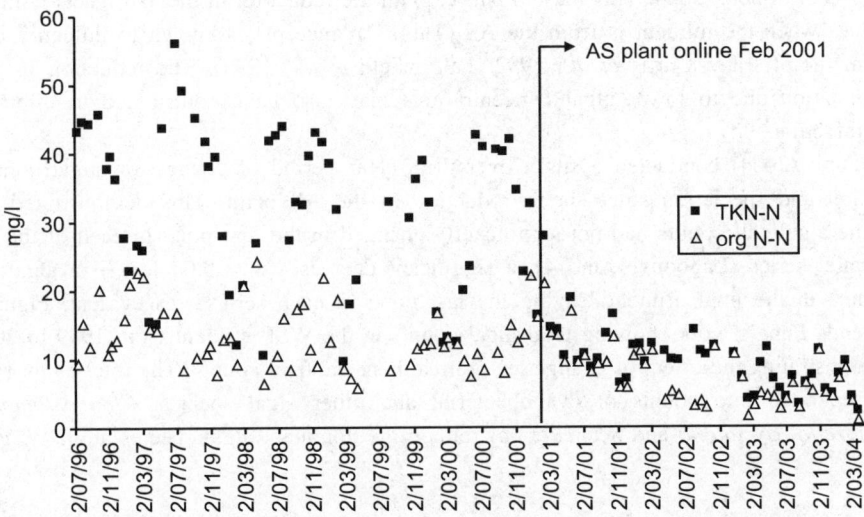

Figure 3 Total Kjeldahl nitrogen and organic nitrogen (mg/L) in Bolivar WSP lagoons 1996–2004

Table 2 Comparison of the treated effluent composition from the waste stabilisation ponds fed either trickling filter (1996–2001) or activated sludge effluent (2001–2004)

Parameter	Jul 1996–Jan 2001 WSP effluent (Median)	Mar 2001–Mar 2004 WSP effluent (Median)
BOD_{tot} (mg/L)	68	18
BOD_{sol} (mg/L)	4	4
COD (mg/L)	271	202
Suspended solids (mg/L)	79	43
Organic N (mg/L)	12	6.4
TKN (mg/L)	33	9.6
NH_4-N (mg/L)	22	0.95
NO_2-N + NO_3-N (mg/L)	3	0.9
Total P (mg/L)	5.6	3.4
Faecal coliforms ($\log_{10}$/100 ml)	3.21	2.15
pH	7.8	8.2

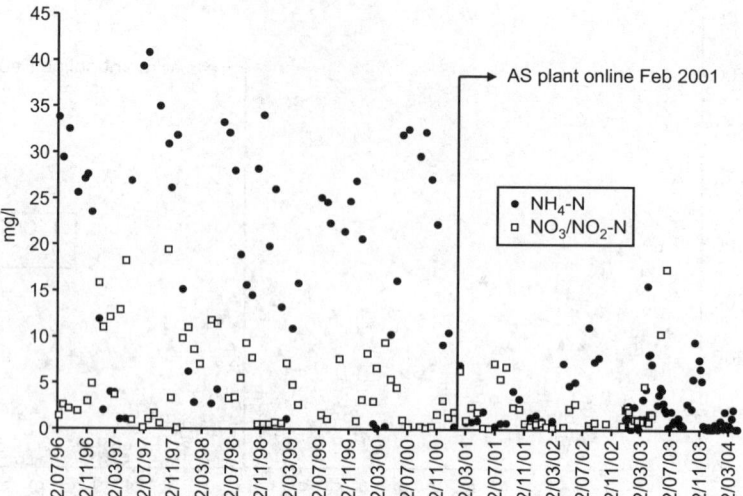

Figure 4 NH_4-N and NO_3/NO_2-N (mg/L) in Bolivar WSP lagoons 1996–2004

fed from the AS plant. For example, the ranges were 7.2–8.5 in 1997/98, compared to 7.1–10.5 in 2003/2004. This may in part explain the reduction in median faecal coliform values when the influent is from the AS (Table 2) since pH is known to influence coliform die-off rate (Curtis *et al.*, 1992; Fallowfield *et al.*, 1996). The reduction in light attenuation due to lower algal concentrations may also have contributed to improved disinfection.

Fung (2004) conducted a study over the 3-year period of changes in algal concentration and speciation since the introduction of the AS plant. This demonstrated that while algal cell counts had not significantly changed in the first pond in each of the two parallel series (Lagoons 1 and 4), a significant decrease ($\alpha = 0.05$ level) in algal cell counts in the final effluent leaving the last pond in each series was evident. Figure 6 extends Fung's work showing total algal counts in the WSP effluent from 1999 to 2004; demonstrating the effect of changes in pretreatment on algal counts. The total count trend is separated into counts of cyanobacterial and other algal species. *Chlamydomonas*, *Chlorella*, *Cyclotella* and *Scenedesmus* remain the dominant algal species at all WSPs at

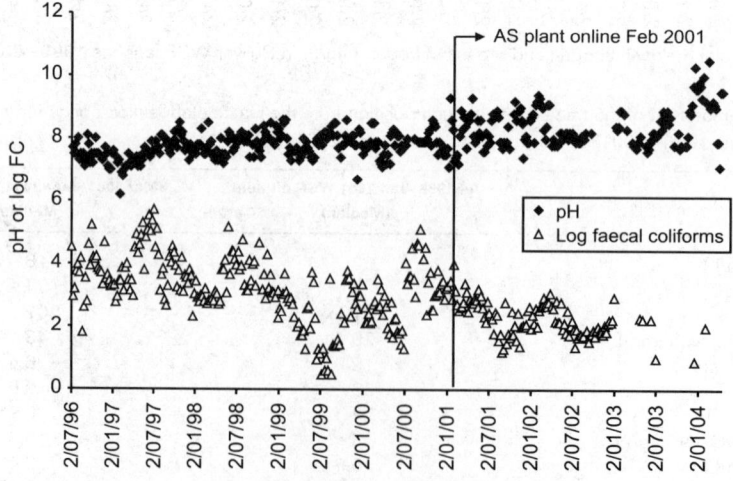

Figure 5 Faecal coliforms ($\log_{10}$/100 mL) and pH in Bolivar WSP lagoons 1996–2004

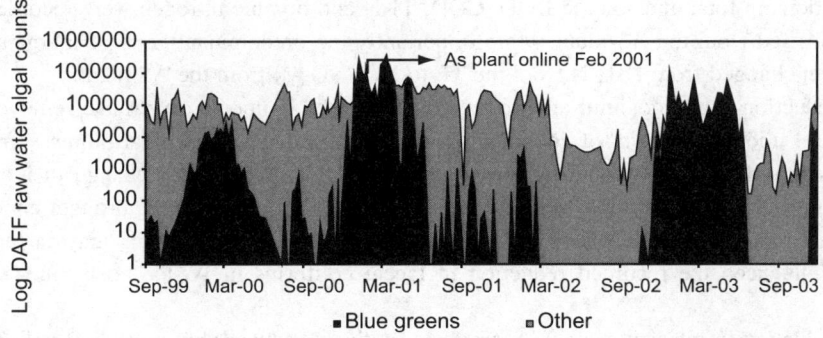

Figure 6 Bolivar DAFF plant raw water algal counts by cyanobacteria (blue greens) or other species

Bolivar WWTP. Some changes in species are evident at the site including increases in *Chlamydomonas*, *Cyptomonas* and *Phormidium*, and a decrease in *Euglena*.

In addition to the overall decrease in algal counts in the final pond effluent subsequent to February 2001, the variability of algal counts at the site has also increased. While cyanobacterial species still predominate in summer, little evidence of them is seen during the winter months, with total counts of all species falling as low as 219 counts/mL. Green algal species account for the majority of the lagoon algal biomass during this time. At such times, visibility through the pond depth increased markedly from several centimetres to, in some cases, over a metre. These algal population crashes may be the result of increasing grazing pressure from zooplankton, and anecdotal evidence of high numbers of these organisms exists, although quantitative data is currently limited. The increase in grazing pressure might be anticipated as a result of the reduction in potentially toxic NH_3-N following introduction of the AS plant. A consequence of the decrease in algal growth and photosynthesis is that the first ponds in the series, while remaining facultative, have on occasion recorded DO values as low as 1 mg/L (Sweeney *et al.*, 2005).

These lower DO values, coupled with nitrate being the predominant form of inorganic nitrogen input from the AS plant, may lead to enhanced nitrogen loss from the water column via denitrification in the sediment. The importance of denitrification in this system is yet to be quantified. However, the data available suggests that the algal population in the WSPs is becoming nitrogen limited. The optimum molar N:P for algal growth is purported to be 16:1 (Redfield, 1958). The N:P status of the Bolivar system was derived using data for the winter period when algal growth is at its lowest and nutrient availability at its highest. Molar N:P ratios of 17:1 and 13:1 were determined for ponds operating on trickling filter influent in the winters of 1997 and 1998 respectively. In contrast, N:P ratios of 10:1 and 4:1 were found in the winters of 2003 and 2004 respectively, in the ponds fed AS influent. The molar ratio calculated for the whole period 1996–2001 (to January) when the ponds were fed trickling filter effluent was 10:1 N:P, whereas for the period 2001 (from February)–2004, when the ponds were fed AS plant influent, the N:P ratio was 1:1. These data suggest that the WSPs are nitrogen limited; a condition which may favour the predominance of nitrogen fixing cyanobacteria, however, such species of cyanobacteria have not yet been recorded in the ponds. The management issues and economic benefits associated with the introduction of the AS plant are discussed by Sweeney *et al.* (2005).

Conclusions

- The water quality in WSP supplied with wastewater pretreated by an activated sludge plant was significantly improved when compared to the same WSP system fed wastewater pretreated using trickling filters.

- Reductions in total and soluble BOD, COD, TKN and organic nitrogen were recorded in ponds fed from the AS plant. More importantly the predominant form of inorganic nitrogen changed from NH_4-N from the TF to NO_2/NO_3-N from the AS plant.
- The reduction in ammonium, and potentially toxic free ammonia, may have removed a control upon the growth of zooplankton, resulting in decrease in algal biomass and photosynthesis, and consequently lower DO in the final WSPs. The change in inorganic nitrogen speciation may also have contributed to an elevated pH, brought about by algal reduction of NO_3-N to NH_4-N for growth. The increase in pH may in turn have influenced the recorded reduction in faecal coliforms in WSPs fed by the AS plant.
- The reduction in nitrogen in the AS pretreated influent reduced the molar N:P ratio in the WSPs from 17–13:1 when fed TF effluent to 10–4:1. The algae within the WSPs may now be nitrogen limited, a condition favouring the growth of nitrogen fixing cyanobacteria.
- The decrease in dissolved oxygen levels may enhance sedimentary denitrification, further driving the system towards nitrogen limitation.

Acknowledgements

Information supplied by Lauren Fung (United Water International) was of benefit in compiling this summary and is very much appreciated.

References

Curtis, T.P., Mara, D.D. and Silva, S.A. (1992). Influence of pH, oxygen and humic substances on ability of sunlight to damage fecal coliforms in waste stabilisation water. *Appl. Env. Microbiol.*, **58**(4), 1335–1343.

Fallowfield, H., Cromar, N. and Evison, L. (1996). Coliform die-off rate constants in a high rate algal pond and the effect of operational and environmental parameters. *Wat. Sci. Tech.*, **34**(11), 141–147.

Fung, L. (2004). *Investigating the changes in lagoon algae species since the Bolivar WWTP upgrades*, Internal report, United Water International, Adelaide.

Redfield, A.C. (1958). The biological control of chemical factors in the environment. *Am. Sci.*, **46**, 205–221.

Sweeney, D.G., O'Brien, M.J., Cromar, N.J. and Fallowfield, H.J. (2005). Changes in waste stabilisation pond performance resulting from the retrofit of activated sludge treatment upstream—Part II: Management and operating issues. *Wat. Sci. Tech.*, **51**(12), 17–22.

Changes in waste stabilisation pond performance resulting from the retrofit of activated sludge treatment upstream: part II – management and operating issues

D.G. Sweeney*, M.J. O'Brien*, N.J. Cromar** and H.J. Fallowfield**

*United Water International, GPO Box 1875, Adelaide SA 5001, Australia
(E-mail: *david.sweeney@uwi.com.au*; *michael.obrien@uwi.com.au*)
**Department of Environmental Health, Flinders University, GPO Box 2100, Adelaide SA 5001, Australia
(E-mail: *nancy.cromar@flinders.edu.au*; *howard.fallowfield@flinders.edu.au*)

Abstract Bolivar Wastewater Treatment Plant (WWTP) was originally commissioned with trickling filter secondary treatment, followed by waste stabilisation pond (WSP) treatment and marine discharge. In 1999, a dissolved air flotation/filtration (DAFF) plant was commissioned to treat a portion of the WSP effluent for horticultural reuse. In 2001, the trickling filters were replaced with activated sludge treatment. A shift in WSP ecology became evident soon after this time, characterised by a statistically significant reduction in algal counts in the pond effluent, and increased variability in algal counts and occasional population crashes in the ponds. While the photosynthetic capacity of the WSPs has been reduced, the concomitant reduction in organic loading has meant that the WSPs have not become overloaded. As a result of the improvement in water quality leaving the ponds, significant cost savings and improved product water quality have been realised in the subsequent DAFF treatment stage. A number of operating issues have arisen from the change, however, including the re-emergence of a midge fly nuisance at the site. Control of midge flies using chemical spraying has negated the cost savings realised in the DAFF treatment stage. While biomanipulation of the WSP may provide a less aggressive method of midge control, this case demonstrates the difficulty of predicting in advance all ramifications of a retrospective process change.
Keywords Activated sludge retrofit; waste stabilisation pond; operating changes; midge flies

Introduction

The waste stabilisation pond (WSP) system at Bolivar Wastewater Treatment Plant (WWTP) in Adelaide, South Australia, comprises the tertiary component in a four-stage process. The main plant was commissioned between 1964 and 1969 and comprised primary sedimentation, secondary trickling filters with recirculation, and waste stabilisation pond (WSP) treatment prior to marine discharge. A quaternary dissolved air flotation/filtration (DAFF) plant was commissioned in 1999 to treat a component of the WSP effluent for horticultural reuse (described in Buisine and Oemcke, 2003).

An activated sludge plant (ASP) was commissioned in February 2001 to replace the secondary trickling filters. This was part of a range of upgrades introduced at the site (including the introduction of the DAFF plant in 1999) as part of an Environmental Improvement Programme (EIP) to reduce odour emission from the site and reduce nutrient levels entering the marine environment. The ASP is of conventional design with intermittent anoxic zones for targeted nitrogen removal.

The WSP system consists of six large ponds arranged in two parallel streams of three ponds each. The system has a combined surface area of 346 ha and nominal depth of 1.4 m. Nominal mean residence time through the ponds is greater than 30 days, at 150 ML/day. Prior to the introduction of the ASP, WSP influent was not typically nutrient limited, and conditions in the first WSP in each stream were facultative. The final pond in each system was mainly used for pathogen removal, with a transition from facultative to maturation behaviour in between. Since the introduction of the WSP, evidence exists that pond operation is, on occasions, nutrient limited (Cromar *et al.*, 2005).

The EIP has been particularly successful in reducing odour emissions and associated complaints from local residents, and improving the quality and decreasing the quantity of treated wastewater discharged to sea. As a result of the EIP, however, a number of operating changes have been required and operating issues of varying severity have surfaced, predominantly due to the changes in the WSPs which have occurred as a result of the plant upgrade upstream of the WSPs.

Changes in operation and performance of Bolivar WWTP WSPS
Improved pond influent quality

Figure 1 shows the change in typical pond influent quality (median values) corresponding to the change in the secondary treatment stage at the WWTP. For the previous trickling filter configuration, effluent entering the pond was considerably higher in total nitrogen, particularly in the form of NH_3-N. Nitrification in the trickling filters occurred only during the warmer months. While phosphorus levels have also declined, the absolute improvement in removal of phosphorus through the ASP compared to the trickling filters is lower. The ASP has also been responsible for a significant reduction in organic loading levels in the pond influent.

Changes in pond effluent quality

Fung (2004) undertook a statistical study of the changes in algal concentration and speciation that have occurred in the ponds since the introduction of the ASP. Analysis of weekly algal counts demonstrated that, overall, algal cell counts in the first ponds of each series have not significantly changed. However, a statistically significant ($\alpha = 0.05$) decrease in algal cell counts in the pond effluent leaving the last pond in each series was evident.

Plant operators have also observed that brief excursions in the DAFF raw water pH occur during the summer months. While typically between 7.8 and 9, since the introduction of the ASP daily pH averages above 10 have been recorded, with the instantaneous pH on some occasions exceeding 11. This may result from the reduction of nitrate to ammonia by algae in the WSPs.

Reduction in post-treatment (DAFF) operating costs

DAFF treatment consists of coagulation and flocculation, flotation, sand filtration and chlorination. Alum is used as the primary coagulant together with a cationic polyelectrolyte coagulant aid.

Improvements in WSP effluent quality have resulted in a reduction in the operating cost of the DAFF plant. The DAFF raw water turbidity (WSP effluent turbidity) and

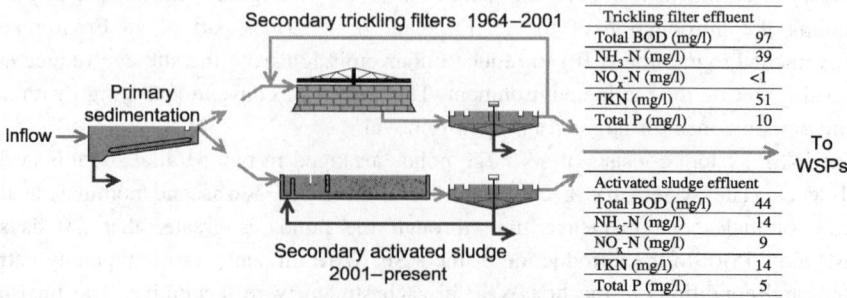

Figure 1 Typical WSP influent quality from alternative pretreatment methods at Bolivar WWTP

filtered water turbidity (outflow from DAFF) are shown in Figure 2. Although seasonal fluctuations in the product water turbidity still occur, the quality of the DAFF product water has improved overall. The optimum coagulant dose in the DAFF plant has decreased from approximately 100 mg/L to 50 mg/L since the ASP was commissioned (Figure 3). Relative treatment costs, based on chemical and energy usage per volume of water treated, are also shown in Figure 3. These are indexed to inflation, and normalised with respect to the highest value recorded at the site, to show the relative cost at other times. It is apparent that the relative treatment cost has decreased by at least 50%, and that the decrease corresponds to the introduction of the ASP.

DAFF plant flowrate, which is driven by demand for irrigation, is seasonal in nature (Figure 4). Increasing yearly demand is also evident. Seasonal peaks in sludge production per volume treated equate to seasonal increases in algal numbers in the WSPs. While limited sludge production data are available for the period prior to the introduction of the ASP, it is evident that sludge production correlates well with the trends of raw water turbidity and optimum alum dose at the site. A reduction in sludge production at the site leads to savings in sludge pumping, digestion, dewatering and disposal costs.

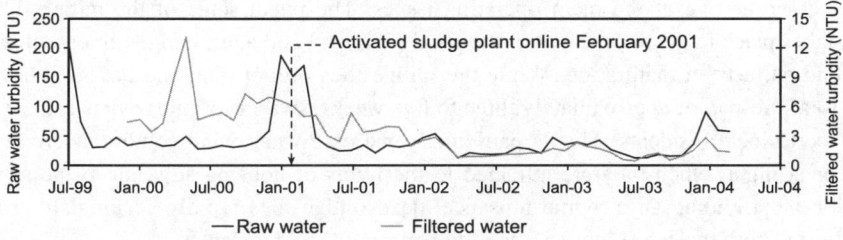

Figure 2 DAFF raw water and treated water turbidity at Bolivar WWTP, from plant commissioning in mid-1999 to early 2004. Date of activated sludge plant commissioning marked

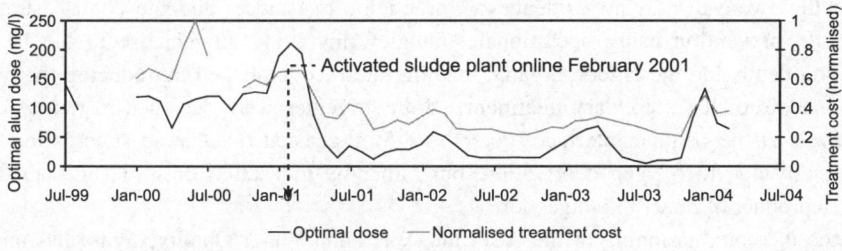

Figure 3 DAFF plant optimal alum dose and normalised treatment chemical cost, from plant commissioning in mid-1999 to early 2004. Date of activated sludge plant commissioning marked

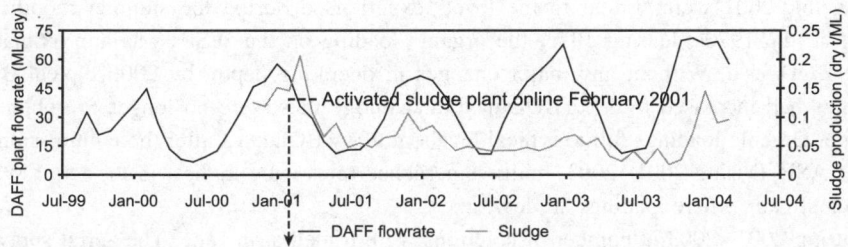

Figure 4 DAFF plant flowrate and dry mass sludge production per volume of wastewater treated at Bolivar WWTP, from plant commissioning in mid-1999 to early 2004. Date of activated sludge plant commissioning marked

Bird numbers at the site

Operational staff have commented on the significant increase in bird life at the site which has accompanied the improvement in water quality in the ponds, particularly in the primary WSP in each series. In addition to the potential for increased pathogens entering the ponds, this has meant that additional expenditure at the DAFF plant has been required for cleaning and the installation of bird netting.

Decreased pond capacity/robustness

With the increased variability in algal populations since the introduction of the ASP, insufficient photosynthetic oxygen production capacity in the WSPs has been identified as a potential problem. While sampling to date has shown that the primary ponds in each series have remained facultative, dissolved oxygen levels in the WSPs have decreased down to as low as 1 mg/L on occasions.

Midge fly nuisance

Since the introduction of the ASP, and the corresponding improvement in pond water quality, the increase in midge fly (*Chironomidae*) numbers at the site during summer months has emerged as one of the main operating issues. The pupal stage of the midge fly life cycle takes place in the upper layer of the pond sediment, and adult midges emerge through the pond surface on maturation. While the midge flies do not bite, and have a relatively short total lifespan of approximately three to five weeks, swarms of midge flies can cause a nuisance to local residents. This is particularly the case when wind conditions are amenable, or at night, when they are attracted to the lights of housing adjacent to the site. In addition to providing an airborne nuisance, dead midge flies rapidly accumulate around buildings at both the site (Figure 5), and the homes of local residents.

Midge fly problems were first reported at the site during the summer of 1967/1968, following commissioning of the WSPs in February 1967 (Peters, 1975). Insecticide dosing using Abate was initially used to control midge numbers. By the early 1970s, the midge flies were displaying evidence of insecticide resistance, and the control strategy shifted to prevention using operational strategies, involving an increase in the lagoon operating depth and the lagoon organic loading rate (accomplished by reducing the recirculation ratio of the secondary treatment). Both strategies were designed to reduce oxygen levels at the benthic interface. Peters (1975) suggested that larvae suppression was not directly due to oxygen deprivation, but rather by the action of some unestablished toxic byproduct of anaerobic digestion.

Since the commissioning of the ASP, the Australian Water Quality Centre has undertaken enumeration of midge larvae and emergent adult midge numbers in all six WSPs at the site (Madden, 2001). An example of the apparatus used for capturing emergent adult midges is shown in Figure 6. The sampling showed that emergent midge fly numbers during mid-2001 were similar to the levels experienced during the summer months by Peters in the 1970s. During 1973, the organic loading on the WSPs was approximately 10 kg BOD/ha d. Without any major changes in operating depth, by 2000 typical BOD loadings had increased to 40 kg BOD/ha d, and midge flies were no longer a problem at the site. Organic loadings fell to typical levels of 20 kg BOD/ha d after the commissioning of the ASP. During 2001–2003, midge emergence rates were highest in the latter WSPs in each stream, where loadings are lowest.

During 2001–2003, a number of solutions were trialed at the site. The aerial spraying of a chemical larvicide and two insecticides was tested, including Abate. Partial bypassing of secondary treatment was again trialed to increase organic loading rate. A number of portable light towers were installed at the site in an attempt to limit the migration of

Figure 5 Accumulation of moribund midge flies at DAFF Plant. From Roder (2003)

Figure 6 Apparatus for sampling of emergent midge flies, with netting open at top. From Roder (2003)

midges outside the plant locality. Of these, only the larvicide Methoprene had a significant impact on emergent midge numbers, and it was concluded that aerial larvicide spraying is the most effective control strategy at the site. While a reduction in emergent midge numbers occurred after the organic loading was increased, it is possible that the reduction was in part due to other factors. Since the introduction of the ASP, no sampling of the WSP benthic interface has been undertaken to test the theory that it has become more aerobic. The increase in midge abundance may more directly result from other factors such as a reduction in ammonia levels, and therefore toxicity, in the WSPs.

Discussion

A considerable change in WSP ecology has occurred as a result of the introduction of the ASP at Bolivar WWTP. The ponds have undergone a shift from a more traditional facultative-maturation configuration to ponds in which are primarily used for pathogen disinfection, and to maintain a hydraulic buffer upstream of DAFF treatment.

The reduction in algal growth in the ponds and subsequent improvement in raw water turbidity entering the DAFF plant has led to significant reductions in both operating costs and sludge production of DAFF treatment. The treated wastewater from the DAFF plant is supplied to a horticultural reuse scheme. Improvements in treated water quality increase the value of this resource, and broaden the scope of uses to which it may be applied. The nature and scale of a number of negative impacts resulting from the changes were, however, not clearly evident prior to the commissioning of the ASP. Operating expenditure for midge control at the site has increased during the summer months. On a yearly basis, this cost of midge management using spraying has been of a similar order of magnitude to the saving in DAFF treatment costs. This is in addition to the expenditure associated with investigating these issues and revising operational practices. Currently, both the costs of operating the DAFF plant and controlling midge flies are subject to a high degree of variability, and from the available data it is not possible to conclude whether or not the lagoons have reached a new "steady state" phase of operation.

In an attempt to realise the full benefits of activated sludge secondary treatment at the site, it may be necessary to make further operational changes in the WSPs. A number of potential solutions have been considered, including biomanipulation of the ponds using shrimp or fish for midge control, or increasing the organic loading by only using a portion of the WSP system. The introduction of the ASP has demonstrated that the net benefits of large-scale changes can be difficult to predict. The impact of further operational changes on algal and zooplankton ecology, and pathogen levels in the WSPs, will have to be clearly understood prior to implementation.

Conclusions

Retrospective modifications of waste stabilisation pond pretreatment stages have the potential to produce significant benefits, as a result of changes in WSP operation. This has been the case at Bolivar WWTP, where the introduction of secondary activated sludge treatment has resulted in a significant improvement in the quality of treated wastewater produced at the site.

While still subject to large seasonal variation, the optimal alum dose for DAFF treatment of the lagoon effluent has reduced by approximately 50% since the introduction of the ASP. A similar reduction in water treatment costs has been achieved, along with reduced sludge production.

A number of operating issues have resulted from the change, the scope and impact of which could not be predicted prior to the change. In addition to capital expenditure associated with control of increased bird numbers on the WSPs, considerable ongoing operating costs are incurred for the management of midge flies at the site.

Biomanipulation of the ponds using aquaculture may provide a potential means to control midge numbers in the pond, however any such further changes which are likely to impact on the algal and zooplankton ecology, and pathogen levels in the pond, will require thorough investigation to ensure that they will be benign.

Acknowledgements

In compiling this summary, the authors appreciate the information and assistance provided by Jason Downard and Liz Roder of United Water International.

References

Buisine, F. and Oemcke, D.J. (2003). Seasonal influence of waste stabilisation pond effluent on DAF/F process performance. *Wat. Sci. Tech.*, **48**(2), 357–364.

Cromar, N.J., Sweeney, D.G., O'Brien, M.J. and Fallowfield, H.J. (2005). Changes in waste stabilisation pond performance resulting from the retrofit of activated sludge treatment upstream—Part I: Water quality issues. *Wat. Sci. Tech.*, **51**(12), 11–16.

Fung, L. (2004). *Investigating the changes in lagoon algae species since the Bolivar WWTP upgrades*, Internal report, United Water International, Adelaide, Australia.

Madden, C. (2001). *Report of Bolivar Lagoon Midge Sampling October 22–23 2001*, Internal report, Australian Water Quality Centre, South Australian Water Corporation, Adelaide, Australia.

Peters, B.C. (1975). *Control of the midge fly at Bolivar Sewage Treatment Works*. Internal report, Engineering and Water Supply Department, Adelaide, Australia.

Roder, L. (2003). *Midge control strategies for the Bolivar WWTP stabilisation lagoons*. Internal report, United Water International, Adelaide, Australia.

Twenty years' monitoring of Mèze stabilisation ponds: part I – removal of organic matter and nutrients

B. Picot*, T. Andrianarison**, J.P. Gosselin*** and F. Brissaud**

*Département Sciences de l'Environnement et Santé Publique, UMR 5569 Hydrosciences, Université Montpellier I, Faculté de Pharmacie, BP 1149, 34093 Montpellier Cedex 5, France
(E-mail: *picot@univ-montp1.fr*)
**Hydrosciences, MSE, Université Montpellier II, 34095 Montpellier Cedex 05, France
(E-mail: *brissaud@msem.univ-montp2.fr; tahina@msem.univ-montp2.fr*)
***Ecosite du Pays de Thau, Parc Scientifique et Environnement, B.P.118, 34140 Mèze, France
(E-mail: *jpgosselin@ecosite.fr*)

Abstract The Mèze stabilisation pond system has been monitored over more than 20 years. Despite the enlargement of the plant, the organic load doubled between the early 1980s and recent years, the removal of organic matter and nutrients has been maintained at the same level for COD and increased for BOD_5, N and P. Combining anaerobic, step-fed aerated and maturation ponds and multiplying the number of cells resulted in a significant improvement in the performances of the plant. Respectively 34, 24 and 23% of the applied COD was eliminated in the anaerobic, the step-fed and the first three maturation ponds, while the figures for BOD_5 were 47, 26, and 19% respectively. 38% of the applied nitrogen was eliminated in the first three maturation ponds. Nitrification and denitrification seem to be a major process of nitrogen removal in warm periods. Most of the phosphorus removal was observed to take place in the two polishing ponds.
Keywords Nitrogen; phosphorus; removal; step-fed facultative pond®; upgrading; waste stabilisation pond

Introduction

Waste stabilisation ponds (WSPs) systems have been widely developed during the last 30 years in France. They account for over 20% of the wastewater treatment plants. They are chosen both for small communities because of their low operation and maintenance costs, low risk of failure and low production of sludge and for bigger coastal towns because of their ability to treat variable loads in tourist zones and to achieve good removal of pathogens when the discharge is located in bathing waters or shellfish breeding areas.

Very few waste stabilisation ponds have been monitored for SS, organic matter, nutrients and microbiological faecal indicators over more than 20 years. The Mèze WSP system which aims to protect the Thau lagoon where more than 10% of the French production of oysters takes place, is an outstanding exception. This paper presents the performance of the plant from the beginning of the 1980s up to 2004. The impacts of pollution load, the ageing of the plant and its enlargement and retrofitting on the removal of organic matter and nutrients are discussed. The specific performance of each stage of treatment is quantified and methods of pollution removal specified.

Material and methods

Site description

The Mèze WSP plant located on the Mediterranean coast (03°35′06″E, 43°25′10″N) was a typical example of the French state-of-the-art when it was constructed in 1980. Made up of three ponds with a total surface of 8 ha, the plant was designed with a treatment capacity of 8,000 p.e. and a design organic load of $50\,kg\,BOD\,ha^{-1}\,d^{-1}$. It received an initial average organic load of only $33\,kg\,BOD\,ha^{-1}\,d^{-1}$ augmented with winery wastewater, particularly in autumn. It discharged into the Thau lagoon. As the population connected to the plant increased, as sludge accumulated and the criteria applying to the treated water quality became more exacting, maintenance and enlargement works had to be undertaken. In October 1994, a 1.5 ha area of the 4 ha primary pond was dredged and 800 tonnes of dried matter were extracted. In 1996 the 3 initial ponds (M_1, M_2, M_3) were supplemented with 2 polishing cells in series (P_1 and P_2), in order to improve the microbiological quality of effluents. Two anaerobic ponds in parallel (A_1 and A_2) were added in 1998 and 4 step-fed facultative ponds (SFP) in series (R_1 to R_4) with recirculation and aeration® in 1999 (Entech patent). Each pond has a surface area of 0.65 ha, is 1.6–1.8 deep and equipped with an 11 kW stirring aerator. Step feeding distribution (50% R_1, 30% R_2, 20% R_3), and recirculation from R4 to R1 were based on a previous study (Sambuco *et al.*, 2002). The total area of the ponds nowadays is 14.4 ha; the load to be treated amounts to up to 1120 kg BOD_5 per day, being equivalent to 19,000 p.e. The present installation can be operated combining new and old ponds either in series or in parallel, the 2 polishing cells serving in both cases (Figure 1). Actually the SFP received a constant flow of about $2500 \, m^3 d^{-1}$, excess of flow at the outlet of A_{1-2} entered directly in M_1.

Sampling and analysis methods

From 1980 to 1992 the WSP was monitored at the inlet and outlet of the 3 initial facultative ponds M_1, M_2 and M_3 fortnightly at first and then every 3 months. Since 2000 the inlet and the outlet have been monitored fortnightly and the inflow and outflow recorded

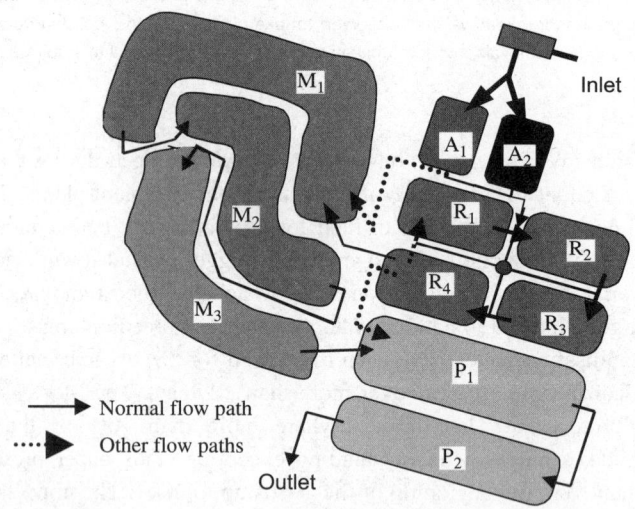

Ponds	Type	Area (ha)	Depth (m)	Start up
A_1, A_2	Anaerobic ponds then surface aerated anerobic ponds	2 × 0.23	3.1	1998 2001
R_1, R_2, R_3, R_4	Step-fed facultative ponds with recirculation (SFP)®	4 × 0.67	1.8	1999
M_1, M_2, M_2	Facultative then maturation ponds	4 + 2 + 2	1.4 – 1.7	1980 1999
P_1, P_2	Polishing ponds	1.9 + 1.2	0.8 – 1.3	1996

Figure 1 Lay-out of the current Mèze waste stabilisation pond system

with an electromagnetic flow meter. During the year 2003–2004 analysing the water quality fortnightly at the inlet and outlet of: (i) the anaerobic ponds in parallel, (ii) the step-fed ponds, (iii) the 3 first maturation ponds, and (iv) the 2 last polishing ponds allowed specification of the role of each type of pond in the overall performance. Composite 24 h samples were taken and analysed daily. The following parameters were measured according to *Standard Methods*: suspended solids (SS), total chemical oxygen demand (COD), filtered COD (Whatman GF/C filters), Kjeldahl nitrogen (TKN), ammonia (NH_4-N), nitrate (NO_3-N), nitrite (NO_2-N), total phosphorus (TP) and orthophosphate (PO_4-P). Biological oxygen demand (BOD) and filtered BOD were measured with a respirometer (Oxytop WTW). *E. coli* and faecal enterococci were analysed using microplate methods. Removal efficiency was calculated in % concentration (inlet concentration – outlet concentration reported to inlet concentration) or in % mass flow (inlet mass flow – outlet mass flow reported to inlet mass flow).

Results and discussion
Comparison of performance at the starting up and prior and after upgrading
During the first 2 years of operation the average overall organic load on the system was 33 kg BOD $ha^{-1} d^{-1}$ and performances were high, 73% COD and 55% N-NH_4 removal, 2.9 log unit faecal coliform and 2.6 log unit enterococci reductions (Table 1).

After 9 years of operation the organic load had increased (65 kg BOD $ha^{-1} d^{-1}$) due to population growth. Meanwhile removal, mainly nitrogen removal, (19%), COD (60%) and faecal coliform (2.3 log unit removal efficiency) had decreased due to both increase in loading rates and accumulation of sludge in the first cell which reduced the hydraulic residence time.

After the enlargement and retrofitting at the end of the 1990s the area of the plant was 8.3 m^2/P.E. The overall improvements to effluent quality from the upgraded WSP have been substantial, particularly for faecal coliform and enterococci indicators (average removal during the last 4 years of 4.1 and 3.4log. units respectively). Brissaud *et al.*, (2005) investigated this improvement in the microbiological effluent quality. Despite the enlargement of the plant, the surface organic loading still doubled between the early 1980s and the present time but the same level of COD has been maintained and even BOD_5 and nutrient removals increased. While influent ammonia concentrations increased with time, effluent ammonia concentrations decreased. Influent phosphorus concentrations decreased with time due to less use of phosphorus in washing products, P removal was the same but effluent concentrations were lower.

It is interesting to investigate the performance of the first two stages of treatment: a surface aerated anaerobic pond (AP) and step-fed facultative pond (SFP). With an area of 1.7 m^2/p.e removal efficiencies were high for SS, COD and BOD_5 (68, 67 and 82% respectively). These first two stages (or one of them) could be used when retrofitting to increase the capacity of overloaded WSPs.

Evolution of performances of the upgraded Mèze WSP
Mean daily inflow increased between 2000 and 2003 from 1875 $m^3 d^{-1}$ to 3127 $m^3 d^{-1}$. In summer the outflow is halved by evaporation (Figure 2). During periods of heavy rain, the plant can have an inflow greater than 10,000 $m^3 d^{-1}$ and the maturation ponds M_{1-2-3} and polishing ponds P_{1-2} have a reserve height of 30 cm and 60 cm respectively to cope with sudden increases in hydraulic charge.

During the same time population increased between by 14000–18800 p.e. Significant seasonal variations of organic matter and nutrients contents and removal were observed, due to tourist population, winery effluents and climate conditions.

Table 1 Comparison of Mèze WSP performance during the first 2 years of operation, after 9 years of operation and after upgrading works for overall plant. Mean, [standard deviation], % removal and log unit reduction for faecal indicators

Years	1981–1982			1988–1990			2000–2003		
Type	3 cells in series			3 cells in series			4 stages-10 cells in series		
Surface area	8 ha			8 ha			14.4 ha		
Population (p.e.)	4400			8700			17300		
Area (m²/p.e.)	18			9.3			8.3		
Surface loading	33 kg DBO/ha.d			65 kg DBO/ha.d			72 kg DBO/ha.d		
	In	Out	%	In	Out	%	In	Out	%
TSS (mg/L)	171			251(109)	136 (69)	46	266 (133)	109(63)	59
COD(mg/L)	430	116 (27)	73	541(409)	216 (91)	60	645 (225)	179(72)	72
fCOD(mg/L)				337 (407)	76 (26)	77 (86*)	262(120)	80 (29)	69 (88*)
BOD(mg/L)	172 (13)	58 (14)	66				405 (136)	50 (23)	88
fBOD (mg/L)							163(72)	9 (6)	94 (98*)
NTK (mg/L)	24 (3)	11 (4)	55	42 (6)	34 (5)	19	46 (13)	15 (8)	67
NH4-N (mg/L)				27 (7)	21 (9)	22	36 (11)	8 (9)	77
TP (mg/L)				14.7 (3.3)	11.3 (3.2)	23	7.9 (2.8)	5.0 (1.7)	37
PO4-P (mg/L)	15 (2/4)	10 (2.7)	35	9.1 (3.2)	5.0 (3.6)	45	4.5 (1.7)	2.8 (1.8)	38
FC (log₁₀/100 ml)	6.2 (0.5)	3.4 (1.2)	2.9	7.0 (0.5)	4.7 (1.0)	2.3	6.8 (0.5)	2.7 (0.7)	4.1
Entero. (log₁₀/100 ml)	5.2 (0.6)	2.6 (0.7)	2.6	6.0 (0.3)	3.6 (0.3)	2.4	6.0 (0.5)	2.6 (0.7)	3.4

% removal was calculated with concentration (in-out/in); fCOD* removal was calculated COD$_{in}$-f COD$_{out}$/ COD$_{in}$; FC and enterococci removals were expressed in mean(log. unit) reduction

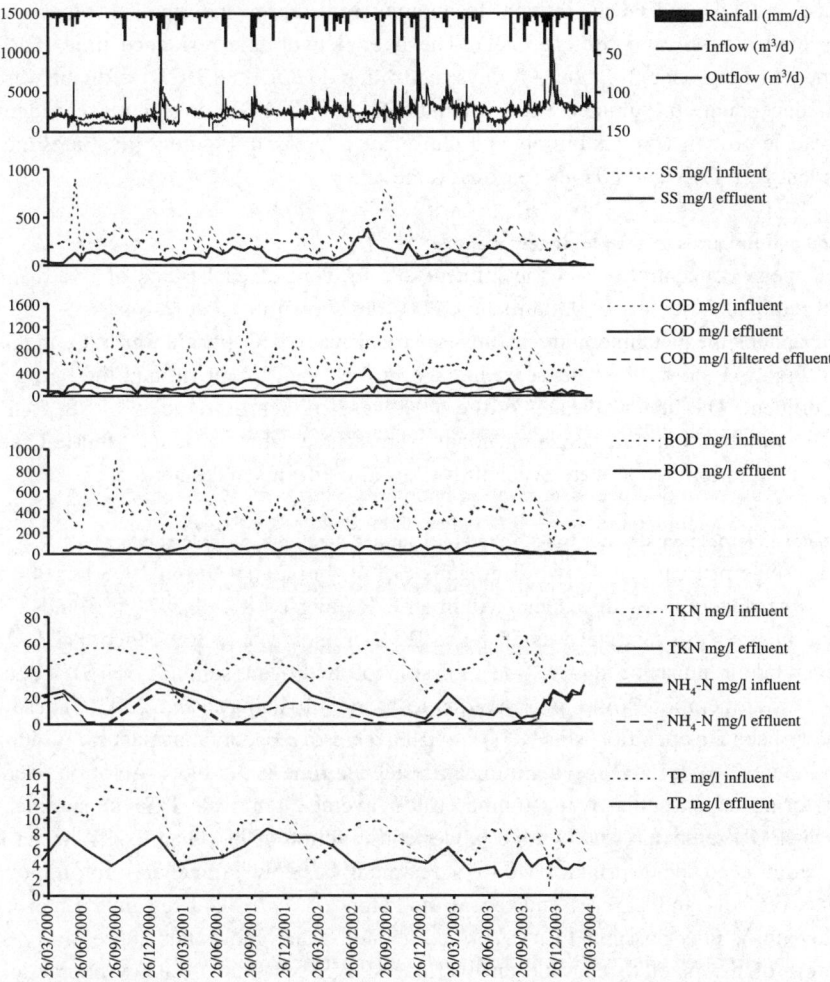

Figure 2 Times series (2000–2004) of influent wastewater and effluent from Mèze WSP for flow, SS, COD, BOD, Kjeldahl nitrogen (TKN), ammonia (NH_4-N) and total-P

Table 2 Annual mean concentrations (standard deviation $n = 23$) and % removal calculated with concentrations or with mass flow in each stage of Mèze WSP from July 2003 until June 2004

	Influent	Anaerobic ponds	SFP	Maturation ponds	Effluent	Overall removal	
						% concentration	% mass flow
SS (mg/L)	256 (85)	158 (99)	81 (47)	54 (39)	70 (62)	73	79
COD (mg/L)	557 (208)	369 (169)	183 (72)	107 (46)	117 (66)	79	81
fCOD (mg/L)	237 (94)	161 (70)	86 (35)	51 (17)	55 (20)	77	77
BOD (mg/L)	347 (133)	187 (82)	63 (22)	32 (26)	26 (24)	92	94
TKN (mg/L)	51 (16)	43 (10)	39 (7)	20 (10)	16 (9)	69	68
NH_4-N (mg/L)	36 (9)	37 (7)	31 (5)	15 (13)	13 (11)	68	73
TP (mg/L)	7.1 (1.7)	6.2 (1.5)	5.6 (1.2)	5.3 (0.7)	4.3 (1.2)	40	45
PO_4-P (mg/L)	4.3 (1.4)	3.5 (1.0)	3.0 (1.8)	4.0 (1.5)	3.0 (1.5	33	36

Peaks in COD and BOD_5 influent in autumn were due to the input of winery wastewaters loaded with high soluble COD. The overall hydraulic residence time (HRT) of effluent was long enough, about 95 days, and filtered COD and BOD_5 effluent were not season dependent. In summer SS, particulate BOD and COD concentrations increased due to algae growth while ammonia and phosphate decreased. Figure 2 illustrates the cyclic tendency of TKN and NH_4-N seasonal removal.

Role and performances of each treatment stage

Annual mean concentrations in the influent and effluent of each stage of treatment and overall removal from July 2003 until June 2004 are shown in Table 2.

The annual mean temperature in the last pond was 17 °C (minimum 5 °C, maximum 28 °C). Figure 3 shows the role of each stage in the overall removal and the part staying in the effluent. The first two stages of the plant were efficient in removal of SS, COD and BOD_5, the maturation and polishing ponds were efficient for nutrients. Step-fed facultative and maturation ponds were efficient for faecal coliforms and enterococci.

First stage: anaerobic ponds later transformed into surface aerated anaerobic ponds

The first stage of treatment is two deep cells (3.1 m) in parallel. From 1998 to 2000 these ponds were anaerobic with a mean volumetric loading of 86 g BOD_5/m^3d and a mean HRT of 4.6 days. SS removal was 55% and BOD_5 removal only 30% (Picot et al., 2003). However odour nuisance due to the emission of hydrogen sulphide (H_2S) led to the addition of an aeration power of 4.4 W/m^3 to each cell (Paing et al., 2003). These aerators have been in operation since 2001 and have resolved odour nuisances. Volumetric loading is now 105 g BOD_5/m^3 d and mean residence time is 3.5 days. Aeration decreases SS performance particularly in summer (39% average removal, 17% in summer) but increases BOD removal (47%) with a power consumption of less than 0.5 kWhr per kg of BOD_5 eliminated. This primary step of treatment has low efficiencies for removal of nutrients (16% N and 12% P) and faecal indicators (E. coli: 0.42 log unit, Enterococci: 0.45 log unit). It is possible to use surface aerated anaerobic ponds successfully at the beginning of the plant to upgrade an overloaded WSP plant with a low energetic cost. (Copin et al., 2004).

Second stage: step-fed facultative ponds with recirculation and aeration® (SFP)

The mean residence time in the 4 step-fed facultative ponds with recirculation and aeration (SFP) is about 25 days. The organic loading applied to the 4 ponds is around

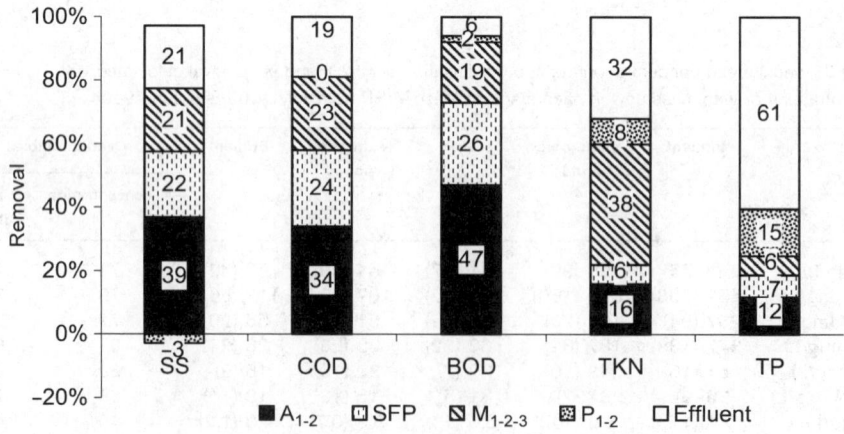

Figure 3 Contribution of each stage of treatment in the overall removal

166 kg BOD ha^{-1} d^{-1} overall. A recirculation of once the inflow was applied after doing tests with ratios of 0.5, 1 and 2 (Sambuco et al., 2002). Four aerators can work if necessary, preferably by night (0.5 kWhr/kgBOD eliminated). This step had very high removal performance SS (50%), COD (50%), dissolved (77%) and total (66%) BOD. It was also efficient to remove microbiological indicators (*E. coli*: 1.62 log unit in summer, 1.19 log unit in winter). This stage had a minimal contribution to the elimination of N (6%) and P (7%). The SFP effluent after these two stages of treatment is below the limit set by the EU Directive 1991 for COD and BOD. Recirculation of effluent from the 4 ponds also has the primary advantage of seeding active algal biomass into the first pond and maintaining aerobic conditions thus increasing pond efficiency. It is therefore possible to more than double the organic load on the first pond in comparison to the usual configuration of stabilisation ponds in series (Shelef and Kanarek, 1995).

During one year (2001) the first stage of treatment was stopped but the SFP still showed good efficiency even though the influent was not pre-treated in the anaerobic pond. However in this case sludge increased in the pond R_1. After 4 years of operation (including one of raw wastewater), sludge heights were still low, 0.30, 0.21, 0.18 and 0.14 m in R1, R_2, R_3 and R_4 respectively, leading to a reduction of volume of 16% R_1, 11% R_2, 10% R_3 and 7% R_4. The SFP is a stage which can replace the facultative pond in the waste stabilisation pond process and produces high effluent quality with decreased land requirements and low operating costs (Copin et al., 2004).

Third stage: maturation ponds

The actual average organic load on the 3 maturation ponds M_{123} is 38 kg BOD ha^{-1} d^{-1} and the hydraulic residence time is about 47 days. This stage of treatment removes the majority of total N (38%) and contributes to faecal indicators removal. The water level is kept low in the 3 ponds during summer to allow them to stock heavy autumn rain and limit its impact on the Thau lagoon.

Fourth stage: polishing ponds

The two last cells P_1 and P_2 are polishing ponds. They were commissioned to stock heavy autumn rain and limit its impact on the Thau lagoon. Loading rate was 26 kg BOD ha^{-1} d^{-1} and the hydraulic residence time was 20 days. Removal was very low in these ponds for all parameters except phosphorus (15%). Increases in SS and particulate COD concentration were even observed due to algae in summer.

Decrease of organic pollution

Respectively 34, 24 and 23% of the applied COD was eliminated in the anaerobic, the step fed and the first 3 maturation ponds, while the figures for BOD$_5$ were 47, 26, and 19% respectively. The K coefficients for degradation, calculated according to the first-order kinetic equation of Marais (complete mixing) were high for the first 2 stages: $K_{COD(AP)} = 0.1$ d^{-1} in summer, 0.16 d^{-1} in winter and $K_{BOD(AP)}$ 0.20 d^{-1} in summer, 0.28 d^{-1} in winter for deep aerated pond and $K_{COD(SFP)} = 0.04$ the same in summer and winter and $K_{BOD(SFP)} = 0.08$ in summer, 0.06 in winter for STPs.

Behaviour of nitrogen and phosphorus

After enlargement and retrofitting, the Mèze WSP system provided high nitrogen removal. Mean influent nitrogen concentration was 51 mg/L, 71% as ammonia. The overall mean removal of total nitrogen was 65% (removal was higher in summer (89%) than in winter (38%). Twenty six percent of influent total nitrogen was discharged into the Thau lagoon in the form of ammonia, 6% as organic-N and 3% as nitrite or nitrate.

The proportion of organic-N in the effluent was higher in summer and the beginning of autumn, than in winter or spring when N-NH$_4$ effluent concentrations were less than 1 mg/L. The first 2 stages of treatment (AP and SFP) removed very little nitrogen. Maturation ponds provided the bulk of the total nitrogen removal.

Mechanisms involved in N removal are: stripping of NH$_3$, nitrification/denitrification and sedimentation (of bacteria, algae and zooplankton) after biological nitrogen uptake. However scientists disagreed over predominant mechanisms. Pano and Middlebrooks (1982) argued that ammonia volatilisation largely explains total nitrogen removal from ponds. On the other hand Zimmo et al., (2003) found that ammonia volatilisation in waste stabilisation ponds accounts for only a small fraction (1.5%) of total N-removal from domestic sewage and suggested that other nitrogen removal mechanisms such as nitrification/denitrification or sedimentation may be more important in overall nitrogen removal. On the basis of the low prevailing nitrate concentrations in pond systems, several studies concluded that nitrification does not take place and consequently denitrification does not play a major role in nitrogen removal (Reed, 1985).

In the Mèze WSP, although average concentrations of the oxidised forms of nitrogen in the effluent were low (annual mean 1.6 mgL^{-1}), nitrate concentration reached 5 mgN L^{-1} in the M$_3$ effluent for more than 15 days in October 2003 and 8 mgL^{-1} in June 2004 even though it was not present in the SFP effluent. Nitrification followed by denitrification could explain the low N-NH$_4$ concentration observed in summer in pond M$_3$ effluent. In October, when temperature was decreasing, denitrification was limited and N-NO$_3$ concentration increased; in November, the water temperature was less than 13 °C and nitrate decreased and ammonia increased. Nitrate was <0.5 mgL^{-1} in winter and spring and increased again in June when temperatures reached 24 °C. For the same pond M$_3$ prior to upgrading, Gomez et al., (1995) demonstrated correlations that were negative between temperature and ammonium and positive between loss of ammonia and the oxidised forms of nitrogen and suggested the existence of a nitrification/denitrification processes.

The annual total phosphorus removal was 39%, higher in summer (63%) when phosphate concentration in effluent was <1 mgL^{-1} and lower in spring (12%). Most of the phosphorus removal was observed to take place in the two polishing ponds, except in summer where it also took place in maturation ponds

Conclusions

The Mèze facility is a successful example of enlargement and retrofitting of an overloaded WSP; though more sophisticated than the initial system and requiring more energy consumption, it remains reliable, easy to operate and maintain, with only a moderate increase in operation costs and improved performances.

The surface aerated anaerobic pond was very efficient in removing SS and organic matter so it could be used to upgrade an overloaded WSP plant with low land requirement and energetic cost. The step-fed facultative pond with recirculation and aeration is an innovative design pond efficient in removing BOD and faecal indicators; its surface organic loading is twice higher than what is usually applied in primary facultative ponds. Maturation and polishing ponds efficiently removed nutrients. They may have a significant part to play in the mitigation of storm events.

The quality of effluent discharged is in accordance with European Directives and could be improved in warmest periods when replacing polishing ponds with rock filters or constructed wetlands that would be more effective for removing SS and particulate pollution.

Acknowledgements

The authors wish to acknowledge the support of the CCNBT, the Conseil Général de l'Hérault, the Agence de l'eau RMC, SATESE 34, ENTECH and Ecosite de Mèze. They also wish to thank to David Druart and Helen Burnett for their kind assistance.

References

Brissaud, F., Andrianarison, T., Brouillet, J.L. and Picot, B. (2005). Twenty years monitoring of Mèze stabilisation ponds II Removal of faecal indicators. *Wat. Sci. Tech.*, **51**(12), 33–41.

Copin, Y., Brouillet, J.L., Rivière, Y. and Brissaud, F. (2004). Waste stabilisation ponds in the wastewater management strategy of a French department. *Presented at 6th IWA Int. Conf. on Waste Stabilisation Pond.*

Gomez, E., Casellas, C., Picot, B. and Bontoux, J. (1995). Ammonia elimination processes in stabilisation and high rate algal pond systems. *Wat. Sci. Tech.*, **31**(12), 303–312.

Paing, J., Picot, B. and Sambuco, J.P. (2003). Emission of H_2S and mass balance of sulfur in anaerobic ponds. *Wat. Sci. Tech.*, **48**(2), 227–234.

Pano, A. and Middlebrooks, E.J. (1982). Ammonia nitrogen removal in facultative WSP. *JWPCF*, **54**, 344–351.

Picot, B., Paing, J., Sambuco, J.P., Costa, R.H.R. and Rambaud, A. (2003). Biogas production, sludge accumulation and mass balance of carbon in anaerobic ponds. *Wat. Sci. Tech.*, **48**(2), 243–250.

Reed, S.C. (1985). Nitrogen removal in stabilisation ponds. *JWPCF*, **57**(1), 39–45.

Sambuco, J.P., Costa, R.H.R., Paing, J. and Picot, B. (2002). Influence of load distribution and recycle rate in step-fed facultative ponds. *Wat. Sci. Tech.*, **45**(1), 33–39.

Shelef, G. and Kanarek, A. (1995). Stabilization ponds with recirculation. *Wat. Sci. Tech.*, **31**(12), 389–397.

Zimmo, O.R., van der Steen, N.P. and Gijzen, H.J. (2003). Comparison of ammonia volatilisation rates in algae and duckweed-based waste stabilisation ponds treating domestic wastewater. *Wat. res.*, **37**, 4587–4594.

Twenty years' monitoring of Mèze stabilisation ponds: part II – removal of faecal indicators

F. Brissaud*, T. Andrianarison*, J.L. Brouillet** and B. Picot***

*Hydrosciences, MSE, Université Montpellier II, 34095 Montpellier cedex 05, France
(E-mail: *brissaud@msem.univ-montp2.fr*; *tahina@msem.univ-montp2.fr*)
**Conseil Général de l'Hérault, DARE-DEMA, rue d'Alco 34087 Montpellier cedex, France
(E-mail: *dema-gestionglobaleeau@cg34*)
***Département Sciences de l'Environnement et Santé Publique, UMR 5569 Hydrosciences- UM1, Faculté de Pharmacie, BP 1149, 34093 Montpellier cedex 5, France (E-mail: *picot@univ–montp1.fr*)

Abstract The WSP system serving Mèze and Poussan (French Mediterranean coast) was constructed in 1980 and enlarged and upgraded from 1994 to 1998. Water quality along the waste stabilisation pond to (WSP) system has been monitored over the years, thus allowing us to assess the influence of enlargement and upgrading works. A significant enhancement of the average microbiological quality of the effluent was observed, with respective *E. coli* and streptococci average abatements of 4.1 and 3.4 log. units. Former seasonal variations of microbiological removal have vanished. The contribution of the different ponds to the disinfection performance of the WSP system was analysed. A microbiological quality model was proposed to evaluate the die-off kinetics related to the different ponds and as a tool for the design and management of WSP systems. Though the relationships between die-off coefficients and environmental factors appeared somewhat frail, this modelling is considered a promising approach for the prediction of WSP microbiological performance.
Keywords Die-off kinetics; disinfection; modelling; waste stabilisation ponds

Introduction

Twenty five years ago waste stabilisation ponds (WSPs) were recommended for the treatment of wastewater to be reused for unrestricted irrigation (WHO, 1989). About the same years, WSP technology was chosen to protect bathing waters and shellfish breeding areas along the Mediterranean coast of Languedoc and Roussillon (Southern France). A prominent implementation of this policy was the construction of WSPs to treat the sewage of all the municipalities of the northern part of the watershed of the Thau coastal lagoon. Since then, disinfection in WSPs has been much studied, at the real scale, on pilot plants and in the laboratory. However, predicting disinfection performance still remains a somewhat tricky aspect of WSP state-of-the-art. This is due to the high number of factors and inter-related processes involved. Disinfection in WSP can be seen as resulting from the combination of pond hydrodynamics, which control water retention times, and micro-organism decay processes. Both, hydrodynamics and die-off processes, are highly dependent on meteorological factors, mainly temperature, solar irradiation and wind. The natural variability of these factors adds to the difficulty of disinfection investigation and prediction (Brissaud *et al.*, 2003).

Therefore, despite the valuable and promising findings of many investigators, from Racault *et al.* (1984) to Shilton and Harrison (2003) on hydrodynamics and from Mezrioui (1987) to Davies-Colley *et al.* (2003), on micro-organism die-off kinetics, there is still much to be learnt from the monitoring of full-scale plants. Monitoring the impact of upgrading and enlargement works on effluent quality may be highly significant for practical purposes (Archer and Donaldson, 2003). Such was the motivation for this paper, trying to take advantage of data collected over several periods, between the beginning of the 1980s and the year 2004, at the Mèze WSPs. We attempted to link the die-off kinetics in the different ponds of the system to meteorological variables in order to work out a model that could be used as a tool for the design and management of WSP systems.

Materials and methods

Mèze WSP system

The main characteristics of the Mèze WSP system are described in Picot *et al.* (2005). The plant was constructed in 1980 in order to protect the shellfish breeding areas of the Thau coastal lagoon. It was a 3-cell system (M_1, M_2, M_3) with a total surface of 8 ha and a treatment capacity of 8,000 p.e. In the early 1980s, the treated load corresponded to 4,400 p.e., augmented with winery effluents in autumn (Figure 1). At that time, no microbiological criterion was set on the effluents discharged into the Thau lagoon. The plant became overloaded a few years later, due to serviced population growth as well as increasing industrial loads. Nowadays, the load to be treated amounts to up to 1,120 kg BOD_5 per day, being equivalent to 19,000 p.e. Furthermore, the new consent conditions require that WSP effluent *Escherichia coli* and enterococci contents are less than 1,000 cfu/100 mL in summer and 10,000 cfu/100 mL in winter. These requirements may be strengthened in the near future. The plant had to be retrofitted in order to cope with these new conditions.

Eight new ponds were added to the 3 old ones between 1996 and 1998, the total pond surface reaching 14 ha. At first, in the early 1980s, a typical French 3-pond system, it is nowadays a sophisticated system of 11 ponds, some of them being aerated, step fed and with possible recirculation. Moreover, the management of the system has been complicated in order to optimise the performance of the step-fed facultative stage and alleviate the impact of storm events on microbiological water quality. A part of the effluent of the anaerobic ponds is deviated to M_1 without passing through the step-fed ponds (Figure 1). Water depth is controlled in every pond through weir level adjustments. Wastewater is stored during storm events in order to increase water detention times and prevent degradation of the effluent microbiological quality. Raising and lowering the water levels entail

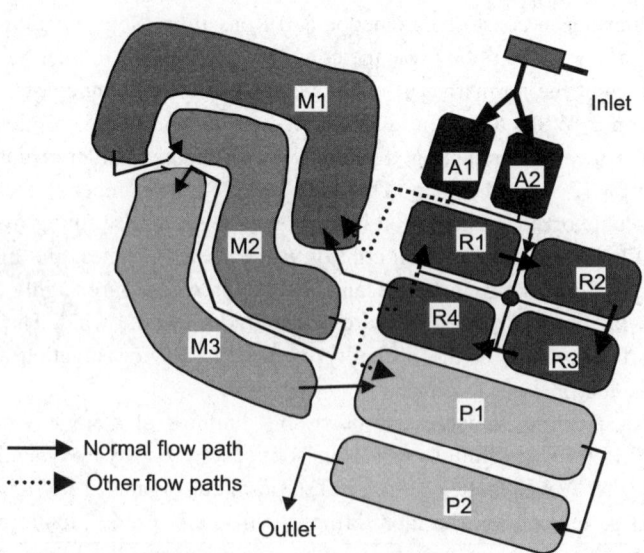

Figure 1 Layout of the current Mèze WSP system

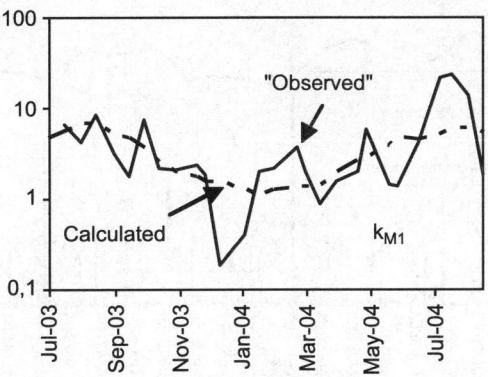

Figure 5 E. coli die-off coefficient in pond M_1 (July 2003 to August 2004). "Observed" values were obtained from Equation 1 while calculated values were derived from Equation 2

a mean value of $0.4 \pm 0.07\,d^{-1}$. E. coli die-off coefficient in ponds $R_{1,2,3,4}$, k_R, showed clear seasonal variations: from $1.1\,d^{-1}$ in July and August, k_R decreased steadily to $0.17\,d^{-1}$ at the end of January, started increasing at the end of April to reach $0.8\,d^{-1}$ at the end of May. Despite abrupt variations, a similar trend was observed in pond M_1, with k_{M1} values around $4.5\,d^{-1}$ in July and August 2003, a steady decrease to a minimum of $0.15\,d^{-1}$ at the end of December. k_{M1} started increasing at the end of March, from 0.17 to more than $20\,d^{-1}$ in July 2004 (Figure 5). The optimisation procedure found variations of k_R and k_{M1} mainly related to water temperature.

Die-off coefficient variations in ponds M_2 and M_3 were hardly matched by calculated values (Figure 6). Both variations observed and calculated from the optimisation procedure using data collected in the period 2003–2004 did not reflect any seasonal effect on E. coli removal. On the contrary, k_{M2} values obtained from Equation 2, using meteorological and SS data of the period 2003–2004 and α, β and χ values derived from the optimisation procedure applied to the data of the periods 1986 and 1988–1989, displayed a clear seasonal trend. The same observation applies to the k_{M3} variations calculated from α, β and χ values optimised over 1988–1989 data. Though differences in operation conditions before and after retrofitting can be evoked, it should be stressed that different monitoring periods provided significantly different sets of α, β and χ values. Therefore, this suggests that Equation 2 failed to report accurately the mechanisms involved in E. coli removal, either because inappropriate representation of the impact of meteorological variables or for other influential factors were not taken into account.

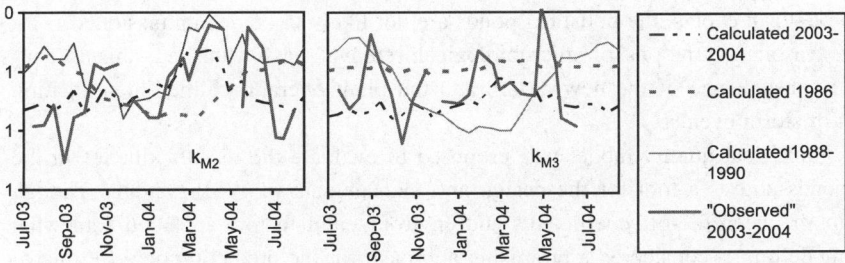

Figure 6 E. coli die-off coefficients in ponds M_2 ad M_3 (July 2003 to August 2004)

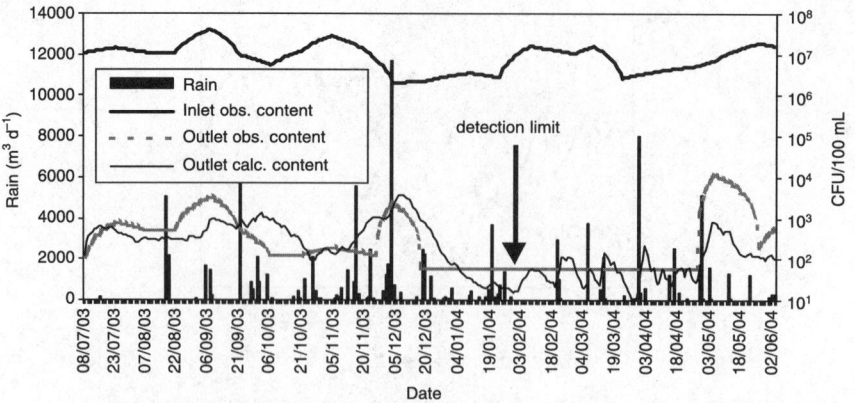

Figure 7 Modelling *E. coli* content in the effluent of Mèze WSP system

Uncertainties related to the amount of R_4 effluent passing directly to P_1 made any relationship between k_{P1} and the meteorological conditions illusive. Die-off coefficient, k_{P2}, appeared to be season dependent with average values of $0.64\,d^{-1}$ from July to September, $0.28\,d^{-1}$ from October to February and $0.78\,d^{-1}$ from March to June.

Effluent quality was calculated through the application of Equation 1 in each cell, one cell after the other, k values being derived at each time step from the relationships established between α, β and χ and water temperature, solar radiation and SS content. Microorganism contents in each cell were thus calculated from the observed microorganism content observed at the inlet of the WSP system. Despite discrepancies between the observed and calculated k values, particularly after the storm of early May, and the above mentioned deficiencies of Equation 2, the simulation was considered satisfactory (Figure 7).

Conclusions

Retrofitting of the Mèze WSP system resulted in a significant enhancement of the average microbiological quality of the effluent, with respective *E. coli* and streptococci average abatements of 4.1 and 3.4 log. units. This improvement was not due to a surface loading decrease but to the subdivision of the plant into a high number of cells and to the stepfed deep aerated pond system. The contribution of the polishing ponds to the disinfection performance was not without ambiguity. Therefore, it might be considered that, owing to the microbiological requirements of the present regulations, the total pond surface could have been significantly reduced. This is considered as a beneficial effect of the design of the first two stages of the retrofitted plant. However, as they play a significant part in the mitigation of storm events, the polishing ponds are not likely to be decommissioned.

Former seasonal variations of microbiological removal have vanished; this is considered the consequence of the new design and the plant operation aimed at mitigating the impact of storm events.

A microbiological quality model was proposed to evaluate the die-off kinetics in the different ponds and as a tool for the design and management of WSP systems. Though relationships between die-off coefficients and environmental factors appeared somewhat frail, this modelling is considered a promising approach for the prediction of WSP microbiological performance and the design of management strategies, provided our knowledge of disinfection mechanisms improves.

Optimisation of the preliminary treatment by supplementary precipitation

If wastewater ponds plants are equiped with a settler as a pre-treatment unit, these precipitations can also with be a temporary method to reduce the organic load of the pond system. Experiments of Barjenbruch *et al.* (2003) show, that with the dosage of $FeCl_3$ or PAX, a COD-elimination ratio up to could 75% be achieved. The additional amount of sludge has to be taken into consideration.

References

ATV-DVWKArbeitsblatt A 201 (2004). *Grundsätze für die Bemessung, Bau und Betrieb von Abwasserteichen für kommunales Abwasser (Dimensioning, construction and operation of wastewater lagoons for communal wastewater)*, Gesellschaft zur Förderung der Abwassertechnik e.V., Hennef.

ATV-DVWKArbeitsblatt A 118 (1999). *Hydraulische Bemessung und Nachweis von Entwässerungssystemen (Hydraulic dimensioning and verification of drainage systems)*, ATV A118, GFA, Hennef.

Barjenbruch, M., Erler, C. and Steinke, M. (2003). *Untersuchungen an Abwasserteichanlagen in Sachsen-Anhalt im Jahr 2003 (Investigation on wastewater lagoons in Saxony-Anhalt in 2003)*, Report for the Ministry of Environment Saxony-Anhalt.

Brockhaus, S., Barjenbruch, M. and Tränckner, J. (2002). Operation experiences with different wastewater ponds in regard to optimisation and extension, *Preprints of 5th international IWA specialist group conference on waste stabilisation pond technology for the new millennium, Auckland.*

Drebes, H. and Grottker, M. (1997): *Einfluss der Vereisung von natürlich belüfteten Abwasserteichan-lagen auf die Ablaufergebnisse und Maßnahmen zur Verbesserung der Reinigungsleistung (Influence of the icing up of naturally aerated wastewater lagoons on the effluent results and measures for the improvement of the cleaning performance)*, 14.1.1997, LANU Schleswig-Holstein.

German Wastewater Directive (2002). Verordnung über Anforderungen an das Einleiten von Abwasser in Gewässer und zur Anpassung der Anlage des Abwasserabgabengesetzes, vom 2.7.2002.

TGL 28722/01 (1982). Natürlich belüftete Abwasserteichanlagen, Anwendung und Bemessung. (Naturally aerated wastewater lagoons, application and calculation), Zentrales Geologisches Institut, Berlin.

Performance of an intensive pond system treating municipal wastewater in a cold region

B. Wang*, P. Qi*, L. Wang**, W. Lu*, S. Liu* and F. Zhao*

*Water Pollution Control Research Center, Harbin Institute of Technology, 202 Haihe Road, Harbin, China, 150090 (E-mail: Baozhen@public.hr.hl.cn)
**Ocean University of China, Qingdao, 260003, China

Abstract A full-scale intensive pond system (IPS) with shorter HRT was designed, constructed and operated in Jining, Inner-Mongolia for the treatment of municipal wastewater, which is a mixed domestic and industrial wastewater characterized by quite high SS and lower BOD_5/COD ratio values or lower biodegradability. Therefore, the pond system was designed as an integrated intensive pond system (IIPS) consisting of settling/anaerobic pond (SAP), intensified anaerobic pond (IAP), facultative pond (FP), and polishing ponds (PPs). In order to improve the performance of the IPS, some intensified measures were made, including inlet and outlet even-distribution systems of each unit pond, package of biofilm carrier in IAP for the increase and even distribution of biomass; overflow waterfalls on the dikes between unit ponds for the increase of DO in pond water, gravel filtration dike (or dam) for removing suspended solids including algae, which have improved the performance of the IPS remarkably in terms of removal of main pollutants, such as SS, COD, BOD_5, TN, NH_3-N, TP and total bacteria. The final effluent from the IPS in warm seasons from May to October were SS 7.2–10.8 mg/L, BOD_5 8.5–19.6 mg/L, COD 44.1–76.5 mg/L, and NH_3-N 1.5–10.2 mg/L, which well meet Chinese national discharge standard (2nd class) of secondary municipal wastewater treatment plants, i.e. BOD_5 and SS 30 mg/L respectively, COD 100 mg/L, and NH_3-N 25 mg/L.
Keywords Biofilm carrier; intensified anaerobic pond; intensive pond system (IPS); settling/anaerobic pond; waterfall aeration

Introduction

The pond system is one of most popular appropriate technologies for municipal and industrial wastewater treatment both in developing and developed countries. Although a pond system has many advantages, such as low capital and operation/maintenance costs, energy saving, ease to operate, and higher removal efficiency for various pollutants present in wastewater, the performance is usually highly dependent on climate and temperature, generally performing well in warm areas or in warm seasons, particularly in the case of intensified or novel pond systems like the advanced integrated pond system (AIPS) developed by Oswald (1991) and high rate pond system (HRPS) developed by the authors (Wang et al., 1996), which performs very well with high removals for various pollutants in all year round operation either in California, USA or in Guangdong, China. However, the operational performance of conventional pond systems in cold regions varies significantly with season and temperature, usually being poor in the winter season, with final effluent not coping with discharge standards. The performance of a pond system operating at low temperature was improved significantly by means of intensified measures such as the package of biofilm carrier in ponds, the formation of surface scum coverage layer in anaerobic ponds and the improvement of hydraulic flow regime in pond system with a final effluent well meeting the discharge standards (Qi et al., 1993; Wang et al., 1995).

The pond system (Figures 1 and 2) designed by the authors for the treatment of mixed municipal and industrial wastewater in Jining City, Inner-Mongolia, is based on the past research results and experience with the intensification of the pond system for its

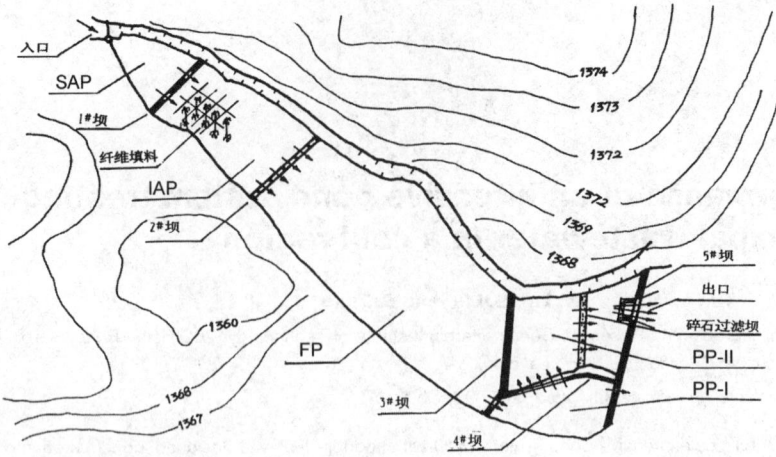

Figure 1 Layout of intensive pond system in Jining, Inner-Mongolia

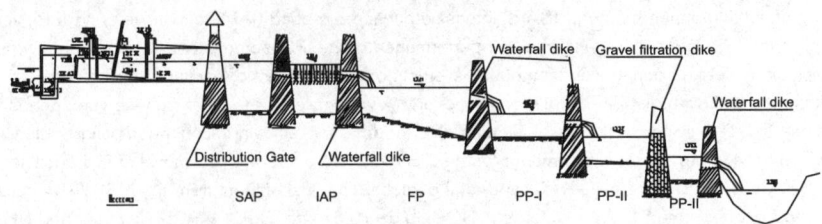

Figure 2 Flowchart of intensive pond system in Jining, Inner-Mongolia

improvement, in the winter season in particular (Wang *et al.*, 1995; Qi *et al.*, 1993). A whole year-round on-site study was carried out and positive results on operational performance were obtained in terms of removal efficiencies for various main pollutants of the wastewater to be treated, their removal variation with operation time, and the removal mechanisms, which are described in this paper.

Methods and materials
Raw wastewater treated by the IIPS

The municipal wastewater treated by the IPS in Jining, Inner-Mongolia is mixed domestic and wastewaters from industries, such as pulp and paper mills, distilleries, wool textile factories, tanneries, slaughterhouses and combined meat processing factories. The raw mixed wastewater was sampled 21 times until February 2000 at the inlet of the settling anaerobic pond or at the head of the IPS and the main pollutants were analyzed by standard methods recommended by Ministry of Construction (Construction standard (Jian bio), 1991). The analytical results and the analytical method are shown in Table 1, from which it was found that the raw wastewater is a typical mixed industrial/domestic wastewater with lower BOD_5/COD ratio of 0.293–0.355 (mean 0.345), whose biodegradability is lower than that of domestic wastewater usually with a BOD5/COD ratio of 0.5–0.6. Besides, the suspended solids content in the raw wastewater was quite high due to industrial discharges. In consideration of the above characteristics, the settling and anaerobic ponds were added in the IPS for the settlement of suspended solids and the improvement of biodegradability of the raw wastewater.

Table 1 Analytical results raw wastewater

Parameter	Concentration		Analytical method
	Range	Mean	
pH	7.80–7.85	7.82	pH-meter
SS (mg/L)	121.1–242.3	206.8	Weighing
COD (mg/L)	176.2–289.2	245.4	Dichromate method
BOD_5 (mg/L)	75.2–12.4	84.6	Dilution seeding method
TN (mg N/L)	35.1–46.8	41.2	Distillation titration
NH_3-N (mg N/L)	21.8–30.4	25.9	Spectrophotometry
TP (mg P/L)	3.5–6.7	4.8	Spectrophotometry
Total bacteria (cfu/ml)	5.1×10^8–2.3×10^9	1.5×10^9	Plate counting

Study site

Jining City is situated in mid-southern part of Inner-Mongolian Autonomous Region at East longitude of 113°03' and North latitude of 40°58' with an average elevation of 1425 m. Its mean annual atmospheric temperature is 3.5 °C with the highest monthly temperature in July (31.4 °C) and lowest in January (-25.6 °C). Its annual average wind velocity is 3.7 m/s, with higher wind speed of 4.1–4.8 m/s in spring and lower one of 2.5 m/s in summer. Annual average precipitation is 298 mm. The annual average solar radiation time is 2983 hours, which is favorable for pond performance as more energy can be obtained from lengthy solar radiation.

Unit ponds

Settling/anaerobic pond (SAP). This is the first unit pond of the pond system, and it mainly functions as a settling pond to remove settlable particular solids and the settled deposit should be removed periodically to prevent abnormal operation due to silt-up on the bottom. An influent diversion and distribution device was built at the inlet of the pond system as shown in Figure 3 to improve the hydraulic flow regime. Anaerobic fermentation and degradation of organic substances also takes place in this pond, which reduces the organic load of the post-positioned ponds.

Intensified anaerobic pond (IAP). Two cell anaerobic ponds in series were designed and built in this pond system: the first one is the settling/anaerobic pond as described above, which is followed by the second one, the intensified anaerobic pond that is packed with biofilm carrier in form of circular plastic plates with synthetic silks, at a package volumetric rate of 10%, for the purpose of the increase of biomass and its even distribution in bulk volume in the front half part of anaerobic pond by the formation

Figure 3 Diversion device at the inlet of the settling/anaerobic pond

and development of biofilm on the carrier surface as well as the biomass in suspension. This can improve the performance of anaerobic pond in terms of organic and nutrient removals or transformation, as indicated by our former studies (Qi et al., 1993; Wang et al., 1996).

The width and length of the intensified anaerobic pond are 170–200 m and 290 m respectively, with a total surface area of 52,000 m^2, total effective volume of 208,800 m^3, the average water depth is 4.0 m, and the HRT is 7.7 days.

Facultative pond (FP). The facultative pond is placed behind the anaerobic pond connected with the later by a common dike with overflow weirs and waterfall for the increase of dissolved oxygen in its influent. The width and length (water flow length) are 200–260 m and 570 m respectively, and the water depth is in the range 1.5–3.5 m. The total area and effective volume of the FP are 116,850 m^2 and 350,550 m^3 respectively, with HRT of 13 days, in which all the anaerobic, anoxic and aerobic processes take place.

Polishing pond I (PP-I). The effluent of the FP flows into the next pond, polishing pond I through a common dike with overflow waterfalls, which make the influent and bulk water in the PP-I maintain aerobic state, and the effluent from the PP-1 meets the standards for agricultural irrigation water, and most of the irrigation water is diverted from this pond, which is thus called treatment pond for agricultural use. The water flow length in the pond is 55–100 m and the area is 11,000 m^2, effective volume 22,000 m^3, and HRT is about 1 day.

Polishing pond II (PP-II). The effluent of PP-I flows into this pond through a waterfall dike between the two polishing ponds as shown in Figure 4. Further purification takes place in this pond to further remove organic matters, nutrients and microorganisms as the last purification step of the IPS and a higher quality of final effluent was obtained.

Improvement of operating conditions by intensified measures

Biomass increase and even distribution by biofilm carrier. Our former studies indicated that the performance of anaerobic pond (AP) and the whole pond system have improved significantly by means of packing biofilm carrier in anaerobic pond due to the increased and even distribution of biomass both in forms of biofilm growing on the carrier's surface and the anaerobic active sludge in suspension and sediments (Qi et al., 1993; Wang et al., 1995). Based on this study result, the anaerobic pond in Jining City was also packed with biofilm carrier. Originally this pond was designed to be filled with biofilm carrier in the whole bulk volume; however, in construction only the front half volume of

Figure 4 The dike at the end of PP-II with overflow outfalls

the AP was filled with this carrier. Nevertheless, the performance of the settling/anaerobic pond plus intensified anaerobic pond in series was quite good with the following removal efficiencies for main pollutants: SS 76.5–84.8%(mean 81.6%); COD 25.4–64.3%(46.6%); BOD_5 45.1–59.4%(52.0%); TN 4.2–8.2%(6.1%); NH_3-N 0–35.2%(21.4%), which are comparable to intensified anaerobic pond in Anda City, Heilongjiang Province (Wang *et al.*, 1995; Qi *et al.*, 1993). The SS, BOD_5 and COD removals were even higher than those of the latter, i.e. 59.8–62.8%, 43.8–51.7% and 43.1–45.1% respectively, which is mainly ascribed to the joint operation of settling/anaerobic pond (SAP) plus intensified anaerobic pond (IAP), which has resulted in a better performance in Jining than the operation of a single intensified anaerobic pond in Anda. Besides, the even distribution and near plug-flow pattern of wastewater in the anaerobic pond by proper design of the pond configuration and their inlet and outlet distribution systems have resulted in better settlement of suspended solids and more adequate contact between pollutants present in wastewater and biomass both in forms of fixed biofilm growing on the carrier surface and activated sludge in suspension (see Figures 5–7).

Improvement of biodegradability of wastewater

The measured BOD_5/COD ratio values in each unit ponds of the IPS are shown in Figure 8, from which it was found that the biodegradability of wastewater was improved to some extent in the intensified anaerobic ponds where there exist some species of anaerobic and facultative bacteria and fungi that are capable of degrading some refractory

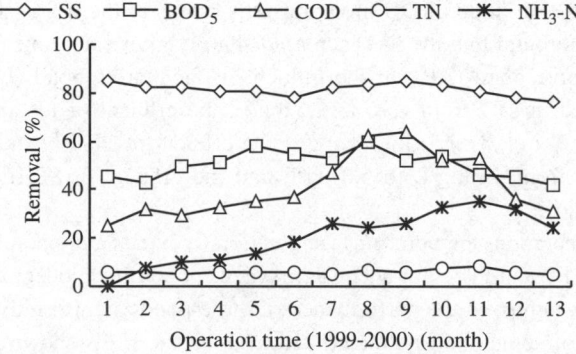

Figure 5 Variation of pollutants removals with operation time in anaerobic ponds

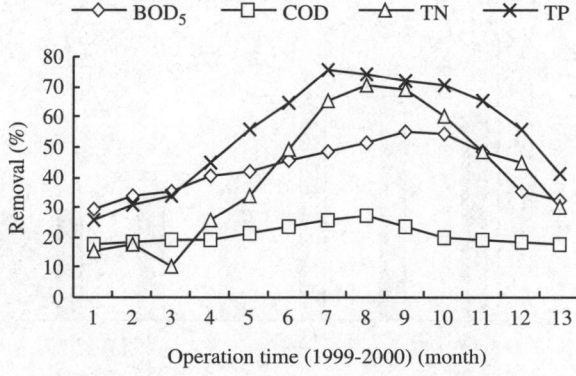

Figure 6 Variation of pollutants removal with operation time in facultative pond

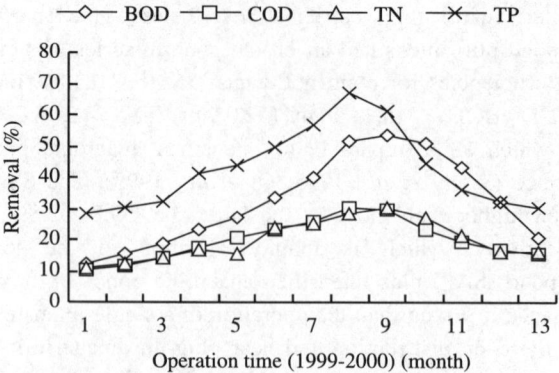

Figure 7 Variation of pollutants removal with operation time in polishing ponds

organic compounds into easily biodegradable ones, which resulted in increase of BOD_5/COD ratio values, which is similar to our earlier study (Wang et al., 1996). In both the facultative and polishing ponds, aerobic process and aerobic bacteria dominated in the treatment. Those are only capable of removing easily degradable organic substances, or BOD_5, thus resulting in a decrease in the BOD_5/COD ratio.

Increase of DO by waterfall aeration. In order to improve the performance of both the facultative and polishing ponds, the influents to these ponds are aerated by overflow waterfalls in their respective dikes by reasonably using the slope of the site where the IPS is situated. The variation of DO in various unit ponds of the IPS is shown in Figure 9, from which it was found that the DO increased sharply after each time of waterfalls. The effluent of anaerobic pond (AP), or the influent of facultative pond (FP) after waterfall aeration, increased its DO from zero to 2.5 mg/L, the effluent of FP, or influent of PP-I increased its DO from 2.8 mg/L to 3.8 mg/L; the effluent of PP-I or the influent of PP-II increased its DO from 4.2 mg/L to 5.0 mg/l; and the effluent of PP-II increased its DO from 5.0 to 6.4 mg/L.

The waterfall aeration, the pond surface aeration by winds and photosynthetic oxygenation provide the FP and PPs with adequate oxygen for aerobic biodegradation and assimilation, which has improved the performance of these ponds significantly.

The variation of some main pollutants removals in the FP is shown in Figure 6, from which it was found that the removal efficiencies of the FP for various pollutants varied with operation time or seasons substantially. In warm seasons from May to October the

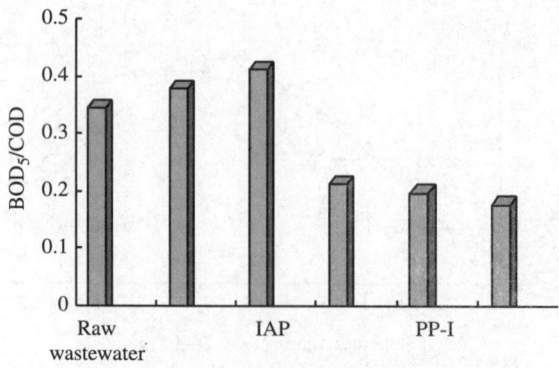

Figure 8 BOD_5/COD ratio values in each unit ponds of the IIPS

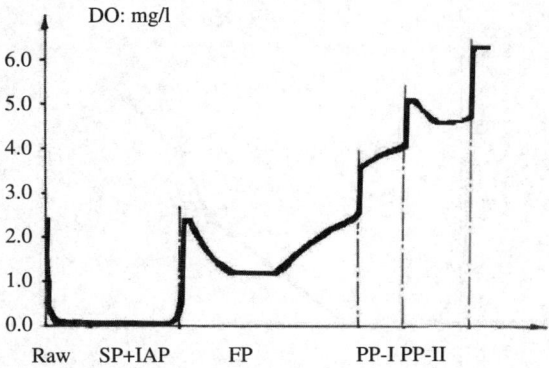

Figure 9 DO variation in the whole IPS

removal efficiencies for BOD$_5$, COD, TN and TP were much higher than those in cold seasons from November through next April. The facultative pond is a major unit pond in an IPS, being in charge of removing most or a substantial part of the organic and nutrient substances with the help of joint function of aerobic, facultative and anaerobic bacteria present in surface upper layer as aerobic zone, middle layer as anoxic zone and lower/bottom layer as anaerobic zone respectively, which degrade organic substrate into final products like CO_2, H_2O, NH_3-N and PO_4-P intermediate products like organic acids, alcohols and formaldehydes by metabolic degradation, while producing new cells of bacteria through metabolic assimilation. Meanwhile the photosynthesis of algae takes place under solar radiation as initial promoting energy, with the uptake of CO_2, NH_4^+ and PO_4^{3-}, for synthesis of new algae cells while generating oxygen through autotrophic assimilation. The recovery and increase of DO in the facultative pond from photosynthetic oxygen generation as well as surface re-aeration is remarkable as shown in Figure 9, and pH also rise significantly with CO_2 consumption by algae photosynthesis as shown in Figure 10.

In addition, the adequate dissolved oxygen promoted the establishment of a more complex eco-system consisting of some longer food chains, such as bacteria/algae → potozoa → metazoa like Rotifier, Daphnia; algae → filtering-feeding fish; algae → benthos like snails and clam that are present in the bottom in the down reach of FP, in which the excess algae and bacteria produced from metabolic assimilation are consumed by predatory organisms as mentioned above partially or sometimes completely, which control the effluent from an increase in suspend solids contributed by excess algae growth. In this respect, the gravel filtration dike (or dam) in PP II plays a very important role in removing SS as well as organic and other pollutants, in which biofilm developed on the gravel surface, which trap, filter out and adsorb suspended algae and bacteria, which made the effluent (filtrate) contain low SS. In the range of 7.2–36.8 mg/L with a mean value of 22.4 mg/L, which is much lower than that of many other pond systems (EPA-625/1-83-015, 1983; Yagoubi *et al.*, 2000). The multi-cell polishing ponds in series also positively contributed to the lower SS in effluent, which conforms to Rich's study (Rich, 1982).

Operational performance of the whole pond system

The intensive pond system (IPS) has performed better in terms of removal efficiencies for various main pollutants in comparison with conventional pond systems because of the application of intensive measures as mentioned above. The variations of influent and

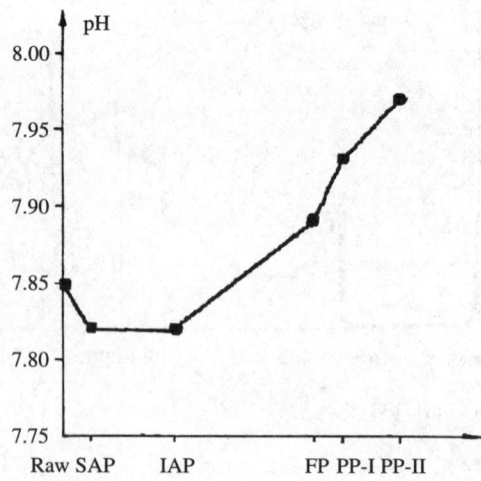

Figure 10 pH variation in the whole IPS

effluent COD and BOD, SS, TP and TN of the whole IPS are shown in Figures 11–13 respectively.

It is evident from Figures 11–13 that the effluent quality varied with season or temperature significantly except for SS. The COD was decreased from influent COD in the range of 176.2–291.1 mg/L to 44.1–112.3 mg/L; the BOD_5 decreased from influent 75.2–171.4 mg/L to effluent 8.5–35.6 mg/L; TN from influent 35.1–46.8 mg/L to effluent 12.6–24.1 mg/L; TP from influent 3.5–5.7 mg/L to effluent 1.1–2.7 mg/L.

All the data showed that the intensive pond system performed well only in warm seasons from May through October, when the effluent quality well complied with the 2nd class discharge standards for secondary wastewater treatment plants. EPA, 1983 [GB-81918-2002], i.e. the effluent SS and BOD_5 should be less than 30 mg/L respectively, COD 100 mg/L, and NH_3-N 25 mg/L. However, in cold seasons from November to April, the effluent did not cope with the discharge standards, which need a further treatment or storage with zero or partial flow discharge according to the self-purification capacity of the receiving water bodies in cold seasons. As there is no storage pond or lagoon, the effluent from polishing pond I (PP-I) or polishing pond II (PP-II) is applied on the nearby farmland for irrigation, thus being as a major irrigation water source from a city with acute water shortage all the year round.

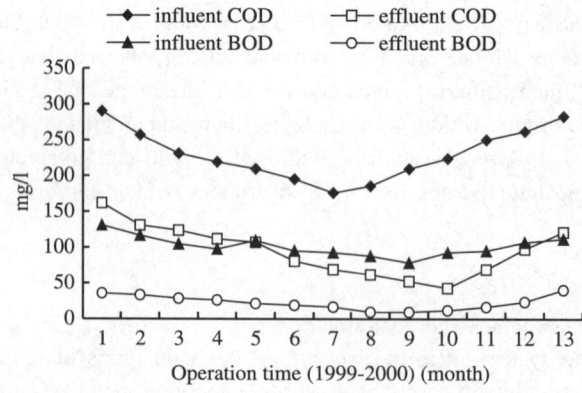

Figure 11 Variation of influent and effluent COD and BOD_5 of the whole IIPS

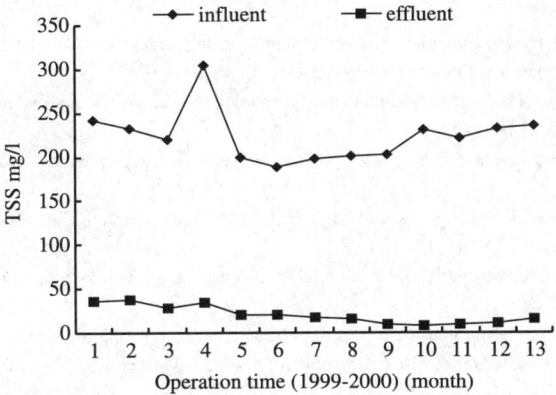

Figure 12 Variation of influent and effluent TSS of the whole IIPS with operating time

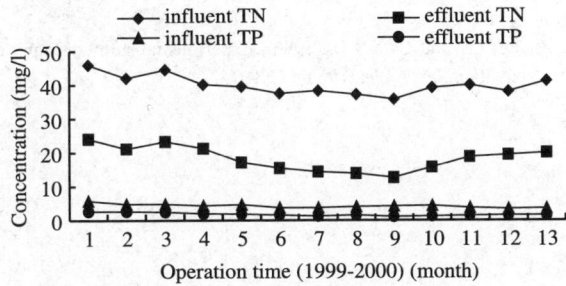

Figure 13 Variation of influent and effluent TN and TP of the whole IIPS

Conclusions

The intensive pond system (IPS) consisting of SAP, IAP packed with biofilm carrier, FP, PPs with gravel filtration dike in Jining, Inner-Mongolia performed quite well, with the final effluent in warm seasons well meeting Chinese national effluent standards (1B or 2nd class) for secondary municipal wastewater treatment plants. However, the effluent in cold seasons from November to April did not meet the standards.

The intensified measures taken in this intensive pond system, i.e. the inlet and outlet distribution systems in the IPS, package of biofilm carrier in the front half volume of IAP, the overflow waterfall dikes installed in AP, FP and PPs, and gravel filtration dike in the polishing pond (PP-I) have improved the performance remarkably, which is better compared with the conventional pond system. If this kind of IPS is built in warm areas, the final effluent all year round will well meet the national effluent standards.

The overflow waterfalls on the dikes in the IPS have increased DO substantially, from 0.8 to 2.5 mg/L after each time of waterfall aeration, which provide the facultative and polishing ponds with adequate oxygen for aerobic process and the establishment and operation of complex eco-system in the pond system.

The anaerobic ponds, the IAP in particular, improved the biodegradability of wastewater remarkably with the increase of BOD_5/COD ratio value, which is mainly contributed by some species of anaerobic and facultative bacteria capable of converting hardly degradable organics into easily degradable ones, or converting part of COD into BOD_5.

In order to make the final effluent of such kind of IPS in cold regions meet the national effluent standards, the IPS should be followed by farmland irrigation in cold seasons both for wastewater storage and water and soil conservation of the farmland.

References

EPA (1983). Design Manual–Municipal wastewater stabilization ponds, EPA-625/-1-83-015 GB-81918-2002, Chinese Integrated National Standard for Wastewater Discharge.

Ministry of Construction (1991). Standard methods for municipal wastewater quality analysis. Jinanbio (construction standard) 1991–551.

Oswald, W.J. (1991). Introduction to integrated pond wastewater ponding systems. *Wat. Sci. Tech.*, **24**(5), 1–7.

Qi, P.S., Wang, B.Z., MA, F., Zhang, J.S. and LI, T.CJ. (1993). Intensification of a pond system by fibrous carriers. *Wat. Sci. Tech.*, **28**(7), 112–123.

Rich, L.G. (1982). Influence of multicellular configurations on algae growth in aerated lagoons. *Wat. Res.*, **16**, 929–931.

Yagoubi, M., Foutlane, A., Bourchich, L., Tellal, J., Wittland, C. and Yaohioni, M.El. (2000). Study on the performance of the wastewater stabilization ponds of Soujaad, Morocco. *J. Wat. Supply Res. Technol. - AQUA*, **49**(4), 203–209.

Wang, B.Z., Qi, P.S, Dai, A.L., Shi, S.X. and Cao, X.D. (1995). Case studies on pond/land systems treating municipal wastewater in northern areas of China. *Proc. of Water Envir. Fed., 68th Annual Conf. Miai Beach, Florida. USA Oct.21–25, 1995 (WEFTEC'as), Vol. 4 Surface Water Quality and Ecology*, pp. 511–522.

Wang, B.I., Dong, W.Y., Zhang, J.L. and Cao, X.D. (1996). Experimental study of high rate pond system treating piggery wastewater. *Wat. Sci. Tech.*, **34**(11), 125–132.

Primary facultative ponds in the UK: the effect of operational parameters on performance and algal populations

K.L. Abis and D.D. Mara

School of Civil Engineering, University of Leeds, Leeds, LS2 9JT, UK (E-mail: *K.Abis@leeds.ac.uk*; *D.D.Mara@leeds.ac.uk*)

Abstract Waste stabilisation pond systems in the UK are used to treat effluents from small rural communities where there are large fluctuations in both BOD load and inflow; the facultative ponds in these systems have a wide range of hydraulic retention times: between 11–86 days. Low hydraulic retention times in UK ponds are sometimes accompanied by a high BOD loading, although some have a low BOD loading due to high inflows of dilute wastewater. It is not certain whether the performance is affected by the short hydraulic retention time or high BOD loading. A pilot-scale experiment tested the effect of hydraulic retention time (20–60 days) on primary facultative pond performance whilst keeping the BOD loading constant at 80 kg/ha d. It was found that no significant loss of performance was experienced at the test range for BOD and ammonia removal; some loss in SS removal was noted at 20 days' retention time. The effect of BOD loading on the maintenance of algal populations during winter (November(February) was tested at loadings of 50 and 80 kg/ha d. Although there was a significant difference in the concentrations of chlorophyll *a* and dissolved oxygen between the two loadings, there was no effect on performance.

Keywords Facultative ponds; hydraulic retention time; performance; temperate climate; United Kingdom; waste stabilisation ponds

Introduction

Waste stabilisation ponds for full treatment of wastewater are used in the UK for small rural communities; these wastewaters are characterised by large fluctuations in both BOD and hydraulic load. Table 1 shows the range of hydraulic retention times (HRT), and the percentage of ponds in each range, for the first pond in UK systems where data are available.

Low hydraulic retention times may be due to high hydraulic loads of very dilute wastewater from infiltration and/or run-off. Alternatively, a low HRT will result if the pond has too small a volume for the population served; in which case it will also be organically overloaded.

In full-scale pond systems it is not possible to test the effect of hydraulic retention time on performance whilst keeping the BOD loading constant. Therefore, it is not certain whether any loss in performance would be due to the low retention time or to high BOD loading.

The UK has a cold temperate climate: the average temperature during the coldest month is 2–4 °C. Some countries with a similar climate to the UK have made recommendations for the hydraulic retention time on the first pond: for example, Germany where the recommendation is ≥20 days (Schleypen, 1997) and France where the recommendation is 15–30 days (CEMAGREF, 1997). In northern states of the USA, a retention time of 60–90 days for the first pond is normally allowed (Reed *et al.*, 1988).

The experimental waste stabilisation pond site located at Esholt Wastewater Treatment Works in Bradford, West Yorkshire, has been operating since July 2000. On the site there are three primary facultative ponds, operating in parallel and fed with screened

Table 1 HRT to first (facultative) pond in UK systems (estimated from volume/design inflow)

HRT (days)	<20	20–29	30–39	40–50	>50	n
Percentage	35%	15%	15%	12%	23%	26

sewage from the host works. During the first trial (July 2000–June 2002), as reported by Abis (2002) and Abis and Mara (2003), the facultative ponds were tested at a range of BOD loadings (60–170 kg/ha d) but the HRT was not controlled. The trial concluded that 80 kg/ha d was the most likely optimal surface loading for UK facultative ponds.

This paper presents the results from the second trial (April 2003 – March 2004) where HRTs of between 20 and 60 days were tested with a fixed BOD loading of 80 kg/ha d. This was achieved by diluting the raw wastewater with tap water at different ratios before it entered the ponds. The surface BOD loading was set by the flow of the wastewater; the HRT was reduced by increasing the flow of tap water.

The results of the first trial found that the algal populations were greatly diminished during mid-winter at all of the test loadings (60–170 kg/ha d) (Abis, 2002). During the winter of this second trial, therefore, a comparison was made between BOD loadings of 50 and 80 kg/ha d with the HRT constant. The aim was to test the United States' recommendation for primary facultative ponds of ~50 kg/ha d for average winter air temperatures of 2–4 °C (US EPA, 1983) and to observe if this loading sustained an algal population.

Methods

The three primary facultative ponds (labelled A, B and C), as described by Abis and Mara (2003), were 40.6, 33.6, and 39.5 m^2 in area respectively, and 1.5 m deep. Each pond received the screened wastewater via a Watson Marlow 604S peristaltic pump. Tap water was supplied from a standpipe on the site; the flow was split three ways and then regulated at each pond inlet using diaphragm valves. The wastewater and tap water were each carried via 16 mm bore plastic tubing; both supplies were fed into the inlet structure. The inlet structure, constructed from two sections of PVC drainpipe connected by an elbow connector, allowed the flows to mix before entering the pond at a depth of 0.75 m.

The BOD of the wastewater at Esholt was very strong (~500 mg/l); to achieve a surface loading of 80 kg/ha d with this wastewater required an inflow rate of ~0.5 m^3/d; the resulting HRT was ~80 days. After dilution with tap water, higher inflow rates could be applied to the ponds thereby reducing the HRT to the required value whilst keeping the BOD loading constant. The schedule for the experimental trial is given in Table 2.

Weekly samples were collected of the influent wastewater, pond column and effluent. The DO of the wastewater and tap water was measured using a YSI model 6820 sonde probe which was also used to measure the DO of the pond profile at 0.25 m depth intervals. Ammonia, chlorophyll *a* and algal identification tests were carried out within 4 hours of sample collection; BOD and SS were analysed after overnight refrigeration. All parameters were analysed as described by Abis and Mara (2003). The free chlorine of the tap water was measured on site using Palintest DPD No.1 chlorine test tablets.

Results and discussion

Unfiltered BOD removal

The results for the BOD removal (based on unfiltered effluent samples) at a surface loading of 80 kg/ha d are given in Table 3. Within the HRT test range of 20–60 days, there was evidence of an effect on BOD removal (median test $p = 0.025$; ANOVA

Table 2 The experimental schedule

Phase	Pond	BOD loading (kg/ha d)	HRT (days)	Wastewater flow (m³/d)	Tap water flow (m³/d)	Approx BOD$_{in}$ (mg/l)
1 Apr–July 03	A	80	45	0.64	0.71	220
	B	80	60	0.53	0.32	280
	C	80	30	0.61	1.20	150
2 July–Sept 03	A	80	40	0.64	0.87	210
	B	80	40	0.53	0.75	230
	C	80	20	0.61	2.08	115
3 Oct–Dec 03	A	50	20	0.40	2.18	80
	B	80	30	0.53	1.15	150
	C	80	20	0.61	2.08	115
4 Jan–Mar 04	A	50	45	0.40	0.89	150
	B	80	40	0.53	0.75	230
	C	80	45	0.61	0.72	220

$p = 0.019$), though this was strongly influenced by the 60 days data set. With the 60 days data set excluded, the median test was $p = 0.859$ and ANOVA $p = 0.330$. The Jonckheere-Terpstra (J-T) test showed a significant increase in removal was related to increasing HRT ($p = 0.007$); but after the exclusion of the 60 days' data, the relationship was not significant ($p = 0.189$). This suggests that the HRT had no significant effect on BOD removal within the range of 20–45 days. There is justification for the omission of the 60 day set (Phase 1, Pond B) because the data were collected only during Phase 1 and had a smaller variance (5–8 times less) and lower chlorophyll a (Figure 1) than the other four groups. Algal-laden effluents, usually experienced during the summer, reduced the calculated BOD removal efficiency. The ponds with lower HRT necessarily had lower influent BOD concentrations, thus their calculated removal efficiencies were more sensitive to high concentrations of algae in the effluent.

Filtered BOD removal

Table 4 shows that, within the range 20–60 days, the HRT had no significant impact on filtered BOD removal. Thus, with the algae removed from the effluent, the 60 days' set could be included with no effect on the analysis. This is evidence that the 60 day anomaly for unfiltered BOD removal was related to the lower concentration of algal solids in the samples.

SS removal

The undiluted wastewater SS concentration was around 800 mg/l; the concentration in the tap water was assumed to be zero. Table 5 shows that there was a significant difference in SS removal across the test range; in particular the removal at 20 days was significantly

Table 3 Unfiltered BOD removals and effluent concentrations at different HRT at a BOD loading of 80 kg/ha d

HRT (days)[a]	BOD mass removal (%)				BOD effluent concentration (mg/l)				n
	Mean	95% conf.	median	quartile range	mean	95% conf.	Median	quartile range	
60	90.5	±2.1	91.2	88.6–93.2	29	±7	27	21–35	11
45	84.6	±3.3	83.3	79.3–92.2	34	±7	36	17–48	23
40	84.8	±2.4	86.6	73.9–89.6	33	±6	29	22–45	34
30	81.6	±4.1	83.7	75.5–89.2	30	±7	27	17–39	21
20	81.9	±4.2	82.2	73.3–88.8	24	±6	23	15–32	20

[a]All sets approximately normal; HRT = 60 days variance small

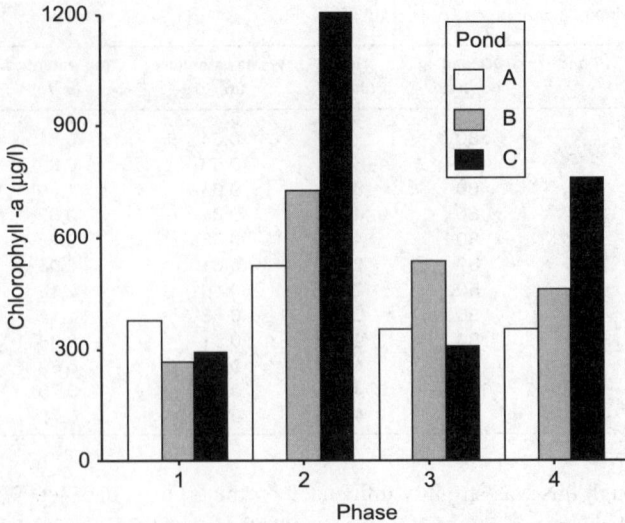

Figure 1 The mean chlorophyll-a concentration in effluent samples

worse than at ≥ 30 days and the removal at 60 days significantly better than the rest. The better removal for the 60 day set was explained in the previous section. The removal data for 20 days' HRT had much greater variability than for the other sets. This may have been a sign of unreliable performance; however, it was strongly affected by the relatively high average concentration of algae in Pond C effluent during Phase 2 (chlorophyll $a = 1254\,\mu g/l$), as shown in Figure 1. These high concentrations of algae in the effluent resulted in SS effluent concentrations of $\sim 100\,mg/l$ during July and August 2003. Like BOD removal, the SS removal calculation was much more sensitive to increases in outlet concentrations at low HRT; the SS effluent concentrations did not vary significantly across the test range.

Ammonia removal

The ammonia concentration in the raw wastewater was $\sim 40\,mg\,N/l$; the ammonia in the tap water was zero. The tap water had a free chlorine concentration of 0.1 mg/l which is insignificant in terms of ammonia-oxidising capacity (7.6 mol of Cl_2 are required to destroy 1 mol of ammonia-N). The results, as shown in Table 6, show that at the test range of HRTs (20–60 days) there was no significant difference in ammonia removal. The ammonia concentration in the effluent reduced relative to the reducing influent concentration.

Table 4 Filtered BOD removals and effluent concentrations at different HRT at a BOD loading of 80 kg/had

HRT (days)	BOD mass removal (filtered effluent) (%)[a]				BOD filtered effluent concentration				n
	mean	95% conf.	Median	quartile range	mean	95% conf.	median	quartile range	
60	96.6	±0.5	96.4	95.8–97.4	11	±1.7	11	8–13	11
45	96.4	±0.7	96.7	95.4–97.7	8	±1.5	8	5–11	23
40[b]	96.4	±0.7	96.9	95.4–97.8	8	±1.5	7	5–10	32
30	96.3	±0.5	96.4	95.4–97.2	6	±1.1	6	4.5–7	21
20	97.3	±0.5	97.3	96.4–98.3	3	±0.6	3	2–4	21

[a]Median test: $p = 0.313$; ANOVA: $p = 0.220$
[b]Data set not normally distributed

Table 5 SS removals and effluent concentrations at different HRT at a BOD loading of 80 kg/ha d

HRT (days)	SS mass removal (%)[a]				SS effluent concentration (mg/l)[b]				n
	mean	95% conf.	median	quartile range	mean	95% conf.	Median	Quartile range	
60	93.0	±2.3	93.3	89.8–96.6	39	±12	37	21–57	11
45	84.8	±4.3	84.6	78.6–93.2	48	±14	47	21–80	23
40	81.6	±3.5	83.3	72.0–88.7	52	±11	49	32–83	36
30	80.8	±4.6	82.2	73.3–89.5	43	±11	35	22–59	21
20	61.5	±10.0	58.5	42.4–86.8	57	±15	62	19–86	22

[a]HRT = 20 and 60 days have unequal variance: median test $p = 0.002$. J-T Test $p < 0.001$
[b]Assumptions of normality and equal variance hold: ANOVA $p = 0.367$

In the previous trial a difference in ammonia removal between the summer and winter months was found (Abis and Mara, 2003). This phenomenon was not noted in this trial ($p = 0.732$): the average removal was around 50% during both summer and winter. Unlike the previous trial, the ammonia concentration in the influent did not vary seasonally. In addition, the DO and chlorophyll a concentrations were higher than in previous winters.

Effect of BOD surface loading on algal populations and DO during winter

The effect of surface BOD loading on the algal population during the winter months, as given by the chlorophyll a concentration, was tested at two loadings: 50 and 80 kg/ha d. The previous trial had showed that the chlorophyll a concentration declined to below 100 μg/l during December, January and February at loadings 60–170 kg/ha d; the rate of decline appeared to be related to BOD loading. At higher loadings, in excess of 100 kg/ha d, the chlorophyll a concentration reduced to below 10 μg/l during these months. However, the pond with the lowest loading (60 kg/ha d) had been affected by blooms of algal predators such as rotifers and *Paramecium*; therefore it was not certain if the reduction in chlorophyll a at 60 kg/ha d during winter was due to the loading or to algal predation (Abis, 2002).

The results of this trial showed that the algal population was more stable, and the DO concentration generally higher, in Pond A loaded at 50 kg/ha d, than in the other two ponds loaded at 80 kg/ha d (Figures 2 and 3). During Phase 3, the ponds appeared to be better oxygenated compared to previous early winters; it was thought that this was perhaps due to higher inflows of of DO-saturated tap water (the wastewater DO was ~20%). Consequently in Phase 4 (January–March 2004) the relative flow of tap water was reduced (see Table 2: the resulting HRTs were 40–45 days). Despite this reduction in tap water flow, the DO in the water column did not reduce (Figure 3).

Table 6 The ammonia removals and effluent concentrations

HRT (days)	Ammonia-N mass removal[a] (%)				Ammonia-N effluent[b] (mg/l)				n
	Mean	95% conf. interval	median	quartile range	mean	95% c.i.	median	quartile range	
60	56	49–64	60	46–63	12	10–14	13	11–14	11
45	53	43–63	52	33–69	7.7	6.0–9.4	8.0	5.0–11	21
40	53	46–60	56	36–71	7.8	6.6–8.9	7.5	5.0–11	36
30	44	37–51	49	34–56	7.4	6.5–8.4	7.0	6.0–9.0	18
20	53	40–66	60	33–72	4.4	3.2–5.7	3.6	2.5–6.5	22

[a]Assumption of normality holds, though HRT = 20 and HRT = 60 have unequal variances: ANOVA $p = 0.537$; median $p = 0.260$; J-T test $p = 0.525$
[b]Assumptions of normality and equal variance hold: ANOVA $p < 0.001$; J-T test $p < 0.001$

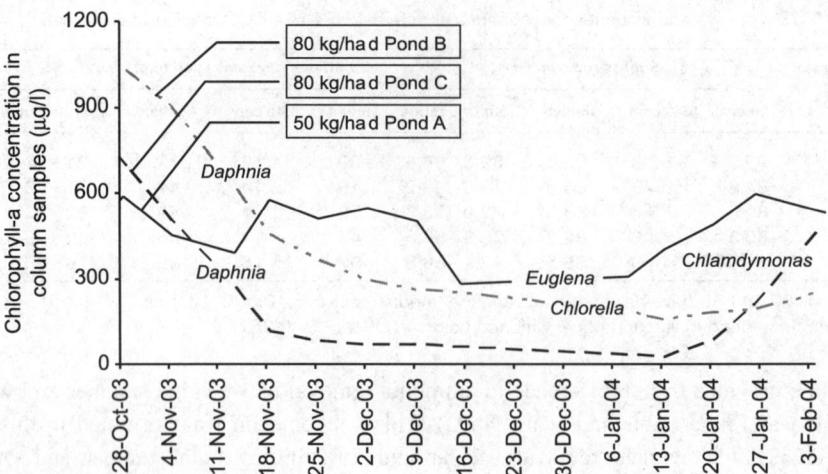

Figure 2 Weekly chlorophyll *a* concentrations in pond column samples during winter 2003/2004. The lines are 2-point moving average trend lines

Daphnia was present in both Ponds B and C during November causing a decline in chlorophyll *a*. The *Daphnia* bloom was most intense in Pond C and consequently the chlorophyll concentration was lower than in Pond B (at the same loading). There were no dominant algal genera in Pond C until the end of January, when *Chlamdymonas* sp. began to flourish. Pond B was dominated by *Chlorella* sp. throughout the winter, the same dominant genus was observed during previous winters at this loading (Abis, 2002). Pond A, loaded at 50 kg/ha d, did not have any *Daphnia* sp. during the autumn and was dominated by *Euglena* sp. throughout the winter. These observations, together with those from previous years, suggest that the effect of BOD loading (especially below 80 kg/ha d) on algal populations may be secondary to the effect of predation.

These results call into question the benefits of maintaining an algal population during the winter in UK facultative ponds. The data suggest that the algae contributed very little to the surface DO between November and January: Ponds B and C had the same surface DO concentration despite having different chlorophyll *a* concentrations (Figures 2 and 3). There was no significant difference in performance between loadings

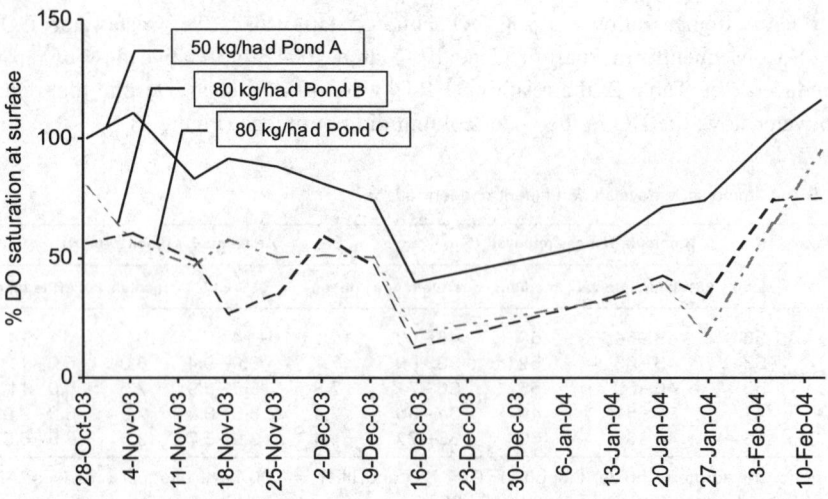

Figure 3 The weekly midday surface DO concentrations. The lines are 2-point moving average trend lines

Table 7 The performance of the ponds loaded at 50 and 80 kg/ha.d during Phases 3 and 4

kg/ha d	BOD removal		SS removal		Ammonia removal	
	50	80	50	80	50	80
mean (%)	88.3	87.2	81.4	83.7	39.6	50.2
median (%)	90.9	88.5	85.7	86.7	46.6	50.8
median test p	0.154		1.000		0.602	

of 50 and 80 kg/ha d during Phases 3 and 4 (Table 7). These findings suggest that the DO introduced from the influent, combined with wind action, was sufficient for the slow rate of aerobic digestion in the cold water; and/or anoxic biological processes played an important role to maintain the high performance of the pond.

Conclusions

The experiments showed there was no loss in performance for BOD and ammonia on reducing the HRT to 20 days. For SS removal there was some loss in performance between 20–30 days' HRT, though this effect was not conclusive due to the effect of algae in the effluent. A significant difference was observed in the DO and chlorophyll *a* concentrations between ponds loaded at 50 and 80 kg/ha d during the winter, though this had no effect on performance. The data from four years' testing suggest that UK primary facultative ponds could be loaded at around 80 kg/ha d and may have a minimum HRT of 20 days. Although algae may not be present during the winter at this loading, this appears to have no detrimental effect. It is not yet known if pond performance can be maintained at an HRT of less than 20 days.

Acknowledgements

The authors gratefully acknowledge the contributions of The Leverhulme Trust and Yorkshire Water to this research.

References

Abis, K.L. (2002). *The Performance of Facultative Waste Stabilisation Ponds in the United Kingdom*, PhD thesis, School of Civil Engineering, University of Leeds, (Available at http://www.leeds.ac.uk/civil/ceri/water/ukponds/publicat/theses/abis/abis.html).

Abis, K.L. and Mara, D.D. (2003). Research on waste stabilisation ponds in the United Kingdom: initial results from pilot-scale facultative ponds. *Wat. Sci. Tech.*, **48**(2), 1–7.

CEMAGREF (1997). *Le Lagunage Naturel: Les Leçons Tirées de 15 Ans de Pratique en France*, Cemagref, Lyon; France.

Reed, S.C., Middlebrooks, E.J. and Crites, R.W. (1988). *Natural Systems for Waste Management and Treatment*, McGraw Hill, New York.

Schleypen, P. (1997). Abwasserteichenlagen. In *Biologiche und Weitergehende Abwasserreinigung ATV Handbuch*, Ernst & Sohn, Berlin, pp. 60–103.

US Environmental Protection Agency (1983). *Design Manual: Municipal Wastewater Stabilization Ponds*, US EPA, Cincinnati, OH, USA.

Implications for physical design: the effect of depth on the performance of waste stabilization ponds

H.W. Pearson*, S.T. Silva Athayde**, G.B. Athayde Jr.*** and S.A. Silva**

*LARHISA, Department of Engenharia Civil, Universidade Federal do Rio Grande Do Norte, Natal, RN. Brazil
(E-mail: howard_william@uol.com.br)
**EXTRABES, Department of Engenharia Civil, Universidade Federal de Campina Grande, Campina Grande, Pb. Brazil (E-mail: sanselmosilva@uol.com.br)
***Department of Engenharia Civil, Universidade Federal da Paraíba, Joao Pessoa, Pb. Brazil

Abstract Studies on experimental primary facultative ponds showed that varying the depth from 1.25 m to 2.3 m had no effect on the rates of BOD removal. In contrast k values for FC removal rates were higher in the shallower (1.25 m) facultative ponds. The risk of odour release via H_2S production was higher in the 2.2 m ponds than the 1.25 m ponds and NH_3 removal was much better in the 1.25 m facultative ponds. A comparison of the efficiency of shallow 5-pond series (1.0 m and 0.61 m deep) with a 2.2 m deep series showed that the shallow systems were more efficient at FC removal, but the deeper series actually saved land area for the same FC final effluent quality under tropical conditions. However, efficient nutrient removal (N and P) only occurred in the shallow series and effluent standards for nutrient concentrations are unlikely to be met by 2.2 m deep 5-pond series in contrast to the norms for pathogen removal.
Keywords Waste stabilization ponds; depth; removal efficiency; FC; BOD; H_2S; NH_3

Introduction

Our knowledge of pond processes for example: the mechanisms of biological disinfection, and the influence of algae on both FC (Pearson *et al.*, 1987; Curtis *et al.*, 1992; Brissaud *et al.*, 2000; Davies-Colley *et al.*, 2000), and BOD removal processes, factors controlling algal growth and oxygen production and the influence of organic loading on pond performance has increased substantially in the last decade or so. Similarly the impact of baffles (Polprasert and Agarwalla, 1995; McLean *et al.*, 2000), inlet and outlet design on the hydraulic regime (Shilton and Harrison, 2002), have been evaluated and the anaerobic processes underpinning sludge stabilization and odour production in ponds are increasingly better understood (Paing *et al.*, 2000).

This paper investigates the impact of depth on treatment efficiency in primary facultative ponds and 5- pond series and discusses the implications for physical design.

Material and methods

Description of the experimental pond systems

The experimental pond systems were located at the Wastewater Treatment Experimental Station (EXTRABES) of the Federal University of Campina Grande (7°13′ 11″ S, 35°52′ 31″ W, 550 m above m.s.l. in the state of Paraíba, N.E. Brazil. The annual mean water temperature was 25 °C. Four identical primary facultative ponds (26 m × 7.5 m) were operated at depths of 1.25 m and 2.30 m. The 5-pond series comprised an anaerobic pond, a secondary facultative pond and then 3 maturation ponds (10 m × 3.35 m) in series. When the system was operated in shallow mode the depth of the anaerobic pond was 1.25 m and the facultative and maturation ponds 1.0 m. When operated in deeper mode all the ponds had a depth of 2.20 m. The innovative system is described in detail elsewhere (Pearson *et al.*, 1995) but basically also comprised 5 ponds in series but with very shallow maturation ponds (0.4–0.9 m deep). Surface organic loadings varied with the experiment.

Sampling and analyses. The effluent of the primary facultative ponds and the two 5-pond series were sampled at 08.00 h which gave values equivalent to the mean effluent concentration over a 24 h period (Silva, 1982). Column samples were used to measure mean effluent quality in the innovative 5-pond series (Pearson *et al.*, 1995). Analyses for BOD, faecal coliforms (FC), ammonia and orthophosphate were as described in *Standard Methods* (APHA, 1995).

Results and discussion
The impact of depth on pond performance

Data for primary facultative ponds designed on the basis of surface organic loading showed that almost doubling the depth had no effect on BOD removal efficiency as shown in Figure 1 where data for 1.25 m and 2.3 m deep primary facultative ponds are plotted as a continuum. Thus the longer hydraulic retention times of the deeper ponds did nothing to improve BOD removal for the same loading range.

However, there was a negative impact of increased depth on the rate of FC removal (Figure 2), since the shallow primary facultative ponds showed significantly higher k values for first-order FC removal compared to the deeper ponds over the same range of surface loadings.

The deeper facultative ponds also produced higher concentrations of H_2S (and total sulphide) in relation to surface organic loading than the shallower ones (Figure 3) and thus the risk of odour release from deeper ponds is potentially higher particularly when the ponds de-stratify and turnover at night.

In contrast to H_2S, NH_3 concentrations were higher in the shallow facultative ponds (Figure 4) even though the total ammonia concentration was higher in the deeper ponds and this was a result of the higher mean pH values recorded for the surface water layers in the 1.25 m deep facultative ponds compared to the deeper ones. Thus ammonia volatilization and hence nitrogen removal will generally be higher from shallow facultative ponds.

Therefore whilst BOD removal efficiency is apparently unaffected by depth at least in the range of 1.25 m to 2.30 m, rates of FC removal and NH_3 volatilization will be better

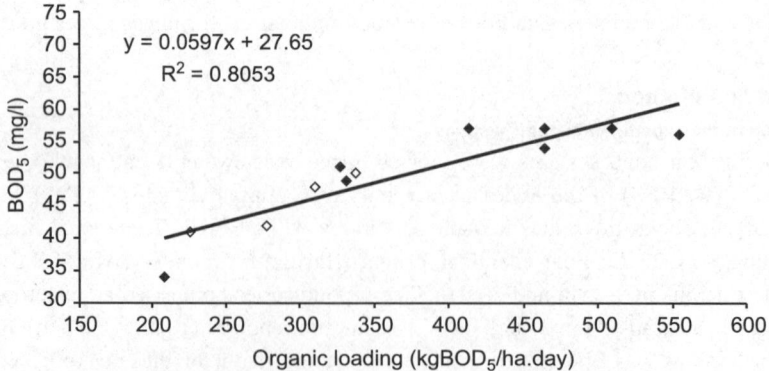

Figure 1 Effluent BOD_5 plotted as a continuum against surface organic loading in the primary facultative ponds. The open diamonds are data for a depth of 2.3 m and the solid diamonds for a depth of 1.25 m. Some solid data points represent values for both 1.25 m and 2.3 m depths

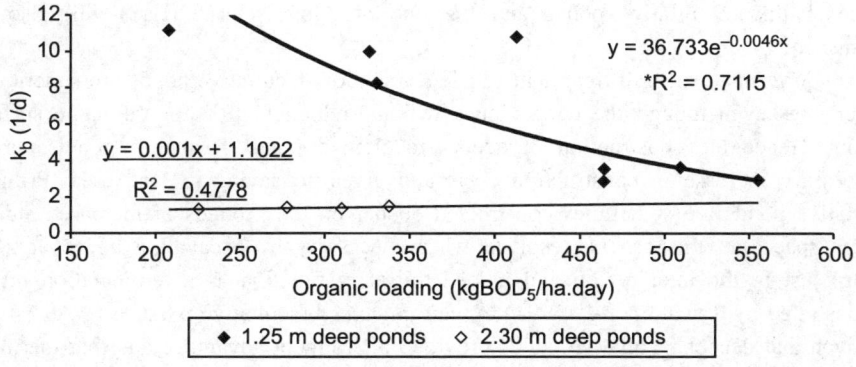

Figure 2 Faecal coliform removal (k_b) plotted against surface organic loading in the primary facultative ponds

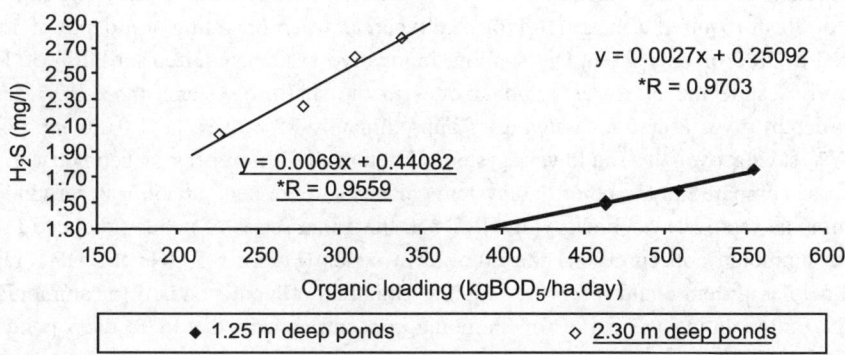

Figure 3 H_2S conc. versus surface organic loading in the primary facultative ponds

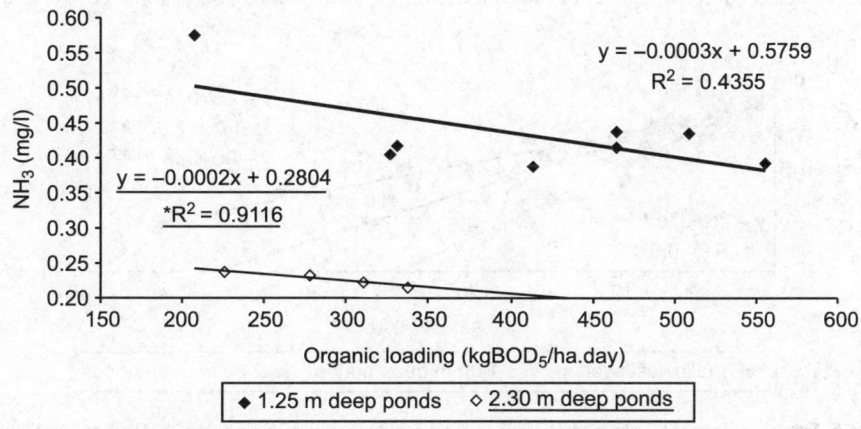

Figure 4 NH_3 plotted against surface organic loading in the primary facultative ponds

in the shallow facultative ponds and the risk of odour release (H_2S) will also be diminished.

Thus in terms of overall treatment efficiency shallow facultative ponds are superior to deeper ones even though the deeper ones will have longer HRT for the same surface loading. Hence the construction of deeper facultative ponds frequently represents an unnecessary increase in construction costs and gives no saving in land area. Primary facultative ponds are sometimes constructed deeper on the grounds of increased sludge storage space but this can be resolved where necessary by excavating a deeper zone (2.5 m) just in the inlet region and indeed this may function as a fermentation pit as demonstrated by Oswald et al., (1994) in their *advanced* facultative pond.

Given that deeper facultative ponds provide no treatment advantage and that the main role of maturation ponds is basically one of pathogen and further nutrient removal then one can think in terms of shallow versus deeper pond series. The data now discussed are for 5-pond series. Figure 5 shows that FC removal was exponential in all series but was more efficient in the shallow pond systems. The shallowest pond series was the one termed the innovative system i.e. the series comprising maturation ponds of various depths (between 0.9 m and 0.3 m) but with a mean depth of only 0.6 m.

To obtain an effluent concentration of 10^3 FC/100 ml the pond series operating at a depth of 2.2 m required a mean HRT of 32 d whereas when operating at a depth of 1.0 m required a HRT of only 21 d. The shallow innovative system required an HRT of 14 d. However, despite the increased removal rates in the shallow systems the deeper series (2.2 m depth) gives a land area saving of approximately 35% over the 1.0 m deep ponds and 17% saving over the shallowest system although the differences in geometry in the latter case compared to the other two systems makes a direct comparison less certain.

Ammonia removal was linear with HRT and the 1.0 m series of ponds produced lower effluent ammonia concentrations than the 2.2 m series (Figure 6). Over the HRT range tested neither system could reach the required Brazilian effluent standard for ammonia of 5 mg/L. Extrapolating the curve for ammonia concentration in the 1.0 m deep ponds in Figure 6, a HRT of approximately 33 days would be required to reach the 5 mg/L standard but in the case of the 2.2 m deep pond series the HRT would be off-scale.

The best removal rates for ammonia (89.6%) and phosphorus (50%) were obtained in the shallow innovative series operating at a reduced organic loading rate on the secondary facultative pond of 250 kgBOD$_5$/ha/d (100 kgBOD$_5$/ha/d less than the normal loading level for 25 °C). In this case the final effluent ammonia concentration was 3 mg/L thus meeting the Brazilian, CONOMA standard and the effluent orthophosphate concentration was 1.1 mg/L. Increasing the organic loading to 770 kgBOD$_5$/ha/d reduced N and P

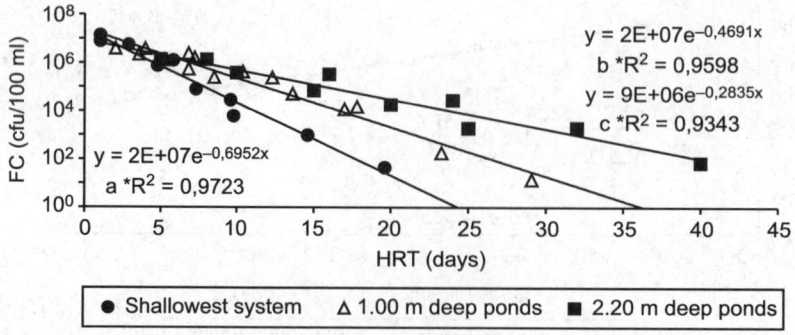

Figure 5 Faecal coliform (FC) concentration plotted against hydraulic retention time (HRT) in the shallowest series (a), 1.00 m deep series (b) and 2.20 m deep series (c)

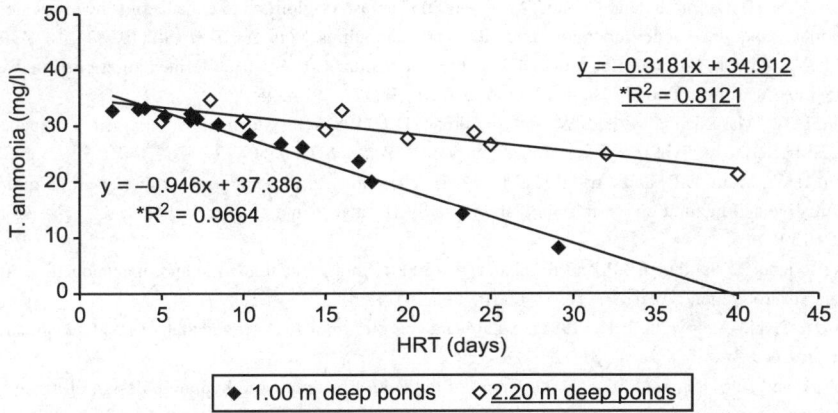

Figure 6 Total effluent ammonia concentration against HRT at the series of ponds 1.00 m and 2.20 m deep

removal efficiency in the system to 67% and 11.5% respectively (but of course also reduced the HRT).

Conclusions

BOD_5 removal was similar in 1.25 m and 2.3 m deep primary facultative ponds but FC and NH_3 removals were better in the 1.25 m ponds.

There is a greater risk of odour release from 2.3 m deep facultative ponds than shallower ones as pH_2S values were higher in the former. Building deeper facultative ponds therefore has no advantage over shallower (1.25 m deep) ones.

Good microbiological effluent quality can be obtained with a series of 2.2 m deep maturation ponds and this actually saves land area compared to shallow systems despite the longer HRT required.

However, knowing that deeper ponds save space for the same HRT, designers may use existing equations to determine the minimum HRT of a system to obtain the desired effluent quality and then increase the depths of the maturation ponds to achieve this HRT to save space without realizing the error (as has happened in the past).

At a mean ambient temperature of $\geq 25\,°C$ under tropical conditions, 2.2 m deep pond series can be used provided that the HRT of the maturation ponds calculated by conventional means is multiplied by a factor of 1.4 to compensate for reduced FC removal efficiencies. This still saves land area. This may be important in effluent reuse situations where FC concentrations must be low but nutrient levels can remain high. However, if maximum removal efficiencies of N and P as well as FC are required then shallow maturation ponds/series should be used in preference to deeper ones.

References

APHA (1995). *Standard Methods for the Examination of Water and Wastewater*, 19th edn., American Public Health Association, Washington, DC.

Brissaud, F., Lazarova, V., Ducoup, C., Joseph, C., Levine, B. and Tournoud, M.G. (2000). Hydrodynamic behavior and faecal coliform removal in a maturation pond. *Wat. Sci. Tech.*, **42**(10–11), 119–126.

Curtis, T.P., Mara, D.D. and Silva, S.A. (1992). Influence of pH, oxygen and humic substances on the ability of sunlight to damage fecal coliforms in waste stabilization pond water. *Appl. Environ. Microbiol.*, **58**, 1335–1343.

Davies-Colley, R.J., Donnison, A.M. and Speed, D.J. (2000). Towards a mechanistic understanding of pond disinfection. *Wat. Sci. Tech.*, **42**(10), 149–158.

McLean, B.M., Baskaran, K. and Connor, M.A. (2000). The use of biofilms to enhance nitrification rates in lagoons: experience under laboratory and pilot-scale conditions. *Wat. Sci. Tech.*, **42**(10–11), 187–194.

Oswald, W.J., Green, F.B. and Lundquist, T.J. (1994). Performance of methane fermentation pits in advanced integrated wastewater pond systems. *Wat. Sci. Tech.*, **30**(12), 287–295.

Pearson, H.W., Mara, D.D., Mills, S.W. and Smallman, D.J. (1987). Physiochemical parameters influencing faecal bacterial survival in waste stabilization ponds. *Wat. Sci. Tech.*, **19**(12), 131–140.

Pearson, H.W., Mara, D.D. and Arridge, H.M. (1995). The influence of pond geometry and configuration on facultative and maturation waste stabilization pond performance and efficiency. *Wat. Sci. Tech.*, **31**(12), 129–139.

Paing, J., Picot, B., Sambuco, J.P. and Rambaud, A. (2000). Sludge accumulation and methanogenic activity in an anaerobic pond. *Wat. Sci. Tech.*, **42**(10–11), 247–255.

Polprasert, C. and Agarwalla, B.K. (1995). Significance of biofilm activity in facultative pond design and performance. *Wat. Sci. Tech.*, **31**(12), 119–128.

Shilton, A. and Harrison, J. (2002). Development of guidelines for improved hydraulic design of waste stabilization ponds. *5TH International IWA Specialist Group Conference on Waste Stabilisation ponds – pond technology for the new millennium.* Auckland, New Zealand, pp. 469–476..

Silva, S.A. (1982). On the treatment of domestic sewage in waste stabilization ponds in N.E. Brazil. PhD thesis, University of Dundee, UK.

Escherichia coli removal in waste stabilization ponds: a comparison of modern and classical designs

C.G. Banda, P.A. Sleigh and D.D. Mara

School of Civil Engineering, University of Leeds, Leeds LS2 9JT, UK
(E-mail: *cen2cbg@leeds.ac.uk; p.a.sleigh@leeds.ac.uk; d.d.mara@leeds.ac.uk*)

Abstract Two PC-based waste stabilization pond design procedures, based on parameter uncertainty and 10,000-trial Monte Carlo simulations, were developed for a series of anaerobic, facultative and maturation ponds to produce ≤ 1000 *E. coli* per 100 ml for both 50% and 95% compliance. One procedure was based on the classical Marais equations and the other on the modern von Sperling equations. For the range of parameter variations selected the classical design procedure required less land area and had a shorter hydraulic retention time than the modern design procedure. For both procedures the design for 90% compliance required substantially more land and a longer retention time than the design for 50% compliance. Regulators and designers should seek a balance between system reliability (as set by the percentage compliance specified or adopted) and system costs, especially (but not only) in developing countries. It is recommended that new waste stabilization pond (WSP) systems be designed for compliance with a given *E. coli* effluent requirement by the classical procedure and that existing overloaded WSP systems be upgraded using the modern procedure.

Keywords Design; *E. coli*; Monte Carlo; removal; waste stabilization ponds

Introduction

The 'classical' design of waste stabilization ponds (WSP) is described in several WSP design manuals and books (e.g., Mara, 1997, 2004; Mara and Pearson, 1998). It essentially uses empirical and first-order equations, the design parameters for which (such as wastewater flow, BOD and *E. coli* numbers, temperature and net evaporation) are the designer's best estimates for the design brief (which is generally to design the WSP for conditions at the end of a 10–20 year planning horizon). However, as pointed out by von Sperling (1996), these best estimates are not generally known with any degree of certainty, and he accordingly introduced the concept of parameter 'uncertainty', in which a range of likely values (rather than a single 'best estimate' value) is selected for each design parameter (although any parameter can be assigned a single fixed value – for example, pond depths). The pond is then designed by a PC-based multi-trial Monte Carlo design procedure, which selects for each trial a random value for each parameter from within its specified range and executes the design for these randomly selected parameter values; it repeats this for the required number of trials (usually 10,000), and then outputs the results of the design based on the median (or, if required, the 95 or any other percentile) value of the required output parameters (such as pond areas, effluent BOD and *E. coli* numbers).

Classical WSP design uses the method of Marais (1974) for *E. coli* removal in WSP, which is satisfactory when the ponds are not overloaded (Pearson *et al.*, 1995, 1996), but often they become overloaded and *E. coli* removal decreases. Under such conditions it may be more appropriate to use the procedure of von Sperling (1999) which is based on the dispersed-flow equation of Wehner and Wilhelm (1956), together with an empirical equation for the dispersed-flow first-order rate constant for *E. coli* removal derived from his survey of 33 WSP systems in tropical and subtropical Brazil.

In this paper we combine von Sperling's uncertainty principle (which he applied only to facultative ponds) and his *E. coli* removal equations to the design of complete WSP systems (i.e., anaerobic, facultative and maturation ponds) to produce effluents suitable for unrestricted irrigation (i.e., ≤1000 *E. coli* per 100 ml; WHO, 2005). We term this 'modern' WSP design, and we compare the results obtained by this design procedure with those from classical pond design. We also apply parameter uncertainty to the classical pond design procedure.

WSP design procedures
Modern WSP design

Anaerobic and facultative ponds. The design of these two ponds is the classical design procedure (Mara, 1997) converted to a 10,000-trial Monte Carlo design procedure (Sleigh and Mara, 2003; Gawasiri, 2003), following the recommendation of von Sperling (1996). The design equations, the notation for which is listed in Table 1, are:

1. Anaerobic pond:

(a) volumetric organic loading:

$\lambda_v = L_i Q / V_a$

For $10 \leq T \leq 20$: $\lambda_v = 20T - 10$

For $20 < T \leq 25$: $\lambda_v = 10T + 100$

(b) retention time (subject to a minimum value of 1 day):

$\theta_a = V_a / Q$

(c) percentage BOD removal:

$\lambda_v = 2T + 20$

2. Facultative pond:

(a) surface BOD loading (Mara, 1987):

$\lambda_{s(f)} = 10 L_i Q / A_f$

$\lambda_{s(f)} = 350(1.107 - 0.002T)^{T-25}$

Table 1 Notation used in WSP design equations

Symbol	Parameter	Subscripts:	
a	$= (1 + 4k_{B(T)}\theta\delta)$	a	anaerobic pond
A	Mid-depth pond area, m^2	e	effluent
B	Pond breadth, m	f	facultative pond
D	Pond depth, m	f/m	facultative or maturation pond
e	Net evaporation, mm d^{-1}	i	influent
$k_{B(T)}$	First-order rate constant for *E. coli* removal, d^{-1}	m	maturation pond
L	BOD, mg l^{-1}; pond length, m	m1	first maturation pond
n	Number of maturation ponds		
N	Number of *E. coli* per 100 ml		
Q	Flow, m^3 d^{-1}		
T	Temperature, °C		
V	Pond volume, m^3		
δ	Dispersion number		
θ	Mean hydraulic retention time, d		
λ_s	Surface BOD loading, kg ha^{-1} d^{-1}		
λ_v	Volumetric BOD loading, g m^{-3} d^{-1}		

(b) retention time (subject to a minimum value of 4 days):

$\theta_f = 2A_f D_f (2Q - 0.001 e A_f)$

(c) effluent BOD:

$L_{e(f)} = L_i / [1 + 0.1(1.05)^{T-20} \theta_f]$

(d) effluent flow:

$Q_e = Q_i - 0.001 e A_f$

E. coli removal. (a) anaerobic pond: *E. coli* removal in the anaerobic pond was estimated from the equations given by Mara (2004):

$N_e = N_i / (1 + k_{B(T)} \theta_a)$

$k_{B(T)} = 2(1.07)^{T-20}$

(b) facultative and maturation ponds: the equations recommended by von Sperling (1999, 2003) were used:

$N_e = N_i [4a/(1+a)^2] \exp[(1-a)/2\delta]$

where $a = (1. + 4k_{B(T)} \theta_{f/m} \delta)^{1/2}$

$\delta = (L_{f/m}/B_{f/m})^{-1}$

$k_{B(T)} = 0.92(D_{f/m})^{-0.88} \theta^{-0.33} (1.07)^{T-20}$

Maturation ponds. (a) retention time: minimum value of 3 days (Marais, 1974), with the retention time in the first maturation pond additionally governed by the constraint that the BOD surface loading on it should not exceed 75% of that on the facultative pond (Mara, 1997). Therefore:

$(\theta_{m1})_{min} = 10 L_{e(f)} D_m / 0.75 \lambda_{s(f)}$

(b) area:

$A_m = 2 Q_i \theta_m / (2 D_m - 0.001 e \theta_m)$

(c) effluent flow:

$Q_e = Q_i - 0.001 e A_m$

Computer program. The Microsoft Excel-based computer program "Modern WSP Series Design" (available as freeware at www.efm.leeds.ac.uk/CIVE/Mcarlo/) was written in Visual Basic. The run time for a 10,000-trial Monte Carlo design simulation is ~45 seconds on a modern PC.

Classical WSP design

Fixed parameter values. (a) Anaerobic, facultative and maturation ponds: the equations given above, except those for *E. coli* removal, were used with the fixed parameter values given in Table 2 (which, in practice, would be the designer's "best estimates"); (b) *E. coli* removal: the equations of Marais (1974) were used as detailed in Mara (1997):

$N_e = N_i / (1 + k_{B(T)} \theta_a)(1 + k_{B(T)} \theta_f)(1 + k_{B(T)} \theta_{m1(min)})(1 + k_{B(T)} \theta_m)^n$

$k_{B(T)} = 2.6(1.19)^{T-20}$

Table 2 Fixed parameter values and percentage parameter variations used in the classical and modern design procedures

Parameter	Value	Variation
Population	100,000	±10%
BOD, g (p.e.)$^{-1}$ d^{-1}	40	±20%
Flow, l (p.e.)$^{-1}$ d^{-1}	120	±20%
E. coli per 100 ml raw wastewater	5 × 10^7	±50%
Temperature, °C	20	±10%
Net evaporation, mm d^{-1}	4	±20%
Pond depths, m:		
Anaerobic pond	3	0
Facultative pond	1.5	0
Maturation ponds	1	0
Minimum retention times, d		
Anaerobic pond	1	0
Facultative pond	4	0
Maturation ponds	3	0

Parameter variation. Here von Sperling's uncertainty principle was applied to the classical WSP design procedure using the computer program "Classical WSP Series Design" (also available as freeware at the above website) and the parameter percentage variations given in Table 2.

WSP design results and discussion

Classical WSP design

Fixed parameter values. The classical design procedure was done to achieve an effluent *E. coli* count of ≤ 1000 per 100 ml. The design output is given in Table 3. The total mid-depth pond area is 18.3 ha and the overall retention time 19 days.

Parameter variation. The design output for an effluent *E. coli* count of ≤ 1000 per 100 ml is given in Tables 4 and 5 for 50% and 95% compliance, respectively. The total

Table 3 Output of classical WSP design procedure with fixed parameter values

Pond	Area (ha)	Retention time (d)	Effluent BOD (mg l^{-1})	Effluent E. coli (per 100 ml)
Anaerobic	0.44	1.1	133	1.3 × 10^7
Facultative	6.30	8.0	74	6.0 × 10^5
First maturation	4.60	3.9	n.c.[a]	5.4 × 10^4
Second maturation	3.50	3.0	n.c	n.c.
Third maturation	3.50	3.0	14[b]	700

[a]Not calculated
[b]Filtered BOD, calculated as 0.3{74/[1 + (0.05 × 3.9)][1 + (0.05 × 3)]^2}

Table 4 Output of classical WSP design procedure with parameter variation for 50% compliance with ≤ 1000 *E. coli* per 100 ml

Pond	Area (ha)	Retention time (d)	Effluent BOD (mg l^{-1})	Effluent E. coli (per 100 ml)
Anaerobic	0.46	1.1	133	1.5 × 10^7
Facultative	6.26	7.9	74	5.5 × 10^5
First maturation	4.56	3.9	19[a]	4.9 × 10^4
Second maturation	3.64	3.0	4[a]	5500
Third maturation	3.61	3.0	3[a]	660

[a]Filtered BOD

Table 5 Output of classical WSP design procedure with parameter variation for 95% compliance with ≤1000 *E. coli* per 100 ml

Pond	Area (ha)	Retention time (d)	Effluent BOD (mg l^{-1})	Effluent *E. coli* (per 100 ml)
Anaerobic	0.56	1.5	178	2.3×10^7
Facultative	8.60	11.5	85	1.1×10^6
First maturation	5.77	4.8	21[a]	1.1×10^5
Second maturation	5.30	4.3	4[a]	1.5×10^4
Third maturation	4.97	4.0	4[a]	2100
Fourth maturation	4.35	3.0	2[a]	560

[a]Filtered BOD

mid-depth pond area for 50% compliance is 18.5 ha and the overall retention time 19 days (i.e., approximately the same as for the design without parameter variation). For 90% compliance an additional maturation pond is required, the total pond area is 29.6 ha and the overall retention time 29 days; these are both ~50% more than for 50% compliance.

Modern WSP design

Fixed parameter values. Initially no parameter variation was selected so as to give a direct (i.e., non-Monte Carlo) comparison between the two design procedures. The output (Table 6) shows that a fourth maturation pond is required in the modern design procedure. The total mid-depth pond area is 24.6 ha and the overall retention time 24 days. These values are 34% and 26% more than for the classical design with fixed parameter values.

Parameter variation. The parameter values were then assigned the variation ranges given in Table 2. Values of 5–1 and 10–1 for the length-to-breadth ratios of the facultative and maturation ponds, respectively, were assumed, so yielding estimated values for δ of 0.2 and 0.1. (In practice, of course, the designer would choose a suitable range for each parameter and appropriate length-to-breadth ratios and not necessarily the ranges and values used here.) The 10,000-trial Monte Carlo program executed the design to give an effluent *E. coli* count of ≤1000 per 100 ml on both a 50-percentile basis and a 95-percentile basis. Each trial calculated the areas, retention times and effluent BOD and *E. coli* numbers for the anaerobic and facultative ponds and for the required number of maturation ponds (i.e., for the whole series) for the parameter values randomly selected for each trial. The design outputs are given in Tables 7 and 8 for 50% and 95% compliance, respectively.

The design for 50% compliance with ≤1000 *E. coli* per 100 ml of final effluent has a total mid-depth pond area of 22.6 ha and an overall retention time of 22 days; these values are similar (as found for the classical WSP design procedure) to those of the non-

Table 6 Output of modern WSP design procedure without parameter variation

Pond	Area (ha)	Retention time (d)	Effluent BOD (mg l^{-1})	Effluent *E. coli* (per 100 ml)
Anaerobic	0.44	1.1	133	1.6×10^7
Facultative	6.32	8.0	74	2.2×10^6
First maturation	4.62	3.9	19[a]	3.1×10^5
Second maturation	4.41	3.8	4[a]	4.6×10^4
Third maturation	4.41	3.8	3[a]	6800
Fourth maturation	4.41	3.8	1[a]	1000

[a] Filtered BOD

Table 7 Output of modern WSP design procedure for 50% compliance with ≤1000 *E. coli* per 100 ml

Pond	Area (ha)	Retention time (d)	Effluent BOD (mg l^{-1})	Effluent *E. coli* (per 100 ml)
Anaerobic	0.46	1.1	133	1.5×10^7
Facultative	6.26	7.9	74	2.1×10^6
First maturation	4.57	3.9	19[a]	3.0×10^5
Second maturation	3.78	3.1	4[a]	4.7×10^4
Third maturation	3.78	3.1	3[a]	6900
Fourth maturation	3.78	3.1	2[a]	1000

[a] Filtered BOD

Table 8 Output of modern WSP design procedure for 95% compliance with ≤1000 *E. coli* per 100 ml

Pond	Area (ha)	Retention time (d)	Effluent BOD (mg l^{-1})	Effluent *E. coli* (per 100 ml)
Anaerobic	0.57	1.5	178	2.3×10^7
Facultative	8.67	11.6	85	3.8×10^6
First maturation	5.81	4.9	21[a]	6.1×10^5
Second maturation	5.00	3.9	5[a]	1.1×10^5
Third maturation	5.00	3.9	4[a]	2.3×10^4
Fourth maturation	5.00	3.9	2[a]	4700
Fifth maturation	4.46	3.3	1[a]	1000

[a] Filtered BOD

Monte Carlo design given in Table 6. However (again as found for the classical WSP design procedure), the design for 95% compliance requires an additional maturation pond, and a much larger area and a much longer retention time: 34.5 ha and 33 days (both ~50% more than for 50% compliance).

The designs based on 95% compliance give high levels of confidence that the required effluent quality (in this case ≤1000 *E. coli* per 100 ml) will be achieved. However, as noted above, the resulting pond system would be considerably more expensive and therefore, especially (but not only) in developing countries, regulators and designers should seek a balance between system reliability and system costs. (Of course, in practice the regulator or designer could choose a percentage compliance lower than 95%: 80% or 90%, for example.)

Conclusions

WSP design procedures based on parameter uncertainty and 10,000-trial Monte Carlo simulations represent a more flexible method of WSP design than those based on fixed parameter values. However, for the ranges of parameter values selected, the modern pond design procedure based on von Sperling's equations required ~34% more pond area than the classical Marais design procedure for 50% compliance with ≤1000 *E. coli* per 100 ml; for 95% compliance the area was ~50% more than that for 50% compliance. The reason for this is that several (perhaps many) of the 33 WSP systems analyzed by von Sperling were overloaded with resultant suboptimal removals of faecal bacteria.

For the design of a new WSP system the classical Marais procedure is therefore recommended. However, when an existing overloaded WSP system is to be upgraded (by, for example, the intelligent incorporation of baffles; Shilton and Harrison, 2003), the modern von Sperling procedure would be more appropriate.

References

Gawasiri, C.B. (2003). *Modern Design of Waste Stabilization Ponds in Warm Climates: Comparison with Traditional Design Methods* MSc thesis, University of Leeds, Leeds, England.

Mara, D.D. (1987). Waste stabilization ponds: problems and controversies. *Water Quality International* (1) 20–22.

Mara, D.D. (1997). *Design Manual for Waste Stabilization Ponds in India*, Lagoon Technology International, Leeds, England.

Mara, D.D. (2004). *Domestic Wastewater Treatment in Developing Countries*, Earthscan Publications, London, England.

Mara, D.D. and Pearson, H.W. (1998). *Design Manual for Waste Stabilization Ponds in Mediterranean Countries*, Lagoon Technology International, Leeds, England.

Marais, G.v.R. (1974). Faecal bacterial kinetics in waste stabilization ponds. *Journal of the Environmental Engineering Division, American Society of Civil Engineers*, **100**(EE1), 119–139.

Pearson, H.W., Mara, D.D. and Arridge, H. (1995). The influence of pond geometry and configuration on facultative and maturation pond performance and efficiency. *Water Science and Technology*, **19**(12), 129–139.

Pearson, H.W., Mara, D.D., Arridge, H., Cawley, L.R. and Silva, S.A. (1996). The performance of an innovative tropical experimental waste stabilization system operating at high organic loadings. *Water Science and Technology*, **33**(7), 63–73.

Shilton, A.N. and Harrison, J. (2003). *Guidelines for the Hydraulic Design of Waste Stabilization Ponds*, Massey University, Palmerston North, New Zealand.

Sleigh, P.A. and Mara, D.D. (2003). *Monte Carlo Program for Facultative Pond Design*, (available at http://www.efm.leeds.ac.uk/CIVE/Mcarlo/index.html).

von Sperling, M. (1996). Design of facultative ponds based on uncertainty analysis. *Water Science and Technology*, **33**(7), 41–47.

von Sperling, M. (1999). Performance evaluation and mathematical modeling of coliform die-off in tropical and subtropical waste stabilization ponds. *Water Research*, **33**(6), 1435–1448.

von Sperling, M. (2003). Influence of the dispersion number on the estimation of coliform removal in ponds. *Water Science and Technology*, **48**(2), 181–188.

Wehner, J.F. and Wilhelm, R.H. (1956). Boundary conditions of flow reactor. *Chemical Engineering Science*, **6**(2), 89–93.

WHO (2005). *Guidelines for the Safe Use of Wastewater in Agriculture*, 2nd edn., World Health Organization, Geneva, Switzerland (in press).

Performance of very shallow ponds treating effluents from UASB reactors

M. von Sperling and L.C.A.M. Mascarenhas

Department of Sanitary and Environmental Engineering, Federal University of Minas Gerais, Brazil (E-mail: *marcos@desa.ufmg.br*)

Abstract Polishing ponds are units conceived for the post-treatment of the effluents from anaerobic reactors, are designed as maturation ponds, and aim at a further removal of organic matter and a high removal of pathogenic organisms. The paper investigates the performance of four very shallow ($H = 0.40$ m) polishing ponds in series, with very low detention times (1.4–2.5 days in each pond), treating anaerobic effluent from the city of Belo Horizonte, Brazil. The system was able to achieve excellent results in terms of BOD and *E. coli* removal, and good results in terms of ammonia removal, allowing compliance with European standards for urban wastewater and WHO guidelines for unrestricted irrigation. The paper presents the values of BOD and *E. coli* removal coefficients, which were much higher than those found in conventional pond systems. No statistically significant difference was found in the effluent *E. coli* concentrations from a pond with low depth and low detention time, and another pond in parallel, with double the depth and approximately double the detention time. The results endorse the applicability of the system composed by UASB reactors followed by very shallow ponds in series, with low detention times.
Keywords BOD; COD; *E. coli*; nitrogen; polishing ponds; UASB reactors

Introduction

Ponds for the post-treatment of the effluent from UASB (Upflow Anaerobic Sludge Blanket) reactors are conceived to give an additional polishing in terms of BOD and COD, if necessary ammonia and phosphorus and, especially, pathogens and indicator organisms. Experience with the application of such systems in Brazil has shown that the ponds can be designed as maturation ponds, without the intermediate stage of a facultative pond. Therefore, considerable savings in land requirements can be achieved. The geometry and the arrangement of the ponds can also follow the same principles of the maturation ponds: cells in series and high length/breadth ratios, in order to increase the efficiency of coliform removal (Cavalcanti *et al.*, 2001; von Sperling *et al.*, 2003b). No detrimental organic overloading in the first cell or compartment has been observed in polishing ponds.

An additional design aspect has also shown to substantially influence the performance of the ponds: depth. As demonstrated by von Sperling (1999) and Cavalcanti *et al.* (2001), shallow ponds are likely to have higher photosynthetic activity and UV radiation throughout the pond depth, higher pH and DO values, which, altogether, contribute to higher coliform and nutrients removal efficiencies. Values of the coliform decay coefficient are higher in ponds with lower depths (von Sperling, 1999). Experimental evaluation of some shallow ponds in Brazil has led to very good results (Cavalcanti *et al.*, 2001; von Sperling *et al.*, 2002, 2003a, 2003b).

As a result of this evidence, it was decided to investigate the performance of very shallow ponds (depth around 0.40 m). Considering that shallow ponds could be associated to large land requirements, it was also decided to keep very low retention times in the ponds, in some cases even lower than the minimum design value of 3 days, suggested by Mara (1997).

The present paper investigates the performance, loading rates and removal coefficients of a system composed by one UASB reactor followed by four shallow polishing ponds. The system was located in the city of Belo Horizonte, Brazil (latitude 20° South), treating actual wastewater from a population around 250 inhabitants. Each pond had the following bottom dimensions: length = 25.00 m, breadth = 5.25 m. Depth varied, as explained in the methodology. The variables investigated were: BOD, COD, SS, nitrogen, phosphorus and *E. coli*. The relevance of the research lies in the novelty of the configuration, in the very good results obtained and in the derivation of design criteria and coefficients for this system.

Methodology

The experiment was conducted in two distinct phases, as detailed in Table 1. A major difference lied in the fact that in Phase 1 the four ponds operated in series, and in Phase 2 they operated as two lines in parallel, each line having two ponds in series. Also, in Phase 1 all ponds operated with very shallow depths (varying from 0.65 m to 0.40 m) and hence short retention times. In Phase 2 the first line (Ponds 1 and 2) operated with double the depth of line 2 (Ponds 3 and 4). As a result, during Phase 2 the retention times in Ponds 3 and 4 were much lower than in Ponds 1 and 2 (not exactly half, because the pond sides were sloped). The very short hydraulic retention times in each of the shallow ponds should be noted, especially in Phase 1 and Phase 2 (set 2). In both phases, Pond 4 operated with duckweed (*Lemna* and *Wolffiella*). Sampling was twice a week in the influent and effluent from all units.

Results

Tables 2 and 3 present the average concentrations of the variables investigated in Phases 1 and 2. In Phase 2, only the first pond in each series (Ponds 1 and 3) are considered, since the presence of duckweed in Pond 4 interferes with the interpretation proposed in this paper. A discussion of the results is presented together with the figures to follow.

Phase 1 (four ponds in series)

BOD and COD. Figure 1 presents the average values of BOD and COD (total, filtered and particulate) along the system, during Phase 1. In terms of BOD, it is seen that the ponds gave a good contribution, and the final effluent had average concentrations of 44 mg/L, which can be considered a good value for ponds. In all ponds, the average filtered BOD was lower than 25 mg/L, which is the limit value for ponds in the European legislation for urban wastewaters (Council of the European Communities, 1991). As expected, the decrease of both particulate and filtered BOD after the first pond was only modest. In terms of COD, the contribution of the ponds was not so effective, and the

Table 1 Pond arrangement and operational phases

Item	Phase 1				Phase 2			
Duration	6 months (Jul–Dec. 2002)				6 months (Jan.–Jun. 2003)			
Arrangement	4 ponds in series				2 sets of 2 ponds in series			
Ponds investigated	Ponds P1, P2, P3 and P4				1st line–P1 and P2		2nd line–P3 and P4	
Pond characteristics	P1	P2	P3	P4	P1	P2	P3	P4
Depth (m)	0.65	0.55	0.40	0.40	0.80	0.80	0.40	0.40
Retention time (d)	2.5	2.1	1.4	1.4	6.2	6.2	2.9	2.9
Average flow (m³/d)	40	40	40	40	20	20	20	20

Table 2 Average effluent concentrations of the main variables analysed in Phases 1 and 2

Variable	Unit	Phase 1						Phase 2		
		Raw	UASB	Pond 1	Pond 2	Pond 3	Pond 4	UASB	Pond 1	Pond 3
BOD total	mg/L	354	108	66	59	50	44	53	28	29
BOD filtered	mg/L	109	43	18	11	9	10	25	10	9
BOD particulate	mg/L	245	65	48	48	41	34	28	18	20
COD total	mg/L	436	188	221	177	178	170	129	122	130
COD filtered	mg/L	188	101	94	94	90	101	81	61	73
COD particulate	mg/L	248	87	127	83	88	69	48	61	57
SS	mg/L	286	86	95	96	104	113	45	55	52
TKN	mg/L	30.2	33.4	24.9	14.7	12.5	10.8	27.6	16.3	14.8
N ammonia	mg/L	24.6	29	19.2	10.2	8	7.3	23.9	13.5	12.3
N organic	mg/L	5.6	4.4	5.7	4.5	4.5	3.5	3.7	2.8	2.5
N nitrate	mg/L	0.36	0.41	0.4	0.39	0.38	0.59	0.27	0.18	0.22
P total	mg/L	3.88	3.75	3.72	3.24	3.46	2.78	1.86	1.93	1.91
E. coli	MPN/100 mL	1.7×10^9	4.5×10^7	8.5×10^5	2.9×10^4	9.2×10^2	3.8×10^2	9.4×10^7	2.4×10^5	1.4×10^5

Table 3 Average concentrations of the main variables measured in situ by sensors (pH, DO, temperature) – Phase 1 (all ponds) and Phase 2 (first pond of each series only)

	Phase 1										Phase 2				
	08:00 a.m.						2:00 p.m.					08:00 a.m.		2:00 p.m.	
	Raw	UASB	Pond 1	Pond 2	Pond 3	Pond 4	Pond 1	Pond 2	Pond 3	Pond 4	Raw	Pond 1	Pond 3	Pond 1	Pond 3
pH	6.47	6.55	6.98	7.69	8.19	8.10	7.44	8.24	8.43	8.87	6.91	7.68	7.98	8.24	8.41
DO (mg/L)	–	–	0.96	1.29	2.63	2.43	1.21	2.27	3.36	4.84	–	2.02	2.50	1.94	3.60
T (°C)	23.94	23.48	21.86	21.96	21.69	21.27	24.61	25.17	25.13	23.98	24.92	24.34	24.00	27.17	26.50

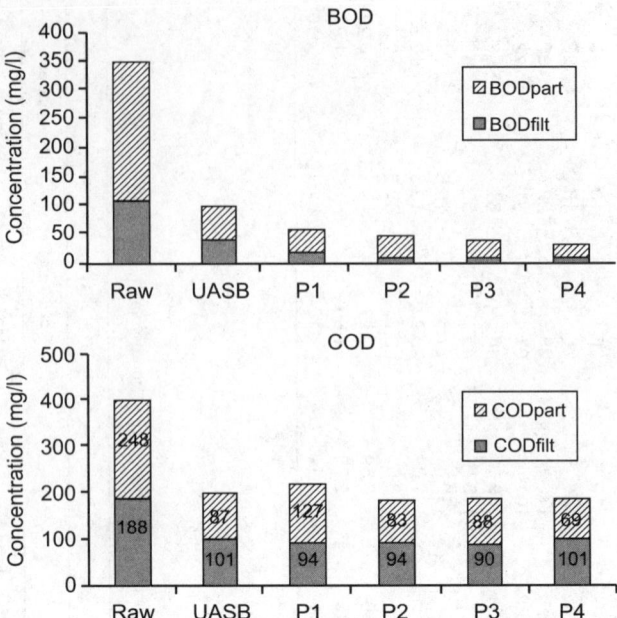

Figure 1 Average BOD and COD concentrations during Phase 1, along the various treatment units

average concentrations remained similar to those at the effluent from the UASB reactor. Even so, compliance with the European standards for filtered COD (125 mg/L) was also achieved for the final effluent average concentration. The average ratio of particulate BOD/SS varied from 0.30 to 0.51 mg BOD_{part} per mgSS, while the average ratio of particulate COD varied from 0.61 to 1.34 $mgCOD_{part}$ per mgSS.

It is worth mentioning that throughout the operational period, there were never problems with the release of malodorous gases. The first pond in the series, even with a retention time as short as 2.5 days, always presented measurable DO concentrations in the top layer, which helped to control any possible foul smell. Even at early morning, the average DO concentration in Pond 1 was 0.96 mg/L (Table 3). The shallow depth helped in creating a predominantly aerobic environment. The average organic surface loading rate in the first pond was 268 kgBOD/ha.d, which is a relatively high value for the regional conditions (southeast Brazil), and even in moments of higher loadings the pond never became anaerobic.

The coefficients of BOD removal in Pond 1 were calculated using the influent total BOD and effluent filtered BOD, and expressed for the temperature of 20 °C. The average values found were: (a) complete-mix model: $K = 1.81\,d^{-1}$; (b) dispersed flow model: $K = 0.88\,d^{-1}$. The temperature coefficient θ used was 1.035 and the dispersion number d was calculated assuming $d = 1/(L/B)$, as done by von Sperling (1999). The K values found are much higher than usual values reported in the literature for secondary ponds (von Sperling, 2002), which are in the order of $0.25\,d^{-1}$ (complete mix) and $0.15\,d^{-1}$ (dispersed flow). The low pond depth and the depth-associated mechanisms may be responsible for the high rate of BOD removal.

Nutrients. Figure 2 presents the profile of nutrients along the system. In the UASB reactor, total nitrogen concentrations increased. This fact had been observed in other UASB reactors, and the possible explanation is the release of nitrogen from the biomass retained in the system for a long sludge age (Aquino, 2004). There was a sequential decrease of total nitrogen along the ponds, caused by the high prevailing pH values

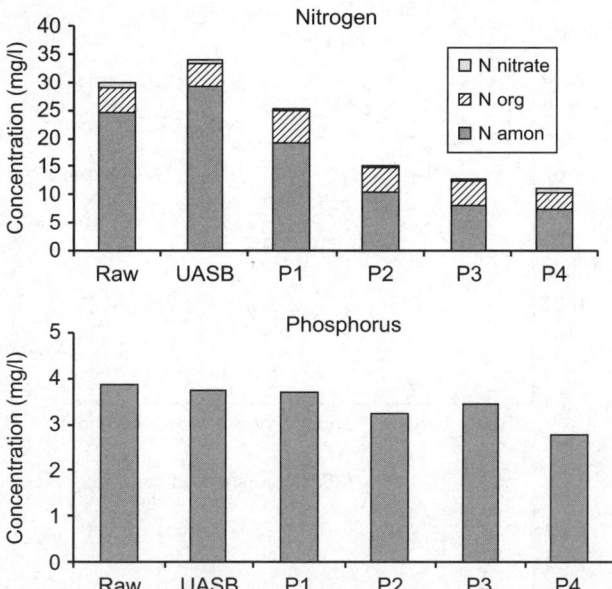

Figure 2 Average nitrogen (ammonia, organic, nitrate) and phosphorus concentrations during Phase 1, along the various treatment units

(ammonia stripping), see Table 2. Nitrification was probably not important, as indicated by the very low nitrate concentrations (nitrate reduction was not expected, because the ponds remained aerobic all the time). For the sake of comparison, total N concentrations were compatible with the discharge standards of 15 mg/L for urban wastewaters from the European Community (Council of the European Communities, 1991). There was some phosphorus removal, although not substantial, highlighting the already known difficulty in achieving good phosphorus removal efficiencies in ponds.

E. coli. Figure 3 illustrates in the form of a box-plot the decay of *E. coli* along the system. The excellent performance of the ponds is clearly seen, and the final effluent had geometric mean concentrations lower than 1000 MPN/100 mL (limit for unrestricted irrigation, according to WHO, 1989). The mean overall removal efficiency was remarkably high (6.42 log units, or 99.99996%), considering the low total retention time in the system (7.4 days). The coliform die-off coefficient K_b (see Figure 3) was much higher than the usual values reported in the literature for maturation ponds (0.4 to $0.7 \, d^{-1}$) (von Sperling, 1999), indicating the very favourable conditions for coliform removal in shallow polishing ponds. Coliform removal depends on the dimensionless product $K_b.t$. The low values of the retention time t were compensated by the high K_b values, giving a final product largely favouring high coliform removals. As expected, K_b in Pond 4 was lower, due to the presence of the duckweeds in the pond surface, hindering sunlight penetration. K_b coefficients for dispersed flow were calculated using the temperature coefficient $\theta = 1.07$ and the dispersion number d was calculated assuming $d = 1/(L/B)$.

Phase 2 (two pond systems in parallel, each with two ponds in series)

This section discusses some aspects of the system performance during Phase 2 (two systems in parallel, each with two ponds in series). The major comparison is between the first ponds in each series (Ponds 1 and 3), since Pond 4 had duckweeds, which of course affect the interpretation of the results within the scope of the present paper.

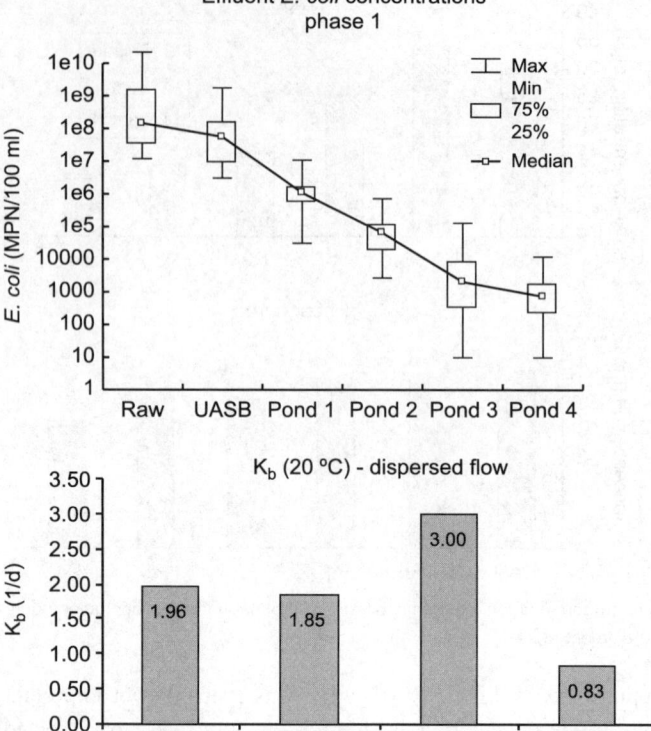

Figure 3 Average *E. coli* concentrations (geometric means) and K_b values (20 °C, dispersed flow) during Phase 1, along the various treatment units

In the comparison, it is important to notice that Pond 3 operated with lower depth ($H = 0.40$ m) and retention time ($t = 2.9$ d), compared with Pond 1, which operated with $H = 0.80$ m and $t = 6.2$ d. Both ponds had the same surface area and received the same flow and concentrations (effluent from the UASB reactor).

BOD removal coefficients. The values of the average BOD removal coefficient, calculated as described for Phase 1, were: (a) Pond 1: complete mix: $K = 0.50\,\mathrm{d}^{-1}$; dispersed flow: $K = 0.26\,\mathrm{d}^{-1}$; (b) Pond 3: complete mix: $K = 1.27\,\mathrm{d}^{-1}$; dispersed flow: $K = 0.62\,\mathrm{d}^{-1}$. Thus, the shallower pond had a higher BOD removal coefficient. The $K.t$ products in both ponds were similar, associated with the fact that both ponds had similar BOD removal efficiencies.

E. coli removal. The median effluent concentration from Pond 3 was lower, but the variability of the results was greater. The average coliform decay coefficients (dispersed flow, 20 °C) obtained were: Pond 1: $K_b = 1.57\,\mathrm{d}^{-1}$; Pond 3: $K_b = 3.73\,\mathrm{d}^{-1}$. Again, it is seen that, the shallower the pond, the higher the coliform decay coefficient. The K_b values obtained are very high, guaranteeing the high removal efficiencies encountered. In order to compare the performance of Ponds 1 and 3, the non-parametric Wilcoxon matched-pairs test was done, comparing effluent *E. coli* from both ponds in Phase 2, and the conclusion was that the concentrations from both ponds were not significantly different ($p = 0.56$). The shallower pond had a lower retention time t but a higher K_b value, and the $K_b.t$ products were thus similar, leading to similar performances. As with

BOD removal, the shallower pond, involving much lower construction costs, was able to lead to a similar performance.

Conclusions

The wastewater treatment system composed by an UASB reactor and four shallow polishing ponds in series was able to achieve excellent results in terms of BOD and *E. coli* removal, and good results in terms of ammonia removal. European standards for filtered BOD, filtered COD and total nitrogen, and WHO guidelines for unrestricted irrigation could be achieved, even with the prevailing short retention time (retention times in each pond varying from 1.4 to 2.5 d; overall retention time of 7.4 days during Phase 1).

Average BOD removal coefficients obtained in the first pond of the series were (20 °C): (a) complete-mix model: $K = 0.50$ to $1.81 \, d^{-1}$; (b) dispersed flow model: $K = 0.26$ to $0.88 \, d^{-1}$. The higher values were associated with the lower depths. The values are much higher than values usually encountered in secondary ponds.

The average ratio of particulate BOD/SS varied from 0.30 to $0.51 \, mg \, BOD_{part}$ per mgSS, while the average ratio of particulate COD varied from 0.61 to $1.34 \, mg \, COD_{part}$ per mgSS.

No organic overloading problems, such as foul odour, were encountered in the first pond of the series, even during periods of high surface BOD load (equal to or greater than 265 kgBOD/ha.d).

The ponds showed good ammonia removal efficiencies (average final effluent concentration of 7.3 mg/L), but phosphorus removal was small in the polishing ponds.

The very shallow pond ($H = 0.40 \, m$) showed no statistically significant difference from a pond with double the depth ($H = 0.80 \, m$) and approximately double the retention time in terms of *E. coli* removal efficiency. This conclusion is in line with other research (von Sperling 1999, 2002), which shows an inverse relationship of the decay coefficient with depth.

The very good performance of the shallow polishing ponds in series, especially with the depth of 0.40 m, which is also associated with lower construction costs, indicates the high potential of this configuration for the post-treatment of the effluents from UASB reactors.

Acknowledgements

Brazilian Research Programme on Basic Sanitation (PROSAB), with the support of FINEP, CNPq and CAIXA.

References

Aquino, S (2004). *Personal communication. Nitrogen release from biomass in high-rate reactors.*

Cavalcanti, P.F.F., Van Haandel, A.C., Kato, M.T., Von Sperling, M., Luduvice, M.L., Monteggia, L.O. (2001). Capítulo 3: Pós-tratamento de efluentes de reatores anaeróbios por lagoas de polimento. In Chernicharo, C.A.L. (coord). *Pós-tratamento de efluentes de reatores anaeróbios*. PROSAB/ABES, Rio de Janeiro, pp. 105-170 (in Portuguese).

Council of the European Communities (1991). Council Directive of 21 May 1991 concerning urban waste water treatment (91/271/EEC). *Official Journal of the European Communities*, No. L 135/40.

Mara, D.D. (1997). *Design manual for waste stabilisation ponds in India*, Lagoon Technology International Ltd., Leeds, UK.

Von Sperling, M. (1999). Performance evaluation and mathematical modelling of coliform die-off in tropical and subtropical waste stabilisation ponds. *Water Research*, 33(6), 1435–1448.

Von Sperling, M., Chernicharo, C.A.L., Soares, A.M.E. and Zerbini, A.M. (2002). Coliform and helminth eggs removal in a combined UASB reactor–baffled pond system in Brazil: performance evaluation and mathematical modelling. *Water Science and Technology*, **45**(10), 237–242.

Von Sperling, M., Chernicharo, C.A.L., Soares, A.M.E. and Zerbini, A.M. (2003a). Evaluation and modelling of helminth eggs removal in baffled and unbaffled ponds treating anaerobic effluent. *Water Science Technology*, **48**(2), 113–120.

Von Sperling, M., Jordão, E.P., Kato, M.T., Alem Sobrinho, P., Bastos, R.K.X., Piveli, R (2003b). Capítulo 7: Lagoas de estabilização. In: Gonçalves, R.F (coord). *Desinfecção de efluentes sanitários*. PROSAB/FINEP, Rio de Janeiro, pp. 277–336 (in Portuguese).

WHO (1989). Health guidelines for the use of wastewater in agriculture and aquaculture. *Tech. Bull. Ser. 77*, WHO, Geneva, Switzerland.

Removal of *E. coli* and helminth eggs in UASB: polishing pond systems in Brazil

M. von Sperling*, R.K.X. Bastos** and M.T. Kato***

*Department of Sanitary and Environmental Engineering, Federal University of Minas Gerais, Brazil
(E-mail: *marcos@desa.ufmg.br*)

**Department of Civil Engineering, Federal University of Viçosa, Brazil (E-mail: *rkxb@ufv.br*)

***Department of Civil Engineering, Federal University of Pernambuco, Brazil (E-mail: *kato@ufpe.br*)

Abstract Ponds following anaerobic reactors, such as Upflow Anaerobic Sludge Blanket (UASB) reactors, have been termed polishing ponds in the literature. The present paper analyses the removal of *E. coli* and helminth eggs in five UASB–polishing pond systems in Brazil. Since there were ponds in series, the total number of ponds was 10. The ponds had average retention times varying from 2 to 21 days, and depths ranging from 0.40 to 2.00 m. The shallow ponds in series, even with low retention times, were able to produce effluents complying with the coliform WHO guidelines for unrestricted irrigation (≤1000 MPN/100 ml). An equation for the coliform decay coefficient was proposed: K_b (dispersed flow) = $0.710\,H^{-0.955}$ (20 °C). The equation highlights the inverse relationship between the pond depth and the decay coefficient. All polishing pond systems were able to produce effluents with helminth eggs concentrations predominantly equal to zero, and satisfying the WHO guidelines for unrestricted and restricted irrigation (≤1 egg/L, arithmetic mean). The approximate range of helminth eggs removal efficiency was predicted satisfactorily.

Keywords Coliforms; *E. coli*; helminth eggs; polishing ponds; UASB reactors

Introduction

UASB (Upflow Anaerobic Sludge Blanket) reactors are gaining widespread use for the removal of organic matter in some warm-climate countries. Ponds following these reactors provide an additional polishing in terms of BOD removal, considering that typical removal efficiencies in UASB reactors are around only 70%. Additionally, ammonia and phosphorus removal is also partially achieved, as a result of the high pH levels that may prevail in these ponds. For these reasons, they are termed "polishing ponds". However, their main objective is the removal of pathogenic organisms. In this case, the ponds can be arranged as maturation ponds, either in series or with baffles, in order to maximize the removal of bacteria and viruses.

The first generation of polishing ponds was designed as facultative ponds. However, field experience showed that there is no need for facultative ponds following UASB reactors, since the complementary BOD removal will take place anyway in the polishing ponds, thus not requiring the large surface areas typical of facultative ponds. Within this concept, the flowsheet can be simply as shown in Figure 1, with the UASB reactors being directly followed by the polishing ponds, therefore allowing considerable land savings. Polishing ponds are thus designed as maturation ponds.

There still are only a few data in the international literature concerning this important treatment system configuration. Catunda *et al.* (1994) discussed the applicability of polishing ponds, while von Sperling *et al.* (2003b) and von Sperling and Chernicharo (2005) presented the general description and design criteria for these ponds. Von Sperling *et al.* (2002, 2003a) presented results of *E. coli* and helminth eggs removal in polishing ponds (one baffled and one unbaffled) located in the city of Itabira, Brazil. However, an

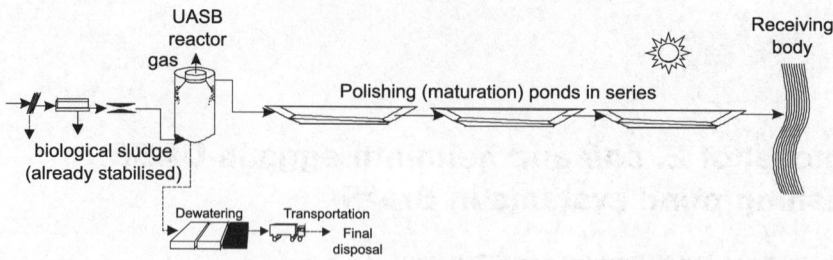

Figure 1 Flowsheet of a system comprised by a UASB reactor followed by polishing ponds in series

overall assessment of different polishing pond systems, in different locations and with different characteristics is yet not available in the literature.

The present paper analyses the removal of *E. coli* and helminth eggs in five UASB–polishing pond systems in Brazil. These ponds have been investigated by three universities (UFMG, UFV and UFPE) as part of the Brazilian Research Programme on Basic Sanitation (PROSAB). Some of the polishing ponds had been designed and built as facultative ponds, whereas the newer polishing ponds were conceived as maturation ponds.

Methodology

Table 1 presents basic data on the treatment systems investigated, including the different operational phases to which most of the ponds have been subjected. In the five systems analysed, the total number of ponds was 10, and the total number of data was 29 (each data represents a long-term average – geometric mean for *E. coli* and arithmetic mean for helminth eggs – of each operational phase). The ponds are located in different regions in Brazil (Northeast and Southeast), between latitudes 8 and 21°S. The total monitoring period, covering different operational phases, was approximately two years, with sampling frequencies varying from 2/week to 1/month.

E. Coli removal

Figure 2 presents box-plots of *E. coli* concentration along the treatment units of the systems that comprised three or more shallow polishing ponds in series. The substantial decrease is clearly seen, together with the fact that the median in both systems is below the reference value of the WHO guidelines for unrestricted irrigation (1000 MPN/100 ml). The performance of the shallow ponds, even with very low retention times, was impressive. Shallow ponds allow the penetration of the bactericide UV radiation throughout the depth, and are able to sustain high levels of photosynthetic activity in its entire volume, thus allowing high DO and pH values. The combination of these factors is responsible for the high coliform removal efficiencies. It was concluded that shallow ponds, even though with shorter retention times, are able to perform similarly to or better than deeper ponds with longer retention times.

The *E. coli* decay coefficient K_b was calculated for all ponds, assuming the dispersed flow model (Equations 1 and 2). The dispersion number d was estimated according to von Sperling (1999), in which $d = 1/(L/B)$ (L = length, B = breadth).

$$N = N_o \cdot \frac{4a e^{1/2d}}{(1+a)^2 e^{a/2d} - (1-a)^2 e^{-a/2d}} \qquad a = \sqrt{1 + 4K_b \cdot t \cdot d} \qquad (1)$$

Table 1 UASB–pond systems investigated

University/place	Plant scale	Treatment system	Oper. phase	Number of ponds in series	HRT in each pond (d)	Total HRT in ponds (d)	Pond depth H (m)
UFPE/Mangueira	Real	UASB–pond (facult.)	1	1	4.7	4.7	1.50
UFV/Viçosa	Pilot	UASB–pond (maturation)	1	3	9.6	28	0.90
		UASB–pond (maturation)	2	3	7.2	21	0.90
		UASB–BAF–pond (matur.)	3	3	7.2–5.5–2.4	15	0.90–0.70–0.30
UFMG/Itabira	Real	UASB–pond (facultative)	1	1	20.9	20.9	2.00
UFMG/Itabira	Pilot	UASB–pond (maturation)	1	2	4.6	9.2	0.60
			2	2	2.9	5.8	0.40
UFMG/Arrudas	Pilot/real	UASB–pond (maturation)	1	4	2.4–2.2–1.6–1.6	7.8	0.65–0.50–0.40–0.40
			2	2	7.3	14.6	0.80
			2	2	3.2	6.4	0.40

HRT: hydraulic retention time; BAF: submerged aerated biofilter

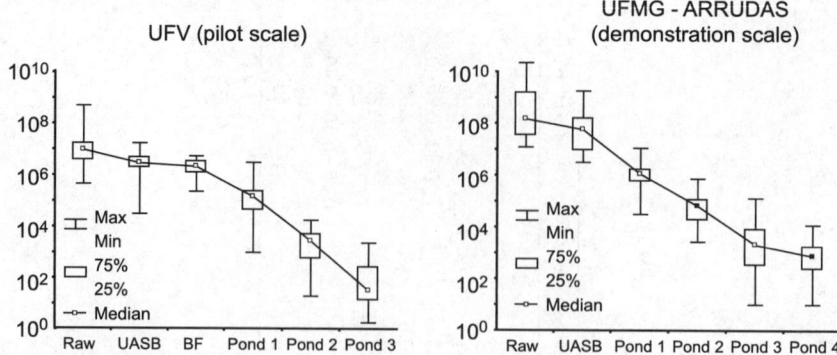

Figure 2 Box-plot of *E. coli* densities along two treatment systems that had three or more shallow ponds in the series

N_o = influent *E. coli* concentration (MPN/100 ml); N = effluent *E. coli* concentration (MPN/100 ml); K_b = coliform decay coefficient (d^{-1}); t = hydraulic retention time (d); and d = dispersion number.

Figure 3 presents the values of the coliform decay coefficient K_b plotted as a function of the pond depth. In order to increase the number of ponds in the regression analysis, data from two other Brazilian UASB–polishing ponds systems were included (Campina Grande, five shallow ponds, Cavalcanti *et al.*, 2001; Juramento, one facultative-type polishing pond, Santos, 2003). The total number of polishing ponds in the analysis was 19, with a total of 45 data on *E. coli* concentrations (each data representing long-term geometric means). It is clearly seen that, the lower the depth, the higher K_b. Based on a regression analysis between K_b and H, an equation was derived (Equation 2). The coefficient of determination for the estimation of K_b was only reasonable ($R^2 = 0.342$), mainly due to the fact that the model was not able to reproduce well the very high K_b values obtained with the low-depth ponds. Figure 4 compares the observed and estimated (using Equation 2 for K_b and the dispersion number $d = 1/(L/B)$) values of effluent *E. coli*. The fitting was very good ($R^2 = 0.925$). It is interesting to comment that applying the equation proposed by von Sperling (1999) ($K_b = 0.917 H^{-0.877} t^{-0.329}$) also led to high R^2 values (0.906) in the comparison of observed and estimated effluent *E. coli* concentrations.

$$K_b(\text{dispersed flow}) = 0.710 H^{-0.955} (20\,°C) \quad (2)$$

Helminth eggs removal

Table 2 presents the descriptive statistics of the helminth eggs concentrations (eggs/L) in the *final effluent* from the systems investigated. It is seen that all systems easily comply

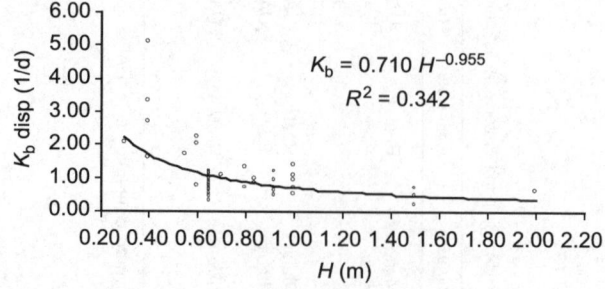

Figure 3 K_b (20 °C) values for dispersed flow (19 ponds, 45 data) estimated according to Eq. 2 and $d = 1/(L/B)$. ○, observed; —, estimated

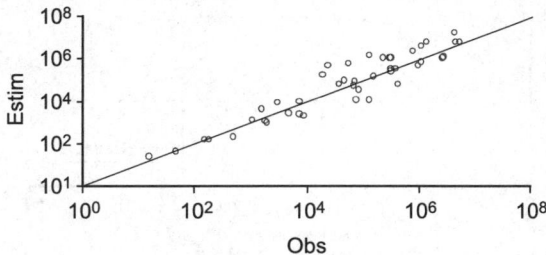

Figure 4 Observed (O) and estimated (−) values of effluent *E. coli* (19 ponds, 45 data) estimated according to Eq. 2 and $d = 1/(L/B)$

Table 2 Summary statistics for helminth eggs in the final effluent

Statistics	UASB reactor–polishing pond systems				
	UFV–Pilot	UFPE–real	UFMG–Itabira real	UFMG–Itabira pilot	UFMG–Arrudas demonstration
Total hydraulic retention time (d)	28	4.7	21	6–9	8
Arithmetic mean	0.0	0.0	0.2	0.4	0.0
Median	0.0	0.0	0.0	0.0	0.0
Standard deviation	0.0	0.0	0.4	1.4	0.0
Minimum	0.0	0.0	0.0	0.0	0.0
Maximum	0.0	0.0	1.3	6.7	0.0

with the WHO guideline for restricted and unrestricted irrigation (arithmetic mean ≤ 1 egg/L).

When analysing individual data from each pond, already in the first pond (in some systems, the only pond) effluent, the eggs concentrations are mostly zero (all medians are equal to zero). Only few samples are greater than zero, or 1 egg/L.

It is worth mentioning that, due to the large variability of the data and the predominance of nil concentrations, arithmetic means do not lead to a good representation of their central tendency. Few or sometimes a single large value lead to a large arithmetic mean. The medians are systematically equal to zero in all systems, after Pond 1. Geometric means cannot be calculated because the existence of a single nil value in the whole series invalidates the calculation of the geometric mean.

The average removal efficiencies in the UASB reactor and in the first pond in the series are shown in Table 3. In most cases, it was not possible to calculate the removal efficiency in the subsequent ponds in the series because they received an influent with zero concentration. It can be seen that the average removal efficiencies in the UASB reactors varied from 0% to 88%, and in the first pond in the series, from 96.5% to 100%.

Table 3 Average removal efficiencies for helminth eggs (%)

Unit	UFV (UASB full scale; ponds pilot scale)	UFPE (full scale)	UFMG (Itabira– full scale)	UFMG (Itabira– UASB demonstration; ponds pilot scale)	UFMG (Arrudas– demonstration scale)
UASB reactor	71	0	88	86	63
First pond	98,1	100.0	98.4	96.5	100

Note: removal efficiencies calculated based on the arithmetic means of the influent and effluent concentrations in each unit

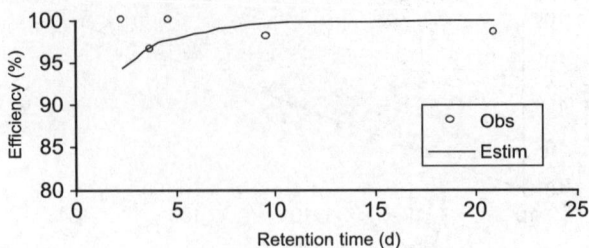

Figure 5 Comparison between the observed and the estimated (Ayres et al., 1992) helminth eggs removal efficiency.

Figure 5 compares the average removal efficiencies observed in the five systems, with those estimated according to the Ayres et al. (1992) model: $E = 100 \times [1 - 0.14 e^{-0.38 \cdot t}]$, where E is expressed in percentages. The model results indicate that the removal efficiency should be greater than 96%, which would be in agreement with the experimental data. However, a fine fitting was not observed, highlighting the difficulty in representing the behaviour of helminth eggs in ponds using a simple model (laboratory analyses are not trivial and arithmetic means do not represent the central tendency of the data) using a simple model. However, the overall removal range was well reproduced.

Table 4 presents the percentage of the samples from the final effluent in each of the five systems investigated according to the following conditions: (a) percentage of samples with concentration ≤ 1 egg/L and (b) percentage of samples with zero concentration. The very high percentages within both criteria clearly indicate the adequacy of the pond systems for the removal of helminth eggs and the production of an effluent that complies with the WHO guidelines for restricted and unrestricted irrigation. The general indication of 8–10 days (WHO, 1989) for effective eggs removal are endorsed by the results obtained. Additionally, it may be said that ponds designed for high levels of coliform removal are very likely to produce effluents that comply with the mentioned guidelines.

Conclusions

Polishing ponds (ponds following anaerobic reactors such as the UASB reactors) in series are able to produce *E. coli* concentrations lower than the WHO guideline value of 1000 MPN/100 ml for unrestricted irrigation.

Shallow ponds performed very well in the removal of coliforms, even with very short retention times. The coliform decay coefficient K_b was very high in these shallow ponds. An equation was derived for the estimation of the coliform decay coefficient (20 °C, dispersed flow): $K_b = 0.710 \, H^{-0.955}$.

The equation shows an approximate inverse relationship with depth. As a result, efforts to increase the coliform removal by increasing the retention time via an increase

Table 4 Percentage of the number of samples complying with the indicated conditions

Condition			Final effluent		
	UFV pilot	UFPE real	UFMG–Itabira pilot	UFMG–Itabira pilot	UFMG–Arrudas demonst.
% of samples ≤ 1 egg/L	100	100	92	91	100
% of samples = 0 egg/L	100	100	80	86	100

in depth will not be efficacious. The retention time will rise, but the K_b coefficient will decrease. Retention time shall be increased by increasing the surface area, keeping the depth at a low value.

All polishing pond systems were able to produce effluents with helminth eggs concentrations predominantly equal to zero, and satisfying the WHO guidelines for unrestricted and restricted irrigation (≤ 1 egg/L, arithmetic mean).

The Ayres *et al.* (1992) equation was able to predict the approximate range of helminth eggs removal efficiency.

The results of helminth eggs removal in the polishing ponds endorse the WHO proposition that retention times of 8–10 days are able to produce effluents that comprise with the irrigation requirements of ≤ 1 egg/L.

Polishing ponds, designed for the removal of coliforms (effluent concentrations ≤ 1000 MPN/10 ml), are expected to produce final effluents with helminth eggs concentrations that comply with the WHO guidelines (≤ 1 egg/L).

Acknowledgements
Brazilian Research Programme on Basic Sanitation (PROSAB), with the support of FINEP, CNPq and CAIXA.

References

Ayres, R.M., Alabaster, G.P., Mara, D.D. and Lee, D.L. (1992). A design equation for human intestinal nematode egg removal in waste stabilization ponds. *Wat. Res.*, **26**(6), 863–865.

Catunda, P.F.C., Van Haandel, A.C. and Lettinga, G. (1994). Post treatment of anaerobically treated sewage in waste stabilization ponds. In: *Anaerobic digestion, 7th Symposium*, Capetown, South Africa, pp. 405–415.

Cavalcanti, P.F.F., Van Haandel, A.C, Kato, M.T., Von Sperling, M., Luduvice, M.L. and Monteggia, L.O. (2001). Capítulo 3: Pós-tratamento de efluentes de reatores anaeróbios por lagoas de polimento. In: Chernicharo, C.A.L (coord). *Pós-tratamento de efluentes de reatores anaeróbios*. PROSAB/ABES, Rio de Janeiro. pp 105–170 (in Portuguese).

Santos, S.E. (2003). *Avaliação do desempenho de um sistema reator UASB – lagoa de polimento. Estudo de caso da ETE Juramento-MG*. MSc dissertation, UFMG (in Portuguese).

Von Sperling, M. (1999). Performance evaluation and mathematical modelling of coliform die-off in tropical and subtropical waste stabilisation ponds. *Wat. Res.*, **33**(6), 1435–1448.

Von Sperling, M., Chernicharo, C.A.L., Soares, A.M.E. and Zerbini, A.M. (2002). Coliform and helminth eggs removal in a combined UASB reactor – baffled pond system in Brazil: performance evaluation and mathematical modelling. *Wat. Sci. Tech.*, **45**(10), 237–242.

Von Sperling, M., Chernicharo, C.A.L., Soares, A.M.E. and Zerbini, A.M. (2003a). Evaluation and modelling of helminth eggs removal in baffled and unbaffled ponds treating anaerobic effluent. *Wat. Sci. Tech.*, **48**(2), 113–120.

Von Sperling, M., Jordão, E.P., Kato, M.T., Alem Sobrinho, P., Bastos, R.K.X. and Piveli, R. (2003b). Capítulo 7: Lagoas de estabilização. In: Gonçalves, R.F (coord). *Desinfecção de efluentes sanitários*. PROSAB/FINEP, Rio de Janeiro. pp. 277–336 (in Portuguese).

Von Sperling, M. and Chernicharo, C.A.L. (2005). *Biological wastewater treatment in warm climate countries*, IWA Publishing, London, UK.

WHO (1989). Health guidelines for the use of wastewater in agriculture and aquaculture, *Tech. Bull. Ser. 74*, WHO, Geneva, Switzerland.

Aerated rock filters for enhanced nitrogen and faecal coliform removal from facultative waste stabilization pond effluents

M. Johnson and D.D. Mara

School of Civil Engineering, University of Leeds, Leeds LS2 9JT, UK (E-mail: cenmlj@leeds.ac.uk and d.d.mara@leeds.ac.uk)

Abstract Facultative waste stabilization ponds in the UK, loaded at 80 kg BOD/ha day, produce effluents which comply with the European Urban Waste Water Treatment Directive (i.e., ≤25 mg filtered BOD/l and ≤150 mg SS/l). However, the Environment Agency of England and Wales typically requires a higher effluent quality of ≤40 mg/l unfiltered BOD and ≤60 mg/l SS, both on a 95-percentile basis. An ammonium-nitrogen requirement might also be applied. Traditionally, maturation ponds and reedbeds have been used to upgrade facultative pond effluents, requiring large land areas. This paper describes and compares aerated and unaerated rock filter performance for BOD, SS, nitrogen and faecal coliform removals, and highlights the land-saving opportunities as maturation ponds and reedbeds become redundant.

Keywords Aeration; ammonium; performance; removal; rock filters; United Kingdom

Introduction

Waste stabilization ponds (WSP) are a natural and sustainable wastewater treatment technology used throughout Europe. Their advantages include low capital, operation and maintenance costs. But their dependency on land area and warmer climates has been a major limitation to their adoption in the UK. Previous research carried out at Esholt Sewage Treatment Works, Bradford, West Yorkshire has shown that facultative waste stabilization ponds, loaded at 80 kg BOD/ha day, produce effluents which comply with the European Urban Waste Water Treatment Directive (i.e., ≤25 mg filtered BOD/l and ≤150 mg SS/l) (Abis and Mara, 2003). However, the Environment Agency (EA) of England and Wales typically requires, for communities of less than ∼2,000 persons, a higher effluent quality of ≤40 mg unfiltered BOD/l and ≤60 mg SS/l, both on a 95-percentile basis. Additionally, an ammonium-nitrogen ($NH_4^+ - N$) requirement may be applied – for example, a summer effluent quality of ≤2 mg/l $NH_4^+ - N$ is required for the Duloe Manor WSP, Cornwall, UK (Mara *et al.*, 1998). To comply with these more stringent effluent quality requirements, the effluent from the facultative pond at Esholt required further treatment. Therefore, work began in 2001 on simple (unaerated) rock filters, and although these were able to remove BOD and SS, they were unable to remove ammonia. Middlebrooks (1995), whilst noting that "the advantages of rock filters on a purely cost basis are dramatic", concluded that "high ammonia nitrogen concentrations in rock filter effluents could limit application of the process". We have therefore investigated rock filter aeration to determine if ammonium-N can be effectively removed in rock filters by nitrification.

Methods and materials

Pilot-scale rock filters

Two pilot-scale rock filters, each measuring 4 × 0.5 × 0.5 m, were constructed above ground, lined with a waterproof liner and filled with 40–100 mm limestone aggregate.

One filter was aerated (using a Jun-Air oil-free compressor at an air flow rate of 20 l/min), and the other was not and thus served as the control. Two perforated vertical sampling tubes were placed at 1.4 m and 3 m along the length of each filter. Facultative pond effluent was pumped into the base of each filter by Watson Marlow peristaltic pumps at an HLR of $0.15\,m^3/m^3$ day from August 2003 to April 2004, and was then increased to $0.3\,m^3/m^3$ day giving a retention time of just over 1.5 days. The filter effluents flowed by gravity from the top of the filter to the nearest drain.

Analytical methods

Rock filter influent and effluent samples were analysed weekly for BOD (*Standard Methods* method 5,210 B; APHA, 1998), suspended solids (method 2,540 D), TKN (method 4,500-N_{org} C), ammonia (method 4,500-NH_3 D), and faecal coliforms (method 9,222 D). A Sonde probe was inserted weekly into the sampling tubes to measure dissolved oxygen, pH and temperature. A Dionex DX500 ion analyser was used for monthly analyses of nitrite and nitrate.

Results

BOD and SS

The results for SS and BOD removal are given in Figure 1. The results show that the facultative pond effluent exceeded the required ≤60:40 mg/l 95-percentile concentrations for SS and BOD. The influent SS and BOD trends suggested that the concentrations of these parameters were largely due to the algal cells in the facultative pond effluent. The aerated filter removed >90% of both SS and BOD; its effluent BOD was consistently <5 mg/l. SS and BOD removals in the control filter were much more variable (70–90% and 45–90%, respectively); nevertheless the control filter achieved the EA requirement for both these parameters.

Nitrogen

Figure 2 shows that the TKN in the effluent of both filters decreased during the winter months (November to February) corresponding to a decrease in algal organic N coming

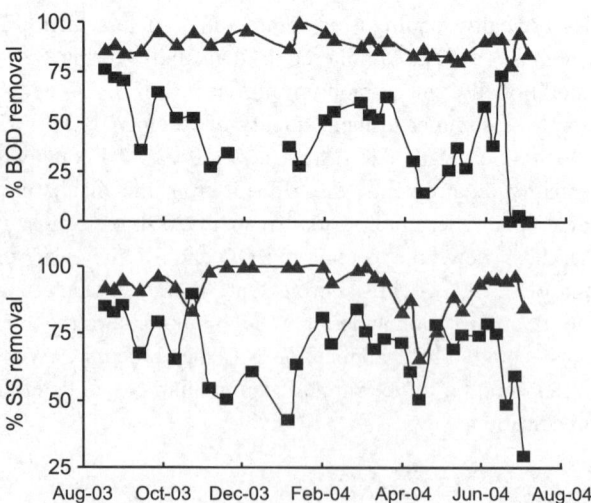

Figure 1 Percentage BOD removals (left) and percentage SS removals (right) in the aerated rock filter (▲) and the unaerated control (■). The table shows rock filter influent and effluent means, standard deviations and 95-percentile values, for SS and BOD in the aerated and control filters

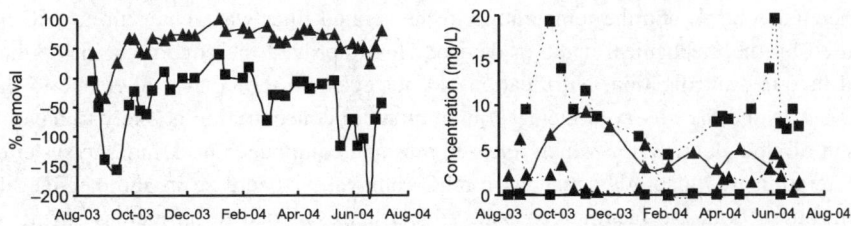

Figure 2 Percentage ammonia-N removal (left) and effluent nitrate-N (–) and TKN concentrations (–) (right) in the aerated rock filter (▲) and the unaerated control (■)

from the facultative pond. This trend is analogous with the BOD and SS trends. Effluent TKN from the aerated filter was consistently <5 mg/l, whereas the control filter frequently failed to reduce TKN. A similar pattern established for ammonium removal. The $NH_4^+ - N$ concentration in the influent to both filters was reasonably similar and varied over the eight month period from 2 to 7 mg/l. The aerated filter effluent consistently removed $NH_4^+ - N$ to <2 mg/l, but the control filter did not remove any $NH_4^+ - N$ – in fact, its concentration generally increased. Nitrate was produced in the aerated filter, but not in the control. Nitrite was below detection levels in both the influent and effluent for both filters.

Faecal coliforms

Figure 3 shows the FC numbers in the effluents of the aerated and control filters. Typical faecal coliform numbers in the facultative pond effluent were 10^5 per 100 ml in winter and 10^3 per 100 ml in summer. They were always >1,000 per 100 ml of the control filter effluent, but in the effluent from the aerated filter they were consistently reduced to <1,000 per 100 ml and often to <100 per 100 ml.

Discussion

All of the parameters measured in the facultative pond effluent required further treatment to comply with the EA discharge consents. BOD and SS were effectively removed in the control filter; and although aeration stabilised the effluent quality, it was an unnecessary improvement. A reduction in influent BOD and SS corresponded with a reduction in influent TKN, yet ammonium concentrations in the influent were reasonably stable, suggesting that the variability in influent TKN was a result of the reduced organic N coming from a declining algal population during the winter months. TKN was removed in the aerated filter but not in the control.

Ammonia was removed in the aerated filter to <2 mg/l. Dissolved oxygen concentrations within the aerated filter averaged 32% over the year (compared to just 4% in the

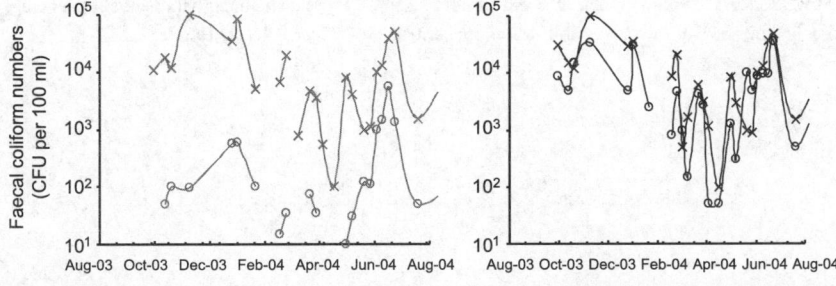

Figure 3 Faecal coliform cound (colony forming units per 100 ml) in the aerated filter (left) and unaerated control filter (right) showing influent (×) and effluent (o) concentrations

unaerated control) and the temperature in the aerated filter was always up to 2 °C higher, caused by the mechanical aeration device. This improved the conditions inside the filter and favoured nitrification. Nitrification did not occur in the control filter due to the low DO levels and the observed increase in ammonium concentration is likely to have been a result of algal degradation which released nitrogen compounds back into the system.

Aeration was also advantageous for FC removal. A decrease in summer faecal coliform numbers in the facultative pond effluent corresponded to an increase in pH in the pond when the pH can be as high as 9.3; pH values >9 are optimal for faecal coliform die-off (Pearson et al., 1987). The high DO levels were probably supporting a healthy predator population (mainly protozoa) that would otherwise barely survive under low DO conditions. The control filter did remove a small amount of FC suggesting that some removal by filtration did occur; however, predation was likely to have been a much more efficient removal mechanism for the reduction of FC numbers to <100 per 100 ml, even during winter.

Aeration has, of course, the disadvantages of increasing operational costs and introducing an electromechanical component into an otherwise natural wastewater treatment system. However, rock filters require less land than maturation ponds and constructed wetlands (~ 1.1 vs. $\sim 5\,m^2$ per person in temperate climates). For small communities in industrialized countries we believe rock filter aeration to be an appropriate tertiary treatment process for facultative pond effluents.

Conclusions

Aerated and unaerated rock filters can achieve effluent requirements of ≤ 40 mg BOD/l and ≤ 60 mg SS/l on a 95-percentile basis at a suitable hydraulic loading rate.

Rock filter aeration induces effective nitrification even in winter. In contrast the ammonia concentration increased in the unaerated filter.

Rock filters can advantageously replace maturation ponds and/or constructed wetlands for the tertiary treatment of facultative pond effluents, with a consequent reduction in land area requirements. The filters should be aerated if low levels of ammonia and/or faecal coliforms are required.

References

Abis, K. and Mara, D.D. (2003). Research on waste stabilization ponds in the United Kingdom – initial results from pilot-scale facultative ponds. *Wat. Sci. Tech.*, **48**(2), 1–7.

APHA (1998). *Standard Methods for the Examination of Water and Wastewater*, 20th edn., American Public Health Association, Washington, DC.

Mara, D.D., Cogman, C.A., Simkins, P. and Schembri, M.C.A. (1998). Performance of the Burwarton Estate waste stabilization ponds. *J. CIWEM*, **12**(4), 260–264.

Middlebrooks, E.J. (1995). Upgrading pond effluents: an overview. *Wat. Sci. Tech.*, **31**(12), 353–368.

Pearson, H.W., Mara, D.D., Smallman, D.J. and Mills, S. (1987). Physicochemical parameters influencing faecal coliform survival in waste stabilization ponds. *Wat. Sci. Tech.*, **19**(12), 145–152.

CFD (computational fluid dynamics) modelling of baffles for optimizing tropical waste stabilization pond systems

A.N. Shilton* and D.D. Mara**

*Centre for Environmental Technology and Engineering, Massey University, Private Bag 11 222, Palmerston North, New Zealand (E-mail: a.n.shilton@massey.ac.nz)
**School of Civil Engineering, University of Leeds, Leeds LS2 9JT, UK

Abstract CFD modelling of the incorporation of two baffles equally spaced along the longitudinal axis of the pond and with a length equal to 70% of the pond breadth, indicated a potential improvement in the removal of E. coli in a 4-day secondary facultative pond at 25°C from 5×10^6 per 100 ml in the effluent from a 1-day anaerobic pond to 4×10^4 per 100 ml; the reduction in an un-baffled pond was an order of magnitude less effective. The addition of a similarly baffled 4-day primary maturation pond reduced the effluent E. coli count to 340 per 100 ml; the reduction in an un-baffled series was two orders of magnitude less effective. Well designed baffles thus have considerable potential for reducing pond area requirements and hence costs in the hot tropics. These very promising results highlight the need for field studies on baffled pond systems to validate (or allow calibration) of the CFD model used in this study.
Keywords Baffles; CFD modelling; design; E. coli; hydraulics; removal; tropical climates; waste stabilization ponds

Introduction

Waste stabilization ponds (WSP) are most efficient in tropical areas where in-pond temperatures are high and decay/reaction rates are fast. The design is typically focused on the reduction of pathogens, as it is not uncommon for the pond effluent to be utilized for crop irrigation. The World Health Organization guidelines for the microbiological quality of treated wastewaters used for this purpose are as follows (WHO, 2005):

(a) Restricted irrigation: $\leq 10^5$ E. coli per 100 ml and ≤ 1 human intestinal nematode egg per litre (reduced to ≤ 0.1 egg per litre if children under 15 are exposed).
(b) Unrestricted irrigation: ≤ 1000 E. coli per 100 ml and the same egg numbers.

The WSP designer's goal is to optimise pond design by minimising cost and land required while maintaining satisfactory treatment. Using CFD modelling, Shilton and Harrison (2003a) showed that the incorporation of baffles could dramatically improve removal of indicator bacteria in primary facultative ponds (Table 1). Their work considered a single primary facultative pond in a temperate region with a design temperature of 14°C, a wastewater flow of $10,000 \, m^3 \, d^{-1}$ and a resultant theoretical hydraulic retention time of 31 days.

WSP in tropical countries have higher design temperatures, faster decay rates and hence shorter retention times than those investigated by Shilton and Harrison (2003a). Unfortunately, this combination of faster decay rates and shorter retention times make the impact of any hydraulic short-circuiting on treatment efficiency much more significant. This implies that use of effective baffling to mitigate short-circuiting is even more important in tropic regions than it is in the temperate ones previously studied. To date, information on the use of baffling for optimizing the design of tropical WSP has been lacking.

Table 1 The effects of baffling on *E. coli* removal in primary facultative ponds at a design temperature of 14°C and for an influent *E. coli* count of 1×10^8 per 100 ml (Shilton and Harrison, 2003a) In practice for different ponds the resultant performance values will depend on the influent concentration, retention time, temperature and many other variables

Number of baffles	*E. coli* per 100 ml of effluent
None	6.2×10^6
1	4.1×10^6
2	6.0×10^3
4	3.9×10^2
6	5.7×10^2
8	10

In this study we have used a CFD model with integrated first-order decay kinetics to evaluate the effect of baffles on *E. coli* removals in secondary facultative and primary maturation ponds in the hot tropics at a design temperature of 25°C.

Methods

The secondary facultative pond was designed for 25°C following standard design procedures (Mara, 1997) for a raw wastewater with a BOD and an *E. coli* count of $300 \, \text{mg} \, l^{-1}$ and 5×10^7 per 100 ml, respectively; these were assumed to have been reduced to $90 \, \text{mg} \, l^{-1}$ and 5×10^6 per 100 ml in a 1-day anaerobic pond. The facultative and maturation pond dimensions were taken as $640 \times 320 \times 1.5 \, \text{m}$ (the same model dimensions used by Shilton and Harrison, 2003). A wastewater flow of $76,800 \, \text{m}^3 \, \text{d}^{-1}$ was used giving a theoretical hydraulic retention time of 4 days.

As was the case in the Shilton and Harrison (2003a) study, this project also used a pond with a length-to-breadth ratio of 2 to 1; a pond depth of 1.5 m; and baffles equally spaced along the longitudinal axis of the pond with a length equal to 70% of the pond width.

Both a facultative pond and a maturation pond were modelled with and without the provision of two baffles. Single baffles were not considered as Shilton and Harrison (2003a) had found them to be generally ineffective (Table 1). The provision of more than two baffles in the facultative pond was not considered in order to avoid significant variation in organic loading across this pond. Another option could have been to use an un-baffled facultative pond followed by a maturation pond with four baffles. However, it can be deduced from the work of Shilton and Harrison (2003a) that this would not be nearly as efficient as the provision of two baffles in each pond.

The PHOENICS CFD model produced by Concentration, Heat and Momentum Ltd (London) was used in this work. The pond models were solved for fluid flow (via the solution of pressure, momentum in three dimensions and turbulence parameters) and first-order *E. coli* die-off at 25°C. For full details of the use and evaluation of the validity of PHOENICS for modelling waste stabilization pond hydraulics refer to Shilton (2001). Shilton and Harrison (2003b) discuss and validate the technique of integrating decay into a CFD model.

Results

The CFD modelling of the facultative pond on its own and the facultative and maturation ponds in series, both with and without two 70%-width baffles in each pond and operating at 25°C, yielded the following results for the effluent *E. coli* count per 100 ml for the five pond systems investigated (see Figure 1).

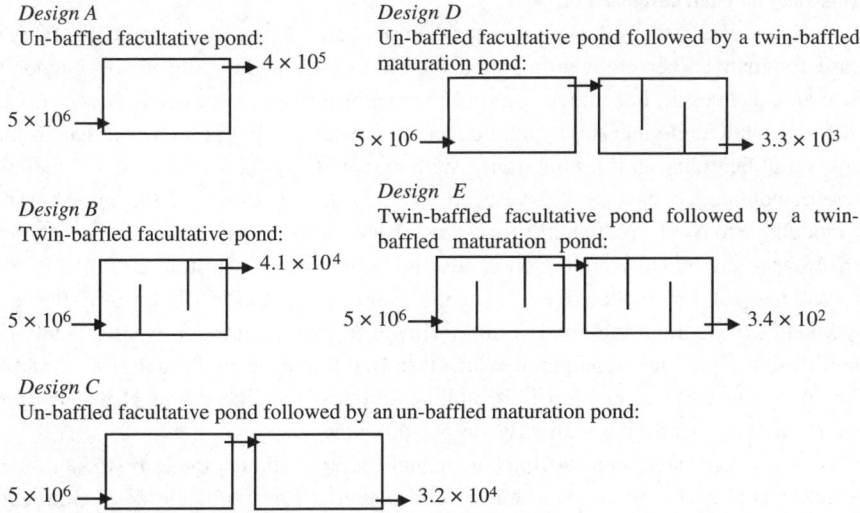

Figure 1 Five pond systems studied

It is possible to calculate the performance of Design A (the un-baffled facultative pond) by simply using the completely mixed flow equation as proposed by Marais (1966). This approach yields a value of 2×10^5. Considering that the results from different designs vary by orders of magnitude, this result is obviously very similar to that predicted by the CFD simulation. As would be expected, the result from the CFD model, which simulates the *actual* hydraulic flow pattern in the pond (and its deficiencies), is somewhat higher than that which is predicted by the completely mixed flow equation which assumes *perfect* mixing.

The incorporation of twin 70%-width baffles in the facultative pond achieves an improvement in *E. coli* reduction of an order of magnitude. The twin-baffled facultative pond produces essentially the same effluent quality as the un-baffled facultative pond and un-baffled maturation pond combined.

Discussion

Practical implications of results

The modelling predicted that restricted irrigation requirements ($\leq 10^5$ *E. coli* per 100 ml) could be met using both Design B (twin-baffled facultative pond system) or Design C (un-baffled facultative pond followed by one un-baffled maturation pond). However, compared with Design C, the use of baffling in Design B reduces the pond area requirement by almost 50%.

Where an effluent suitable for unrestricted irrigation is required, the modelling showed that Design E (twin baffled facultative pond followed by a twin-baffled maturation pond) reduces effluent quality to ≤ 1000 *E. coli* per 100 ml. To achieve this effluent quality without any baffling, the area required would be almost doubled.

The use of four baffles in the primary maturation pond was originally considered, but after obtaining the results given above this was clearly not warranted.

Installing baffles is not, of course, without cost. However, given the above findings that the use of baffles in tropic climates can dramatically reduce pond area requirements, it can be expected that the cost of baffling would be more than offset by the savings in land and pond construction costs.

CFD modelling for pond design assessment

CFD models of WSP attempt to represent a complex and dynamic environment. Just as is the case for many other engineering models, CFD pond models are not be intended to produce an exact result, but rather as a tool for exploring various designs alternatives.

Like all practical engineering equations and models, the CFD pond model has to make various simplifications and assumptions. With regard to its ability to predict hydraulic behaviour, confidence has been developed by previous studies into the application of CFD modelling to WSP, particularly in terms of its ability to assess large 'step changes' in performance that result from different design configurations (Shilton, 2001).

A weakness of the model does lie in the simple assumption of uniform first-order decay kinetics throughout the pond volume. However, this assumption is still widely used in pond design equations. Compared with these traditional design equations, the advantage of using the CFD simulation is its ability to assess the likely impact that hydraulic modifications, such as baffles, can have on performance.

While only extensive field-testing can conclusively validate these results, it would nevertheless appear that there is certainly large potential for significant cost optimization to be achieved by the intelligent incorporation of baffles in facultative and maturation ponds in tropical countries.

Conclusions

This modelling has shown that the provision of two 70%-width baffles in secondary facultative and primary maturation ponds at 25°C significantly increases *E. coli* removals in these ponds.

With respect to the particular pond configuration/parameters used in the CFD modelling undertaken for this study, the results showed that the combination of a 1-day anaerobic pond and a 4-day twin-baffled facultative pond could produce an effluent suitable for restricted crop irrigation. Additionally it was shown that an effluent suitable for unrestricted crop irrigation could be achieved by addition of a 4-day twin-baffled primary maturation pond.

These findings show that there is significant potential for size reduction and cost optimization to be achieved by the incorporation of properly designed baffles in ponds in tropical climates.

References

Mara, D.D. (1997). *Design Manual for Waste Stabilization Ponds in India*, Lagoon Technology International, Leeds, UK.

Marais, G.v.R. (1966). New factors in the design, operation and performance of waste stabilization ponds. *Bulletin of the World Health Organization*, **34**(5), 737–763.

Shilton, A.N. (2001). *Studies into the Hydraulics of Waste Stabilization Ponds* PhD thesis, Massey University, Palmerston North, New Zealand.

Shilton, A.N. and Harrison, J. (2003a). *Guidelines for the Hydraulic Design of Waste Stabilization Ponds*, Institute of Technology and Engineering, Massey University, Palmerston North, New Zealand.

Shilton, A. and Harrison, J. (2003b). Integration of coliform decay within a CFD model of a waste stabilisation pond. *Water Science and Technology*, **48**(2), 205–210.

WHO (2005). *Guideline s for the Safe Use of Wastewater in Agriculture*, World Health Organization, Geneva, Switzerland (in press).

Virus removal in a pilot-scale 'advanced' pond system as indicated by somatic and F-RNA bacteriophages

R.J. Davies-Colley, R.J. Craggs, J. Park, J.P.S. Sukias, J.W. Nagels and R. Stott

National Institute of Water and Atmospheric Research Ltd, P. O. Box 11-115, Hamilton, New Zealand
(E-mail: *r.davies-colley@niwa.co.nz*)

Abstract Advanced pond systems (APS), incorporating high-rate ponds, algal settling ponds, and maturation ponds, typically achieve better and more consistent disinfection as indicated by *Escherichia coli* than conventional waste stabilisation ponds. To see whether this superior disinfection extends also to enteric viruses, we studied the removal of somatic phages ('model' viruses) in a pilot-scale APS treating sewage. Measurements through the three aerobic stages of the APS showed fairly good removal of somatic phage in the summer months (2.2 log reduction), but much less effective removal in winter (0.45 log reduction), whereas *E. coli* was removed efficiently (>4 logs) in both seasons. A very steep depth-gradient of sunlight inactivation of somatic phage in APS pond waters (confined in silica test tubes) is consistent with inactivation mainly by solar UVB wavelengths. Data for F-RNA phage suggests involvement of longer UV wavelengths. These findings imply that efficiency of virus removal in APS will vary seasonally with variation in solar UV radiation.

Keywords APS; WSPs; disinfection; virus; phage; solar radiation; UV radiation

Introduction

Waste stabilisation ponds (WSPs) generally achieve good overall disinfection as indicated by *Escherichia coli*, however performance can be inconsistent and sometimes poor because of the lack of hydraulic and process control in these (passive) natural treatment systems (Davies-Colley, 2004). Conventional WSPs are much used in New Zealand where there is now increasing pressure to upgrade treatment, with one promising option being 'advanced' pond systems (APS, based on AIWPS, Oswald, 1991) that can provide more consistent treatment including disinfection.

Davies-Colley *et al.* (2003) reported excellent removal of *E. coli* in a pilot-scale APS treating domestic sewage in New Zealand. They attributed the high efficiency and consistency of disinfection to sunlight exposure of different ponds separated into three different aerobic modules: (1) a (gently stirred, shallow) *high-rate pond* (HRP) that is strongly exposed to sunlight and promotes the growth of algae, (2) a deep *algal settling pond* (ASP) promoting sedimentation of algal solids, so clarifying the surface water, and (3) *maturation pond* (MP) cells designed for effluent 'polishing' including further disinfection. All three stages contributed significantly to disinfection, although the greatest contribution overall occurred in the HRP. Fallowfield *et al.* (1996) and Bahlaoui *et al.* (1997) have also reported excellent faecal indicator bacterial removal in HRPs, despite short residence times (c. 5 d).

We wanted to know whether the efficient faecal indicator bacterial removal in APS extends to enteric viruses, so we monitored various stages of a pilot-scale APS for bacteriophages (viruses that infect bacteria and that are useful as faecal indicators and enteric virus models – Havelaar *et al.*, 1993). This paper reports the findings from two short monitoring campaigns (summer versus winter), and a preliminary experiment designed to elucidate processes of virus removal in APS.

Methods

The pilot-scale APS used for this study treats human and laboratory wastewaters (flow $\sim 2\,m^3\,d^{-1}$) from Ruakura Research Centre, near Hamilton, New Zealand (37°47'S, 175°19'E). A meteorological station, measuring solar radiation among other variables, is located on the Ruakura campus. The Ruakura APS consists of an anaerobic digester followed by three aerobic stages: an HRP (8 d hydraulic residence time, HRT), an ASP (4 d HRT) and a 2-cell MP (10 d total HRT).

Two types of bacteriophage occurring naturally in domestic sewage were used as convenient indicators and models of enteric virus behaviour. Somatic phages are DNA-viruses that attach by 'tail' structures to sites on host bacteria cell walls (e.g., Sinton et al., 2002), while F-RNA phage are small, round, RNA-viruses that attach to 'F-pili', the hair-like structures on surfaces of some bacteria. Both phages were enumerated using the single agar layer plating method with particular strains of E. coli host bacteria. Somatic phage was assayed by the method of Grabow and Cobrough (1986) and F-RNA phage by the method of Debartolomeis and Cabelli (1991).

Phage monitoring was carried out in two separate (twice-weekly) sampling campaigns focussed on seasonal extremes: 'summer' (7 occasions from 4–27 February) versus 'winter' (7 occasions from 30 June–21 July). The raw sewage influent and the anaerobic digester output were both sampled as well as the aerobic modules (HRP, ASP, MP). Preliminary sampling (summer) suggested that F-RNA phage is naturally low in the Ruakura sewage, so monitoring was restricted to somatic phage.

Experiments were conducted on the depth-dependence of sunlight inactivation in both the HRP and the ASP with both somatic phage and F-RNA phage. Phage numbers were increased by 'spiking' HRP water with primary-treated sewage from Hamilton City, which has consistently higher F-RNA phage than the Ruakura sewage. The spiked pond water was confined in test tubes made of pure silica, which is transparent to all wavelengths in sunlight. Up to 8 test tubes were exposed for up to 4 h near solar noon in summer at each of three different depths in the uppermost 50 mm of the pond water column. Tubes (25 mm diam.) were fixed horizontally with rubber bands on metal frames designed to minimise shading, and sampled at intervals of 30 minutes to 2 hr for measurement of surviving phage and E. coli. Penetration into pond water of visible radiation (400–700 nm), UVA (360 nm) and UVB (311 nm) was measured with a 3-channel radiometer (Macam Photometrics Ltd, Livingston, Scotland).

Results and discussion

Figure 1 shows that somatic phages were fairly efficiently removed in summer months (2.3 log removal) from digester to MP, although appreciably less so than E. coli (4.0 log removal). However, in winter somatic phage removal was much lower (0.45 log difference in medians from digester to MP, a 5-fold difference) despite similarly efficient E. coli removal (4.1 log units). A small *increase* in somatic phage between influent and digester in both seasons (Figure 1) may be attributed to lysis of infected coliform bacteria and release of phage particles in the digester, and it is possible that this process also occurs to some extent further downstream in the APS.

Sinton et al. (2002) showed that somatic phage are inactivated primarily by short-wave UV radiation in sunlight, so we hypothesised that the seasonality in somatic phage removal in the APS reflected seasonality in the solar UV climate. An experiment on depth-dependence of phage inactivation provided a partial test of this hypothesis.

Experiments with sunlight-exposed silica tubes yielded survival curves (not shown) that were well fitted by first-order rate equations as has been reported in earlier work (Davies-Colley et al., 1999; Sinton et al., 2002). Both phages had strong depth gradients

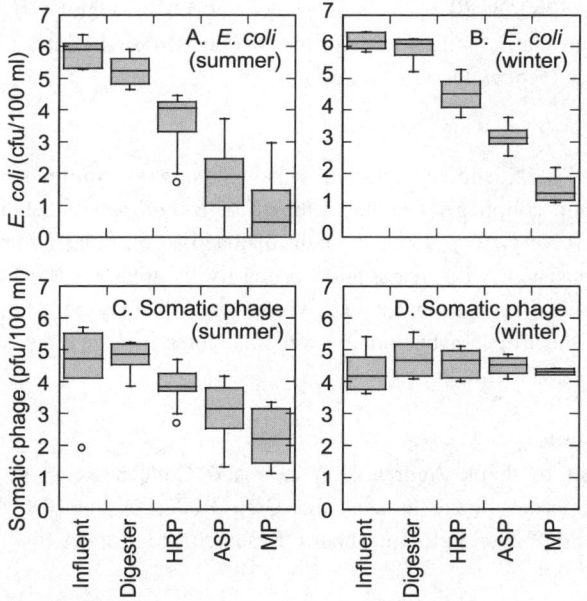

Figure 1 Disinfection through the pilot-scale APS at Ruakura, Hamilton as indicated by box plots of summer and winter monitoring data (log scales) for *E. coli* (Panels A and B) and somatic coliphage (Panels C and D). Boxes contain the middle 50% of the data

of sunlight inactivation rates, particularly the somatic phage for which first-order rate coefficients (m^2/MJ) paralleled UVB radiation in the pond water (Figure 2), suggesting inactivation primarily by solar UVB radiation. The F-RNA phage was a little more resistant (slower inactivation rate) and its less steep depth-gradient of inactivation suggests some involvement of longer UV wavelengths. These inferences about solar UV wavelengths causing phage inactivation are consistent with the findings of Sinton *et al.* (2002) using optical filters.

The solar UV model of Bodeker and McKenzie (1996) predicted an average daily erythemal (skin-burning) solar UVB dose of 3.7 kJ/m^2 (cf. 5.1 kJ/m^2 under clear skies) at the Ruakura Climate Station for the summer (4–27 February) monitoring period,

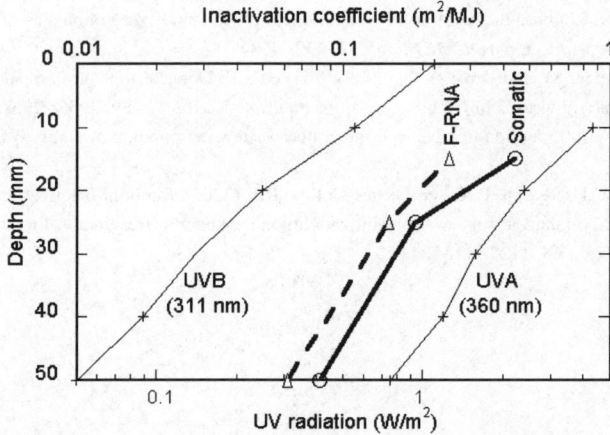

Figure 2 Depth dependence of inactivation of phages compared with the depth-gradient of UVB radiation. Data are for an experiment in the surface waters of the algal settling pond (ASP) of the Ruakura pilot-scale APS on 6 March 2003 (late summer). First-order sunlight inactivation rate coefficients (m^2/MJ) at three different depths are plotted for both somatic and F-RNA phages

compared with $0.52\,kJ/m^2$ ($0.7\,kJ/m^2$ under clear skies) in winter (30 June to 21 July), a 7-fold difference that is consistent with the marked seasonal contrast in somatic phage removal efficiency (Figure 1).

Conclusions

Monitoring of an APS showed marked seasonal contrast (summer versus winter) in removal of somatic coliphage, whereas removal of *E. coli* was excellent (>4 log units) in both seasons. A very steep depth gradient of inactivation rate of somatic coliphage in pond water is consistent with inactivation primarily by solar UVB. Our monitoring and experimental findings suggest that enteric viruses in APS may be efficiently removed in summer (and in the tropics), but not so well in winter in temperate latitudes owing to seasonality in solar UV climate.

Acknowledgements

The authors wish to thank Andrea Donnison and Colleen Ross of AgResearch Ltd, Ruakura, for numerous phage assays, and Greg Bodecker and Richard McKenzie of National Institute of Water and Atmospheric Research Ltd, Lauder for UV model data.

References

Bahlaoui, M.A., Baleux, B. and Troussellier, M. (1997). Dynamics of pollution indicator and pathogenic bacteria in high-rate oxidation wastewater treatment ponds. *Wat. Res.*, **31**, 630–638.

Bodeker, G.E. and McKenzie, R.L. (1996). An algorithm for inferring surface UV irradiance including cloud effects. *J. Appl. Meteorol.*, **35**, 1859–1877.

Davies-Colley, R.J. (2004). Chapter 6. Pond disinfection. In *Waste Stabilisation Pond Treatment Technology*, Shilton, A. (ed.), IWA, London, UK.

Davies-Colley, R.J., Craggs, R.J. and Nagels, J.W. (2003). Disinfection in a pilot-scale "Advanced" pond system (APS) for domestic sewage treatment in New Zealand. *Wat. Sci. Tech.*, **48**(2), 81–87.

Davies-Colley, R.J., Donnison, A.M., Speed, D.J., Ross, C.M. and Nagels, J.W. (1999). Inactivation of faecal indicator micro-organisms in waste stabilization ponds: interactions of environmental factors with sunlight. *Wat. Res.*, **33**, 1220–1230.

Debartolomeis, J. and Cabelli, V.J. (1991). Evaluation of an *Escherichia coli* host strain for enumeration of F male-specific bacteriophages. *Appl. Environ. Microbiol.*, **57**, 1301–1305.

Fallowfield, H.J., Cromar, N.J. and Evison, L.M. (1996). Coliform dieoff rate constants in a high rate algal pond and the effect of operational variables. *Wat. Sci. Tech.*, **34**(11), 141–147.

Grabow, W.O.K. and Coubrough, P. (1986). Practical direct plaque assay for coliphages in 100 mL samples of drinking water. *Appl. Environ. Microbiol*, **52**, 430–433.

Havelaar, A.H., Olphen, M. and Drost, Y.C. (1993). F-specific RNA bacteriophages are adequate model organisms for enteric viruses in fresh water. *Appl. Environ. Microbiol.*, **59**, 2956–2962.

Oswald, W.J. (1991). Introduction to advanced integrated wastewater ponding systems. *Wat. Sci. Tech.*, **24**(5), 1–7.

Sinton, L.W., Hall, C.H., Lynch, P.A. and Davies-Colley, R.J. (2002). Sunlight inactivation of fecal indicator bacteria and bacteriophages from waste stabilization pond effluent in fresh and saline waters. *Appl. Environ. Microbiol.*, **68**, 1122–1131.

'Active' filters for upgrading phosphorus removal from pond systems

A. Shilton*, S. Pratt*, A. Drizo**, B. Mahmood***, S. Banker*, L. Billings*, S. Glenny* and D. Luo*

*The Centre for Environmental Technology and Engineering, Institute of Technology and Engineering, Massey University, Private Bag 11222, Palmerston North, New Zealand

**Plant and Soils Science Department, University of Vermont, 105, Carrigan Drive, Burlington, VT 05405 Canada

***Unitec New Zealand, School of Built Environment, Engineering Discipline, Private Bag 92025, Auckland, New Zealand

Abstract This paper investigates limestone and iron slag filters as an upgrade option for phosphorus removal from wastewater treatment ponds. A review of 'active' filter technology and the results from laboratory and field research using packed columns of the different media is presented. It is shown that both limestone and iron slag can remove phosphorus but highlights that different types of limestone give markedly different performance. Filter performance appears to be improved by increasing temperature and by the presence of algae, presumably because of its tendency to elevate pH. Performance is related to hydraulic retention time (HRT), but this relationship is not linear, particularly at low HRTs. Importantly for future research, the results from field-testing with pond effluent show significant differences compared to those obtained when using a synthetic feed in the laboratory. For the iron slag filter, higher performance was observed in the field (72% in field vs. 27% in laboratory, at a 12 hour-HRT), while the opposite was observed for the limestone (64% in laboratory vs. 18% in field, at a 12-hour HRT).

Keywords Active filters; iron slag; limestone; nutrients; phosphorus; waste stabilisation ponds

Introduction

Water pollution from domestic and agricultural effluents has become a problem of great concern over recent times. Discharge of nutrients, particularly phosphorus (P), leads to eutrophication, algal blooms and general deterioration of receiving water quality (Harper, 1995). The need for new regulations to improve water quality has been recognised around the world and solutions for P removal from wastewater are currently being researched across a range of different wastewater treatment technologies (Ribaudo, 2003).

The use of active filters for P removal appears to offer a promising 'appropriate technology' for upgrading wastewater treatment systems for smaller communities and farms. The term 'active' is defined here to mean that in addition to physical straining, the filter media has chemical properties that support further treatment mechanisms. To date the research literature has predominantly reported on their use as an integral part of 'wetland' treatment systems, however, they potentially also offer an excellent method of upgrading 'pond' treatment systems which are typically poor at removing P.

Research in the constructed wetlands area has indicated that the principal P removal/retention mechanisms are adsorption and/or precipitation and complexation with Fe, Al or Ca rich substrate (Richardson, 1985; Steiner and Freeman, 1989; Richardson and Craft, 1993) and therefore the selection of the material is a crucial factor for filter design (Kadlec and Knight, 1996; Mann, 1997; Johansson, 1999; Drizo et al., 1999; Drizo et al., 2002). Ideally, the material needs to be cheap and locally available, but should also have a suitable combination of physical and chemical properties such as hydraulic conductivity, porosity, granulometry, specific surface area, Fe, Al or Ca content (Steiner and Freeman, 1989; Zhu et al., 1997; Kadlec and Knight, 1996; Mann, 1997; Drizo et al., 1999; Johansson, 1999; Liénard et al., 2001; Drizo et al., 2002). Over the past 10 years, various materials have been tested for their ability to retain P around the world (Drizo et al., 2002; Molle et al., 2003) including gravel, various sands, fly ash, light expanded clay aggregates, shale, bauxite, zeolite, half-burnt dolomite, laterite, wollastonite, marble, calcite, crushed concrete, and blast furnace slag. In Quebec, a thorough investigation was carried out comparing the performance of 57 locally available materials (Forget et al., 2001). Of those, electric arc furnace slag (EAF slag) had one of the highest P retention capacities and was selected for further examination in a full-scale system. Other authors also observed a high affinity of slag material for P at both low (5–10 mg/L, Yamada et al., 1986; Baker et al., 1998; Johansson, 1999; Crolla and Kinsley, 2002) and high (40–400 mg/L) loading rates (Drizo et al., 2002; Naylor et al., 2003; Lospied, 2003).

Filter materials rich in Fe, Al and Ca are known to have pH values that are usually either lower or higher than neutral and as such are detrimental for aquatic plant growth (Gersberg et al., 1986; Brix, 1994; Tanner, 2001). It is noted that this problem can be overcome by employing a multistage wetland design consisting of two or more flow cells, where some cells are planted (to enhance organic matter, suspended matter and nutrient removal via plant and microbial uptake) and others left unplanted, containing only the reactive material to enhance P retention (Brix, 1994; Drizo et al., 1999; Platzer, 2000; Liénard et al., 2001; Drizo et al., 2002). While these unplanted wetland sections may be very similar in appearance to what would be used for upgrading a pond, it must be remembered that the wetland and pond environments do have several different characteristics. Of particular relevance in ponds is the consumption of carbon dioxide by high algal concentrations that can result in significant elevation of pH. To date there has been very limited testing of the 'active filters' for pond applications.

The majority of the investigations so far have been conducted in laboratory batch experiments, while the longer-term studies in columns, pilot and full-scale systems have been limited (Mann, 1997; Baker et al., 1998; Johansson, 1999; Drizo et al., 1999, 2002; Gruneberg and Kern, 2001; Naylor et al., 2003; Molle et al., 2003).

This paper reports on a range of studies undertaken using two promising filter materials (iron slag, as discussed above, and limestone which is Ca rich and widely available throughout New Zealand). The primary objective of the paper is to present a range of preliminary findings to form a platform for further research into the application of active filters for upgrading ponds for P removal. Secondly the paper compares and contrasts performance from laboratory experiments feed on synthetic wastewater against those obtained under real field conditions so as to provide insight into methodology development for future research.

Laboratory and field-scale experiments
This section of the paper firstly reports on a series of shorter batch studies and then presents data obtained from columns run continuously for in excess of a year in the laboratory and for half a year in the field using both limestone and iron slag media.

Batch experiments
In these studies the work focused on limestone alone and was conducted in beakers batch fed with a synthetic P solution. A range of variables were assessed including media type and characteristics, temperature and the influence of algae to assess the sensitivity of these parameters.

Six kinds of limestone were obtained from different quarries throughout New Zealand. Density, porosity and calcium carbonate components were measured. The composition of these varied significantly as did performance. As might be expected, higher performance appeared to be linked to higher porosity. The limestone selected for further use in this study was sourced locally from the Tararua region in the North Island of New Zealand. It had a bulk density of $1.34\,g/cm^3$, an individual density of $2.68\,g/cm^3$, a bulk porosity of 50% and a $CaCO_3$ component of 75%. It was noted to become slightly crumbly after soaking as opposed to other limestones, which had very flat and hard surfaces. It is perhaps important to note here that some previous researchers have simply reported the testing of 'a limestone'. Clearly the characteristics of different limestones do vary widely as does their performance within an active filter. It is believed that this might explain the relatively poor results obtained it the limestone used by Strang and Wareham (2002).

A number of simple batch experiments were conducted to assess if temperature had any influence at temperatures ranging from 4 °C up to 35 °C. It was found that temperature did have a significant effect with performance seen to vary by 60% over the temperature range tested. While more detailed analysis of the reaction kinetics is needed, this preliminary finding draws attention to the need to account for temperature in future experimentation. It also highlights that this technology may be particularly appropriate in warmer climates such as tropical regions where ponds are already extensively used.

A key point of difference in conducting a laboratory experiment using a synthetic P solution as compared with a field experiment using real pond effluent, is the high algae concentration contained within a pond. As mentioned previously, algal activity can elevate the pH in a pond. Because pH can have a significant impact on mechanisms such as precipitation this requires further consideration. To study this, batch testing was conducted using two solutions with and without the presence of algae and clearly indicated that the presence of the algae did correspond with enhanced phosphate removal efficiency.

While the batch experiments were effective for preliminary testing it was felt that the work should be continued using columns where a deeper media bed is used with a continuous through flow of wastewater. While variables such as temperature and P concentration can be held constant in the laboratory, the apparent influence of the algae pointed to the importance of also undertaking experimentation in the field.

Continuous flow column testing – laboratory
In these experiments both the Tararua limestone and an iron slag (of similar size) sourced from the electric arc furnace at the BHP steel mill in Waiuku, New Zealand were tested. This iron slag media had a bulk density of $1.37\,g/cm^3$, an individual density of $3.04\,g/cm^3$, and a bulk porosity of 45%. Columns of 150 mm diameter were constructed and filled to a depth of approximately 1 m with the media. Pumps then delivered a continuous solution containing 10 mg/l of P from a holding tank up through the flooded column. The column testing was undertaken in a constant temperature laboratory at 25 °C.

A column of the limestone was run for approximately 10,000 hours at a hydraulic retention time (HRT) of 12 hours. One slag column was tested at HRTs of 3 and 48 hours and another slag column was tested at HRTs of 12 and 72 hours. The total testing time for the slag columns was approximately 5000 hours at each of the two different HRTs.

At the comparable HRT of 12 hours it was found that the limestone filter performed far better than the iron slag filter giving an average P removal of 64% compared to 27% for the slag. It is noted that a trend for both filter types was for a start-up effect where the P removal efficiencies were initially high, then dropped before settling to a more steady-state performance. This phenomenon was also observed in a similar experiment in Canada (Drizo *et al.*, 2002).

Understanding the relationship between filter performance and hydraulic residence time is important for filter sizing. The effect of HRT on the performance of the iron slag filters for HRTs of 3, 12, 48 and 96 hrs is shown in Figure 1. It can be seen that the performance of the slag filter was shown to increase with an increase in HRT, presumably because the contact time for adsorption, precipitation and complexation was enhanced. It is noted that the relationship is not linear, particularly for low HRTs.

Continuous column testing – in the field

The field experiment was conducted at a local waste stabilisation pond system (Ashhurst, NZ) on the final effluent from a two-stage facultative pond system. Two columns (identical to those use in the laboratory experiments) were established, again using the Tararua limestone and iron slag, and continuously fed with the final pond effluent. Both were operated at a 12-hour HRT. These experiments were conducted over a period of approximately 4400 hours.

The results were surprising. In a reversal from the laboratory work, the efficiency of the slag filter was markedly improved achieving an average performance of 72% (up from 27% in the laboratory).

Conversely, the performance of the limestone filter now decreased to an average of only 18% significantly down from the value of 64% obtained under laboratory conditions. Interestingly, this behaviour of the limestone is similar to that reported by Arias *et al.*, (2003) who also observed that Ca rich filter material (calcite) had higher performance in laboratory experimentation as compared to the field.

Continuous column testing – comparison of laboratory and field results

Based on the findings of the batch experiments, the presence of the algae in the field pond would have been expected to improve the performance in the field experiment, as was the case for the iron slag. Alternatively because the temperatures at the field were certainly much lower than those of the constant temperature room at the laboratory, this would have been expected to have the effect of lowering the performance in the field, as was the case for the limestone.

While this result was unexpected it is certainly interesting and indicates that:
1. The predominant mechanisms acting to remove the P within the filter may be different between these two materials
2. The use of laboratory experiments to research active filters may not be entirely representative of the full-scale systems they are meant to represent.

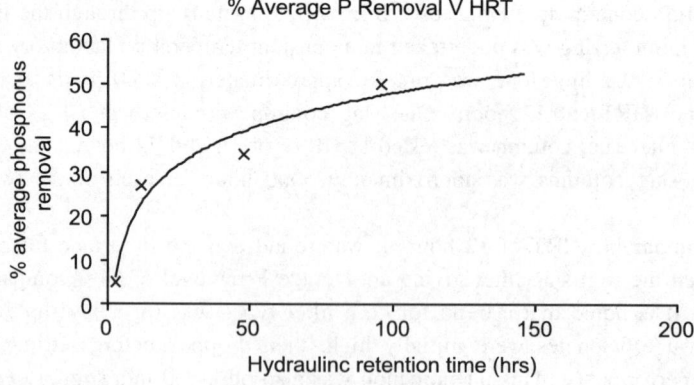

Figure 1 Iron slag filter performance in the laboratory as a function of hydraulic residence time (HRT)

Conclusions

While there is considerably more literature on active P removal filters in the wetlands field, it is noted that differences between the wetland and pond environments require consideration.

The performance of limestone filters varies markedly with the type of limestone used. While the Tararua limestone was reasonably effective in the laboratory, it performed poorly in the field.

Filter performance appears to improve with increasing temperature. The consequence of this being that active filters may be more effective in warmer climates.

Algae activity in ponds appears to improve filter performance, potentially due to the consequential elevation of pH that, in turn, improves precipitation reactions.

Increasing HRT also improves removal efficiency, but this relationship is non-linear with the degree of improvement gained reducing with respect to time.

Iron slag proved to be an effective media in field conditions achieving an average P removal of over 70% at a 12-hour HRT.

Laboratory based research using synthetic wastewater is not effective at representing full-scale systems.

Considering the expensive alternative of chemical dosing, we must investigate more 'appropriate' technologies for upgrading P removal to ensure that pond technology remains a viable treatment option in coming decades. The work presented here can only be considered a start, but given the good results produced by the iron slag filters, it seems clear that with further research this technology offers a very promising solution for upgrading ponds to achieve improved P removal in the future.

Acknowledgements

Acknowledgments are made to the Palmerston North City Council for their financial support of this project and for the access to their ponds for the field research; and to Steelserv for supply of the iron slag material.

References

Arias, C.A., Brix, H. and Johansen, N.-H. (2003). Phosphorus removal from municipal wastewater in an experimental two-stage vertical flow constructed wetland system equipped with a calcite filter. *Wat. Sci. Tech.*, **48**(5), 51–58.

Baker, M.J., Blowes, D.W. and Ptacek, C.J. (1998). Laboratory development of permeable reactive mixtures for the removal of phosphorus from onsite wastewater disposal systems. *Environ. Sci. Technol.*, **32**, 2308–2316.

Brix, H. (1994). Functions of macrophytes in constructed wetlands. *Wat. Sci. Tech.*, **29**(4), 71–78.

Crolla, A. Kinsley, C.B (2002). The Performance of a Free Water Surface Constructed Wetland in the Treatment of Farmstead Runoff. National Conference on Agricultural Nutrients and Their Impact on Rural Water Quality Proceedings, held in Waterloo, Ontario, April 29– 30, 2002.

Drizo, A., Forget, C., Chapuis, R.P. and Comeau, Y. (2002). Phosphorus removal by EAF steel slag – A parameter for the estimation of the longevity of constructed wetland systems. *Environ. Sci. Tech.*, **36**.

Drizo, A., Frost, C.A., Grace, J. and Smith, K.A. (1999). Physico-chemical screening of phosphate-removing substrates for use in constructed wetland systems. *Wat. Res.*, **33**, 3595–3602.

Forget, C., Drizo, A., Comeau, Y. and Chapuis, R.P. (2001). Élimination du phosphore d'effluents de pisciculture par marais artificiel à substrat absorbant (Elimination of phosphorus from aquaculture effluents by constructed wetland with absorbent substrate). In *Americana 2001, Réseau Environnement, Montreal, 28–30 March*.

Gersberg, R.M., Elkins, B.V., Lyon, S.R. and Goldman, C.R. (1986). Role of aquatic plants in wastewater treatment by artificial wetlands. *Wat. Res.*, **20**, 363–368.

Gruneberg, B. and Kern, J. (2001). Phosphorus retention capacity of iron-ore and blast furnace slag in subsurface flow constructed wetland. *Wat. Sci. Tech.*, **44**(11-12), 69–75.

Harper, D. (1995). *Eutrophication of Freshwaters*, 2nd ed., Chapman and Hall, London.

Johansson, L. (1999). Blast furnace slag as phosphorus sorbents – Column studies. *Sci. Total Environ.*, **229**, 89–97.

Kadlec, R.H. and Knight, R. (1996). *Treatment Wetlands*, Lewis Publishers, Chelsea, MI.

Liénard, A., Guellaf, H. and Boutin, C. (2001). Choice of the sand for sand filters used for secondary treatment of wastewater. *Wat. Sci. Tech.*, **44**(2–3), 189–196.

Lospied, C. (2003). *Évaluation des capacités et des conditions d'enlèvement du phosphore dissous par les scories d'aciéries* (Evaluation of capacities and conditions of dissolved phosphorus removal by steelworks slag). M.Sc. Thesis Ecole Polytechnique, Montreal, Canada, (2003).

Mann, R.A. (1997). Phosphorus adsorption and desorption characteristics of constructed wetland gravels and steelworks by-products. *Austr. J. Soil Res.*, **35**, 375–384.

Molle, P., Lienard, A., Grasmick, A. and Iwema, A (2003). Phosphorus retention in subsurface constructed wetlands: investigations focused on calcareous materials and their chemical reactions. *Water Science & Technology* **48**(5), 75–83

Naylor, S., Brisson, J., Labelle, M.A., Drizo, A. and Comeau, Y. (2003). Treatment of freshwater fish farm effluent using constructed wetlands: the role of plants and substrate. *Wat. Sci. Tech.*, **48**(5), 215–222.

Platzer, C (2000). Development of reed bed systems – A European perspective. *Proceedings of the 7th International Conference on Wetland Systems for Water Pollution Control, University of Florida, Lake Buena Vista, Florida, 11–16 Nov., 2000*; pp. 23–27.

Ribaudo, M. (February 2003). Managing Manure; New Clean Water Act Regulations Create Imperative for Livestock Producers. *Economic Research Service USDA. Amber Waves*, **1**(Issue **1**).

Richardson, C.J. (1985). Mechanisms controlling phosphorus retention capacity in freshwater wetlands. *Science*, **228**, 1424–1427.

Richardson, C.J. and Craft, C.B. (1993). Effective phosphorus retention in wetlands, fact or fiction? In Moshiri, G.A. (ed.), *Constructed Wetlands for Water Quality Improvement*. Lewis Publishers, Boca Raton, FL, pp. 271–282.

Steiner, G.R. and Freeman, R.J. (1989). Configuration and substrate design consideration for constructed wetlands wastewater treatment. In Hammer, D.A. (ed.), *Constructed Wetlands for Waste Water Treatment: Municipal, Industrial and Agricultural*, Lewis Publishers, Boca Raton, FL, pp. 499–508.

Strang, T.J. and Wareham, D.G (2002). Phosphorus removal in a waste stabilization pond system with limestone rock filters. *Conference Papers of the 5th IWA International Specialist Group Conference on Waste Stabilisation Ponds. Auckland, New Zealand*.

Tanner, C.C. (2001). Plants as ecosystem engineers in subsurface-flow treatment wetlands. *Wat. Sci. Tech.*, **44**(11–12), 9–17.

Yamada, H., Kayama, M., Saito, K. and Hara, M. (1986). A fundamental research on phosphate removal by using slag. *Wat. Res.*, **20**(5), 547–577.

Zhu, T., Jenssen, P.D., Maehlum, T. and Krogstad, T. (1997). Phosphorus sorption and chemical characteristics of lightweight aggregates (LWA) potential filter media in treatment wetlands. *Wat. Sci. Tech.*, **35**(5), 103–108.

Performance of a pilot-scale high rate algal pond system treating abattoir wastewater in rural South Australia: nitrification and denitrification

R.A. Evans, N.J. Cromar and H.J. Fallowfield

Department of Environmental Health, Flinders University, Adelaide, GPO Box 2100, Adelaide 5001, Australia (E-mail: *richard.evans@flinders.edu.au; nancy.cromar@flinders.edu.au; howard.fallowfield@flinders.edu.au*)

Abstract As part of a study examining the efficacy of high-rate algal pond treatment of high-strength abattoir wastewater, the impact of pond configuration and loading rate on nitrification was determined. The extent of nitrification in all ponds was consistent with mass balance estimates of oxygen demand and availability. Deeper ponds were more stable nitrifying systems, with shallow ponds displaying greater variation in response to changes in nitrogen loading. In a separate experiment the pond system was modified by covering a part of an in-series HRAP to exclude light, providing conditions suitable for denitrification. Specific denitrification rates were often within the range typical for endogenous carbon sources, with mass balance calculations indicating removals of up to 95%.

Keywords Denitrification; high rate algal pond; nitrification; wastewater

Introduction

The high rate algal pond (HRAP) wastewater treatment system was first developed in the United States during the 1950s as a refinement of the waste stabilisation pond and offers great potential as a means to treat both low and high strength domestic and industrial wastewaters (Oswald, 1988a, b). HRAP raceways (tanks) are generally less than one metre in depth and are mechanically mixed, usually by paddlewheel, resulting in higher algal productivity due to an improved light climate. In mixed wastewater treatment systems such as HRAPs, naturally derived bacterial populations facilitate oxidation of organic material and nutrient removal using oxygen provided by photosynthetic algae. Wastewater from the meat processing industry frequently contains high concentrations of both organic and inorganic forms of nitrogen, most commonly, ammonia and proteins/amino-acids (Johns *et al.*, 1995). Whilst ammonium is a readily available nitrogen source, assimilated in preference to nitrate or nitrite in many algae (Kaplan *et al.*, 1986), the non-ionic form, 'free ammonia', is often toxic at relatively low concentrations (Abeliovich and Azov, 1976; Azov and Goldman, 1982; Veenstra *et al.*, 1995) and is the predominant form found at the high pH typical of HRAP systems. A solution to this problem is to increase retention time (reduce volumetric loading rate) and allow ammonia loss, assimilation and conversion processes to prevent the concentration of NH_3 rising to potentially toxic levels (Abeliovich, 1986). Longer retention times also favor the development of a bacterial biomass with the potential for nitrification (Metcalf and Eddy, 1991), a process which also benefits from low free ammonia concentration. Inhibition of nitrifiers can occur at ammonia concentrations as low as 10 mg/L (*nitrosomonads*) and 0.1–6 mg/L (*nitrobacters*) (US EPA, 1993; Surmacz-Górska *et al.*, 1997). A nitrified effluent may be further treated via denitrification, a process in which certain genera of bacteria use modified aerobic metabolic pathways requiring an organic or inorganic electron donor such as nitrate and release nitrogen gas rather than carbon dioxide, a process more correctly termed anoxic rather than anaerobic (US EPA, 1993). In many species of these bacteria the presence of minimal amounts of oxygen permits normal aerobic pathways to be used, preventing nitrogen removal. Numerous studies have shown that the rate of denitrification decreases linearly to zero as the DO concentration increases to 1 mg/L (Metcalf and Eddy, 1991). As a result it is desirable to provide conditions which limit the inhibitory impact of oxygen (Metcalf and Eddy, 1991), typically a separate 'anoxic' zone. In this paper we examine the potential for nitrogen removal via combined nitrification/denitrification in a modified pilot-scale HRAP system treating high-strength abattoir wastewater.

Methods

The pilot treatment system consisted of four high-rate ponds, Table 1, and a dedicated anaerobic pretreatment tank. Wastewater derived from stockyard washing and the evisceration stage of carcass processing was passed through a rendering plant where coarse solids and fats were removed. Approximately 9 m^3 of this wastewater was pumped each day into the anaerobic pretreatment tank (retention time of 15 days). Each HRAP was mixed by an 8-bladed stainless steel paddlewheel operated at a rotation rate of 3 rpm, resulting in a water velocity of 0.2 ms^{-1}. Ammonia concentrations in excess of 130 mg/L suggested a minimum retention time of approximately 10 days in order to limit the phyto-toxic impact of free ammonia upon the algal biomass and permit substantial levels of nitrification (Albeliovich, 1986). Oxygen transfer coefficients were calculated prior to the establishment of a biomass by measurements of the reaeration rate of potable water upon addition of anhydrous sodium sulphite (Metcalf and Eddy, 1991; Fast et al. 1999). These data were used as the basis for estimates of atmospheric oxygen diffusion into the treatment system throughout the trial (Evans et al., 2003). Chemical parameters were measured according to procedures in *Standard Methods for the Examination of Water and Wastewater* (APHA, 1992) and included chlorophyll a, COD, SCOD, SS, VSS, nitrate, nitrite, ammonia and TKN (Fallowfield et al. 2001). Online measurements of dissolved oxygen (Danfoss™ 2100 oxygen probes), temperature (PT-100) and PAR (Licor™) were recorded every 15 minutes for the duration of the trial by datalogger (Datataker™ 500).

Nitrification experiment

In order to determine the optimal configuration for nitrification, a trial was run with ponds in parallel, with configuration differing with respect to depth, loading rate, and as a consequence, retention time (rt), (Table 1). Anaerobic effluent was pumped to the four HRAPs via separate timer controlled pumps on a twice daily basis. Pond depth was

Table 1 Pilot-plant configuration, January–December 1999 (nitrification experiment) and August–December 2000 (Denitrification experiment)

Pond	Depth (m)	Area (m^2)	Volume (m^3)	Pumping time (minutes)	Retention time (days)
Nitrification experiment (parallel configuration)					
1	0.6	8.6	5.2	8	21
2	0.6	8.8	5.3	4	44
3	0.3	8.7	2.6	4	22
4	0.3	8.4	2.5	8	11
Denitrification experiment (in-series configuration)					
1 (standard HRAP)	0.6	8.6	5.2	8	21
2 (denitrification pond)	0.4	8.8	3.5	*	θ_{den}
3 (standard HRAP)	0.3	8.7	2.6	–	θ_{p3}

*determined by DO controlled pump. θ_{den}(denitrification pond retention time) = function of pond 1 effluent volume (constant) and number of daily pump events. θ_{p3}(pond 3 retention time) = function of denitrification pond retention time

controlled by an adjustable PVC outflow pipe situated in the floor of each pond. A detailed description of the parallel configuration is available in Evans *et al.* (2003).

Denitrification experiment

For in-series operation, ponds were connected directly with in-floor PVC pipes (capped during in-parallel operation). Influent entered pond 1 by timer controlled pump on a twice-daily basis and exited via pond 3. Pond 2 was converted into an denitrification zone by covering the surface to exclude light. The existing paddlewheel provided sufficient mixing, however, constant diffusion of atmospheric oxygen via the paddlewheels required either that mixing was intermittent or that additional anaerobically pretreated wastewater (BOD/COD) be provided to reduce average dissolved oxygen of the 'anoxic' pond to appropriate levels. The latter option had the benefit of maintaining the biomass in suspension and providing carbon for denitrification. The addition of wastewater was controlled by use of a datalogger (Datataker™ 500) and custom built relay which activated a pumping event into the denitrification pond when dissolved oxygen levels exceeded 0.5 mg/L. When this limit was exceeded an alarm was activated, upon which another reading was taken after 30 minutes delay. If the alarm condition was still satisfied the pump was triggered for 2.5 minutes. This delay was incorporated into the program to prevent unnecessary pumping events resulting from DO transients occurring as a result of water movement past the probe membrane. Accurate knowledge of the quantity of carbon rich anaerobic wastewater added to balance paddlewheel derived oxygen permitted more precise determination of the various rate constants of denitrification and BOD/COD removal. The third pond in the series, a conventional HRAP, with depth of 0.3 m was intended to limit carryover of COD/BOD or ammonia resulting from the addition of anaerobic wastewater into the anoxic zone. The literature concerning the biology and design criteria necessary to achieve denitrification is extensive with respect to activated sludge, however the authors are not aware of any attempts to achieve nitrogen removal by combined biological nitrification/denitrification using HRAPs. As a consequence, estimates of rates of denitrification in this system were based on a time series mass balance of nitrate removals (Fallowfield *et al.*, 2001) and Equation 1 of Metcalf and Eddy (1991).

$$U'_{DN} = U_{DN} \times 1.09^{(T-20)}(1 - DO) \tag{1}$$

where

U'_{DN} = overall denitrification rate
U_{DN} = specific denitrification rate, mass NO$_3$-N/ mass MLVSS day
T = wastewater temperature, °C
DO = dissolved oxygen in wastewater

Results and discussion

Nitrification experiment

A box and whisker plot of percentage nitrification during the trial period is shown in Figure 1. The box represents the inter-quartile range containing 50% of the data, the whiskers are lines that extend from the box to the highest and lowest values, excluding outliers, with the line across the box indicating the median. Pond 3, with the same retention time and volumetric loading rate as Pond 1, differing only with respect to depth, demonstrated more variation in terms of nitrification. Pond 2, whilst displaying the peak nitrification capacity would have required twice the land area of Pond 1 to treat a given volume of wastewater. The HRAP with the highest volumetric loading rate and shortest retention time, Pond 4, displayed the lowest median nitrification. A summary of HRAP

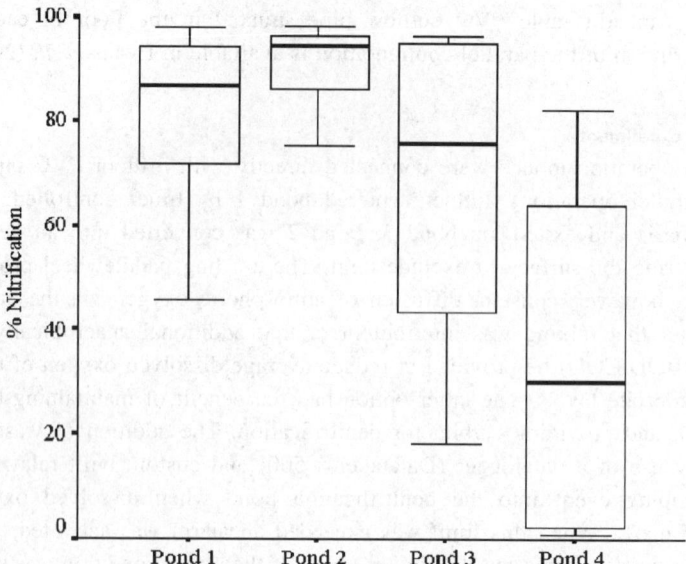

Figure 1 Percentage nitrification, January–December 1999. Pond 1, depth = 0.6 m, Rt = 21 days. Pond 2, depth = 0.6 m, Rt = 44 days. Pond 3, depth = 0.3 m, Rt = 22 days. Pond 4, depth = 0.3 m, Rt = 11 days

effluent concentrations with respect to nitrogen species is shown in Table 2. Substantial losses of nitrogen, with median value of 30–40% over the trial period, were identified by mass balance calculations. Experiments identified ammonia volatilisation as an important mechanism for this nitrogen loss (Fallowfield et al. 2001).

Changes in the C:N ratio are known to influence the efficiency of nitrification (US EPA, 1993). Specifically, wastewater with lower carbon concentration may favor slow growing autotrophic nitrifiers. Therefore, differences with respect to the extent of nitrification between ponds, are likely to be a function of influent COD and ammonia volumetric loading rate as well as retention time and dissolved oxygen concentration. This is illustrated by the factor analysis shown in Figure 2.

A factor analysis may allow correlations amongst a number of variables to be summarised as being mainly due to two or more components (Masters, 1995). The components that are 'factored out' are, by definition, independent of each other. If variables aggregate at opposite ends of a component scale it is an indication that they are inversely correlated with respect to that component. If variables form a group this is indicative that they may be impacted upon by the component(s) in a similar way. With respect to component 1 in Figure 2, pond DO concentration and Pond 4 percentage nitrification form a group which is inversely correlated with respect to a grouping of COD and ammonia volumetric loading, suggesting that oxygen limitation was largely responsible for the reduced nitrification in Pond 4. Nitrification in Ponds 1 and 3 is weighted most strongly upon component 2, indicating that variation in the applied loading rate, and therefore DO concentration was unimportant for nitrification in these ponds, whilst being enhanced in Pond 2 at higher loading rates.

Table 2 Nitrogen speciation, HRAP effluent, mean ± standard deviation, mg/L

Nitrogen form	Pond 1	Pond 2	Pond 3	Pond 4
Ammonia as N (mg/L)	9.0 ± 10	3.1 ± 4.6	12 ± 16	46 ± 48
Nitrite as N (mg/L)	2.6 ± 4.7	1.9 ± 3.5	2.9 ± 4.6	3.3 ± 4.7
Nitrate as N (mg/L)	68 ± 31	63 ± 34	60 ± 43	21 ± 21

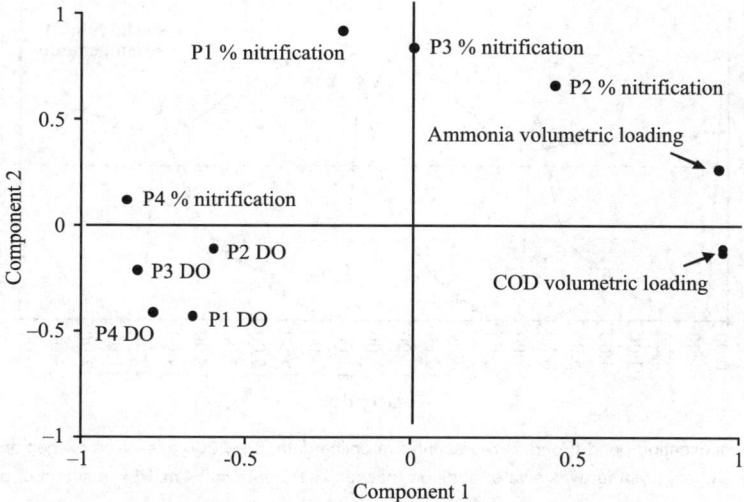

Figure 2 Factor analysis, component plot in rotated space. Pond % nitrification, dissolved oxygen, COD and ammonia volumetric loading rate (mg/L/day). January–December 1999. Pond 1, depth = 0.6 m, Rt = 21 days. Pond 2, depth = 0.6 m, Rt = 44 days. Pond 3, depth = 0.3 m, Rt = 22 days. Pond 4, depth = 0.3 m, Rt = 11 days

The enhancement of nitrification in Pond 2 at higher loading rates may be understood with reference to Figure 3. All ponds displayed a pattern of improved nitrification, with COD loading, to an 'optimal' level after which nitrification capacity declined. At the lowest COD (carbon) loading it is possible that nitrifying bacteria were carbon limited, whilst the highest loading rates favoured the growth of heterotrophs and in Pond 4 limited the amount of oxygen available for nitrification. In summary, a deep pond with intermediate retention time, such as Pond 1, provided the most stable nitrifying configuration. This configuration was used for the denitrification experiment.

Denitrification experiment

Representative data illustrating the change in dissolved oxygen and temperature in the denitrification pond (Pond 2) are shown in Figure 4. Ingress of dissolved oxygen from mixing and the nitrification pond (Pond 1) resulted in non-optimal DO levels for

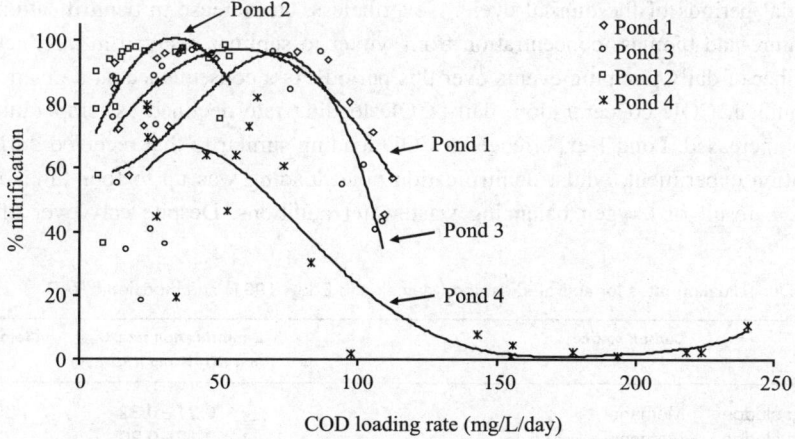

Figure 3 Pond % nitrification as a function of COD volumetric loading rate (mg/L/day). January–December 1999. Pond 1, depth = 0.6 m, Rt = 21 days. Pond 2, depth = 0.6 m, Rt = 44 days. Pond 3, depth = 0.3 m, Rt = 22 days. Pond 4, depth = 0.3 m, Rt = 11 days

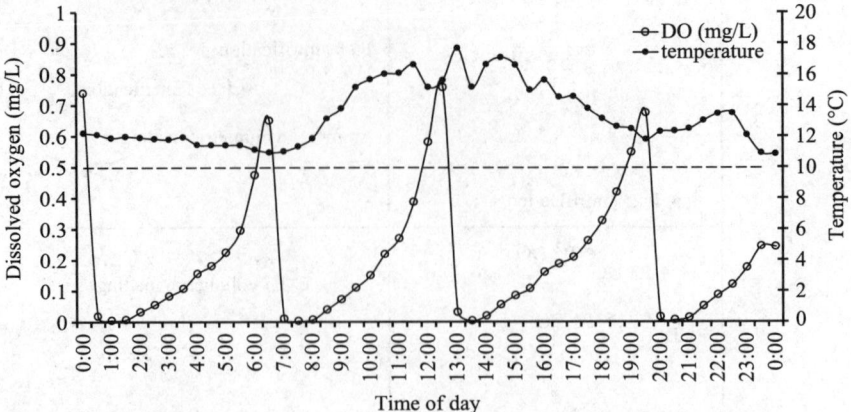

Figure 4 Denitrification pond (Pond 2) representative online data, 8-9-2000, dissolved oxygen and temperature. DO set point for wastewater addition (dashed line) depth = 0.4 m, Rt = function of pond 1 effluent volume (constant) and number of daily pump events

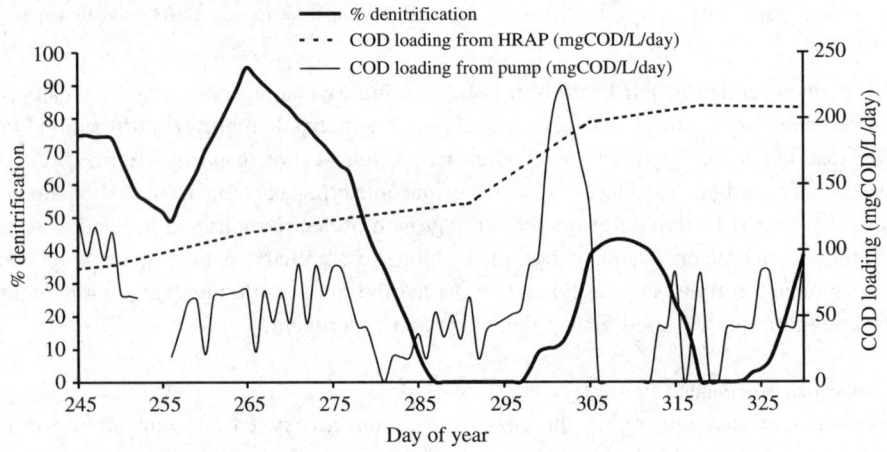

Figure 5 Denitrification pond (Pond 2), % denitrification and COD loading

substantial periods of the diurnal cycle. Nevertheless, an increase in denitrification pond temperature and biomass concentration from winter to summer resulted in a reduction in the number of daily pumping events over this period. As a consequence, at a given wastewater influent COD concentration, daily COD loading rate declined as temperature and biomass increased. Pond 1 experienced a COD loading similar to that received during the nitrification experiment, whilst denitrification pond loading was up to four times that of Pond 1, a result of oxygen balancing wastewater additions. Despite carryover of COD

Table 3 Denitrification rates for activated sludge (Metcalf and Eddy, 1991) and modified HRAP

Process	Carbon source	Denitrification rate, U_{DN}, (mg NO_3-N/mg VSS/day)	Temperature (°C)
Activated sludge	Methanol	0.21–0.32	25
Activated sludge	Methanol	0.12–0.90	20
Activated sludge	Wastewater	0.03–0.11	15–27
Activated sludge	Endogenous metabolism	0.017–0.048	12–20
HRAP	Wastewater and endogenous metabolism?	0–0.025	10–21

from the denitrification pond (Pond 2) the HRAP at the end of the series (Pond 3) displayed similar effluent soluble COD concentration to that observed during the nitrification trial.

Specific denitrification rates calculated as $mgNO_3$-N/mgMLVSS/day (Table 3), were often within the range typical for denitrification using endogenous carbon sources, with mass balance calculations indicating nitrate removal as high as 95%, Figure 5. The pattern of denitrification suggested a lack of available COD and/or low temperature reduced the denitrification potential of the pond during some periods. It follows that COD loading to the denitrification pond, a result of biomass exiting Pond 1, was not readily available to the denitrifying bacteria. Since DO concentrations in the denitrification pond were generally above those considered optimal for denitrification it is likely that the highest rates occurred during periods immediately after addition of wastewater, declining as DO levels increased and non-endogenous sources of carbon were removed.

Conclusions

The nitrification trial (in parallel configuration) showed that differences between ponds, with respect to the extent of nitrification, were a function of depth, retention time, loading rate and in the pond receiving the highest loading, oxygen availability. Pond depth was important in determining the extent of nitrification, as at a constant areal loading rate, increasing pond depth reduced the mass of oxygen required to oxidise daily additions of ammonia. As a consequence deeper ponds were more stable nitrifying systems, with shallow ponds displaying greater variation in response to increases in nitrogen loading. The denitrification trial (in series configuration) demonstrated that, with the correct configuration, HRAP treatment systems have the potential to produce a biomass capable of denitrifying high strength wastewater. Whilst denitrification rates were often of a similar order to those found in other treatment systems, simple modifications to the physical configuration of this proof of concept experiment would limit ingress of oxygen and permit higher rates of denitrification.

Acknowledgements

This research was supported by Meat and Livestock Australia. The authors would also like to thank Briony Vickery and Ben Pol for technical assistance.

References

Albeliovich, A. (1986). Algae in wastewater oxidation ponds. In *CRC Handbook of Microalgal Mass Culture*, Richmond, A. (ed.), CRC Press, Inc., Boca Raton, Florida.

Albeliovich, A. and Azov, Y. (1976). Toxicity of ammonia to algae in sewage oxidation ponds. *Appl. Environ. Microbiol.*, **31**, 801–806.

Azov, Y. and Goldman, J.C. (1982). Free ammonia inhibition of algal photosynthesis in intensive cultures. *Appl. Environ. Microbiol.*, **43**(4), 735–739.

Evans, R.A., Fallowfield, H.J. and Cromar, N.J. (2003). Characterisation of oxygen dynamics within a high-rate algal pond system used to treat abattoir wastewater. *Wat. Sci. Tech.*, **48**(2), 61–68.

Fallowfield, H.J., Cromar, N.J. and Evans, R.A (2001). The use of high rate algal ponds in the treatment of abattoir wastes. Final report for Meat and Livestock Australia, RPDA 501.

Fast, A.W., Tan, E.C., Stevens, D.F., Olson, J.C., Qin, J. and Barclay, D.K. (1999). Paddlewheel aerator oxygen transfer efficiencies at three salinities. *Aquacult. Eng.*, **19**, 99–103.

Johns, M.R., Harrison, M.L., Hutchinson, P.H. and Beswick, P. (1995). Sources of nutrients in wastewater from integrated cattle slaughterhouses. *Wat. Sci. Tech.*, **32**(12), 53–58.

Kaplan, D., Richmond, A.E., Dubinsky, Z. and Aaronson, S. (1986). *CRC Handbook of Microalgal Mass Culture*, Richmond, A. (ed.), CRC Press, Inc., Boca Raton, Florida, pp. 147–198.

Masters, T. (1995). *Neural, Novel and Hybrid Algorithms for Time Series Prediction*, John Wiley and Sons, Inc., Brisbane.

Metcalf and Eddy (1991), Tchobanoglous, G. and Burton, F.L. (1991). *Wastewater Engineering, Treatment, Disposal and Reuse*, McGraw-Hill, Sydney.

Oswald, W.J. (1988a). The role of microalgae in liquid waste treatment and reclamation. In *Algae and Human Affairs*, Lembi, C.A. and Waaland, J.R. (eds), Cambridge University Press, pp. 255–281.

Oswald, W.J. (1988b). Micro-algae and waste-water treatment. In *Micro-algal Biotechnology*, Borowitza, M.A. and Borowitza, L.J. (eds), Cambridge University Press.

Standard Methods for the Examination of Water and Wastewater (1992). AWWA/WEF/APHA, Washington, DC.

Surmacz-Górska, J., Cichon, A. and Korneliusz, M. (1997). Nitrogen removal from wastewater with high ammonia nitrogen concentration via shorter nitrification and denitrification. *Wat. Sci. Tech.*, **36**(10), 73–78.

U.S. Environmental Protection Agency (1993). *Process Design Manual for Nitrogen Control*, Office of Technology Transfer, Washington, D.C.

Veenstra, S., Al-Nozaily, F.A. and Alaerts, G.J. (1995). Purple non-sulfur bacteria and their influence on waste stabilisation pond performance in the Yemen Republic. *Wat. Sci. Tech.*, **31**(12), 141–149.

Anaerobic reactor/high rate pond combined technology for sewage treatment in the Mediterranean area

F. El Hafiane and B. El Hamouri

Wastewater Treatment and Reuse Unit, Department of Rural Engineering, Institut Agronomique et Vétérinaire Hassan II (IAV), BP 6202, Rabat-Instituts, Morocco (E-mail: b.elhamouri@iav.ac.ma)

Abstract Two high-rate, anaerobic/aerobic units were used to treat the sewage of the Institut Agronomique st Vétérinaire Hassan II (Morocco) campus in a 1,100 m^2-plant designed for 1,500 e.p. and receiving 63 m^3 per day. The anaerobic pre-treatment consisted of a two-step up-flow anaerobic reactor (TSUAR) comprising two reactors and one external settler all in series. The aerobic line, or post-treatment, consisted of a high-rate algal pond (HRAP) and one maturation pond in series. The system totalized a hydraulic retention time (HRT) of 9 days. A gravel filter (GF) was constructed behind the TSUAR to trap low-density particles. The TSUAR removed 80% of COD and 90% of SS within 48 h. Solids retention time in the reactors averaged 32 d with a specific sludge production of 0.28 g SS g^{-1} COD removed. Almost 93% of the sludge evacuated from the settler was stabilized. Specific biogas production from both reactors was 0.25 m^3 kg^{-1} COD removed. Used in this configuration, the HRAP lost its BOD removal activity and increased its nutrients and pathogens removal capabilities (tertiary treatment). Results showed that 85% of total nitrogen and 48% of total phosphorus were removed by the HRAP. Land area requirement of this combination was less than 1 m^2 per capita and filtered final effluent was of excellent quality (COD, 82 mg/l; TKN, 8.3 mg/l; total P, 2.7 mg/l, faecal coliforms, 2.4 10^3/100 ml and zero helminths eggs).
Keywords First-order reaction rate constant; high rate algal pond; Mediterranean area; nutrients removal; pathogens removal; sewage treatment; two-step up-flow anaerobic reactor

Introduction

High rate anaerobic reactors have been actively investigated for sewage treatment over the last two decades. Large-scale plants were implemented in tropical climates mainly in India and Latin America (van Haandel and Lettinga, 1994; Hulshoff Pol et al., 1997). Sewage pre-treatment in these systems presents many advantages. They can achieve high removal rates of organic matter and SS in relatively short HRTs, they produce less sludge than equivalent aerobic systems, require small land areas and have moderate operation and construction costs. Other advantages of anaerobic reactors are biogas recovery and use for energy purposes (Malina, 1962; McCarty, 1964; Lettinga et al., 1980).

Pilot-scale units implemented in temperate climates did not give satisfactory results. Investigations have shown that the low temperatures slow down the hydrolysis rate and favour the accumulation of an impairing mass of SS in the reactor (Wang, 1994). To overcome this limitation, it was proposed to switch to two-step reactors, with the first unit mainly operating as a trap and as a "hydrolyser" for SS (Van Lier et al., 1996; Zeeman and Lettinga, 1999; Elmitwalli et al., 2002).

The post-treatment is a stage where N, P and, in some instances faecal coliforms, are removed. Many systems can fulfil this function going from intensive to extensive units depending on the particular conditions of projects (van Haandel and Lettinga, 1994; Lettinga et al., 1980; Zeeman and Lettinga, 1999). The high-rate algal pond (HRAP) associated with MPs could play such a role. The HRAP has been tested for sewage and farm effluent treatment (Oswald and Glueke, 1959; Azov and Shelef, 1982; Picot et al., 1992; Green and Oswald, 1993; El Hamouri et al., 1994; Craggs et al., 2003; Evans et al., 2003). The HRAP was presented as an alternative system to facultative ponds to minimize the land area requirement and construction cost (El Hamouri et al., 2003). Basically, the HRAP has a three-fold role: i) biological degradation of organic matter using algae evolved oxygen (secondary treatment), ii) removal of nitrogen and phosphorus, and iii) exacerbation of faecal pathogen die-off conditions. These three functions are taking place simultaneously, making the overall treatment performance be dictated by the slowest function.

This paper describes a new approach to treat domestic sewage. The two step up-flow anaerobic reactor (TSUAR), developed for organic matter degradation and SS removal, is associated with an HRAP, in which organic matter degradation (secondary treatment) is dropped off while the rates of nutrients and pathogens removals (tertiary treatment) are increased.

Materials and methods

Average temperatures during the reporting period were 14 °C and 24 °C respectively for the cold and the hot season. The average wastewater flow was $63 \text{ m}^3 \text{ d}^{-1}$. The plant included a pre-treatment line based on a duplicated TSUAR line receiving $31.5 \text{ m}^3 \text{ d}^{-1}$ each. The post-treatment line included a high-rate algal pond (HRAP) and two maturation ponds (MP), all in series.

The TSUAR included two reactors (R_1 and R_2), a settler (S) and a gravel filter (GF) (Figure 1 and Table 1). Biogas was collected using external cupola-shaped covers made of acid-resistant glass fibre. The base of the covers was inserted in a (40 cm width × 40 cm depth) channel surrounding the reactors and filled with treated effluent to act as a water seal, preventing odour and biogas release.

The post-treatment unit included a 790 m^2-HRAP with an HRT of 5.2 d and two maturation ponds (MP_1 and MP_2) with dimensions of 17 m length, 5 m width and 1 m depth for an HRT of 0.7 d each. Configuration I (TSAR + HRAP + MP_1 + MP_2) had an overall HRT of 9 days, while configuration II (TSAR + GF + HRAP + MP_1) was obtained by introducing a gravel filter (GF) behind S and by-passing MP2.

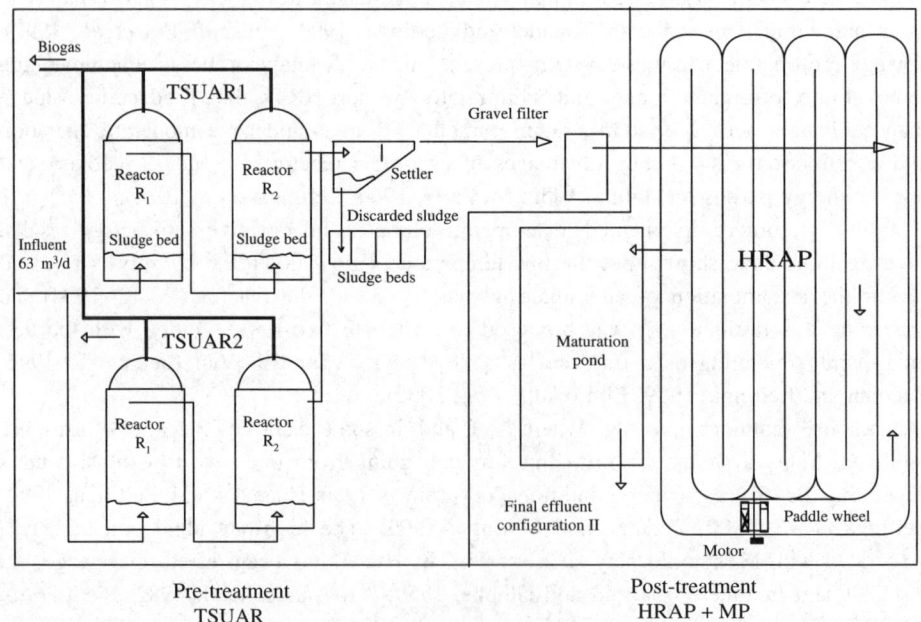

Figure 1 Layout of the treatment plant

Table 1 Dimensions and operation parameters of reactors R_1 and R_2

	Unit	Reactor R_1	Reactor R_2
Depth	m	5.30	5.00
Diameter	m	3.0	3.0
Effective volume	m^3	33	31
Average HRT	hour	24	23
Volumetric loading rate	kg COD $m^{-3} d^{-1}$	0.76	0.4
Up flow velocity	$m h^{-1}$	0.1–0.6	0.1–0.6

The settler S had dimensions of 2 m length, 0.7 m width and 1 m depth and was operated at an overflow rate of $1.5 m h^{-1}$. Trapped sludge in S was removed daily to the sludge drying beds using hydrostatic pressure. The GF consisted of a 1-mm PVC film lined basin operated at a hydraulic loading rate of $1 m d^{-1}$. The role of the GF was to remove low-density sludge particles escaping S.

24-hour composite samples were taken biweekly for main chemical characteristic analysis following Standard Methods (APHA, 1989) while daily *in situ* recording of temperature, pH, electrical conductivity (EC) and dissolved oxygen (DO) were also carried out. Settled COD (CODst) represented the fraction of CODt which did not settle down in 30 min in a 2-litres cylinder. Chlorophyll-a (Chl-a) was analysed following the method described by Pearson *et al.* (1987). The sludge velocity index (SVI) was determined following APHA (1989) and sludge granulometry of particles following the method of Laguna *et al.* (1999). Faecal coliforms (FC) were counted on grab samples using the MPN method (APHA, 1989) and helminth eggs were counted on composite samples following the flotation method described by Arther *et al.* (1981).

Results and discussion

Pre-treatment, anaerobic reactor unit

Performance of the TSUAR presented here are to be analysed under the conditions generally prevailing in the wastewater treatment facilities of small communities, which are characterized by highly varying hydraulic and organic loads. Standard deviations (SD) shown on Table 2 might be explained in this way. For instance, the flow had maximum and minimum values of 110 and $14 m^3 d^{-1}$ respectively. Also, large variations were recorded within the day. Half the daily flow ($30 m^3$) was received within six hours, precisely between 8:00 and 14:00.

On the other hand, the deliberate choice not to remove any excess sludge manually for operational simplicity, forced us to operate the reactors on "maximum sludge hold up" mode (van Haandel and Lettinga, 1994). Washout periods were followed by periods of sludge accumulation during which the sludge washout was at its minimum. The completion of a washout/accumulation cycle took 3 months with an average solid retention

Table 2 TSUAR performance

Parameter	Influent	Reactor R_1		Reactor R_2		Settler		Global RR
		Value	RR	Value	RR	Value	RR	
CODt (mgO$_2$/l)	800 ± 202	530 ± 220	34	380 ± 180	28	310 ± 145	18	61
CODst (mgO$_2$/l)	–	285 ± 105	–	159 ± 79	–	159 ± 81	–	80[1]
CODs (mgO$_2$/l)	420 ± 117	270 ± 97	36	120 ± 71	56	120 ± 67	–	71
BOD$_5$ (mgO$_2$/l)	390 ± 139	200 ± 88	49	150 ± 75	25	120 ± 67	20	69
SS (mg/l)	330 ± 85	300 ± 120	9	280 ± 117	7	230 ± 102	18	30
VSS (mg/l)	190 ± 57	150 ± 51	21	160 ± 41	−7	105 ± 41	34	45

[1]RR in % is calculated as follows (CODt − CODst) * 100/CODt; CODs: soluble COD

time of 32 days. Measurements of sludge bed thickness showed a permanent bed of at least 1 m at the bottom of each of the two reactors even during intensive SS washout periods. The average SS concentration was 23,000 mg/l, with a VSS/SS ratio of 0.79, in R_1 and 14,000 mg/l and 0.70 for R_2 (Figure 2). Specific sludge production in the TSUAR was estimated to be 0.28 g SS g^{-1} COD removed. Based on the extensive and continuous sampling analysis program achieved (136 sampling campaigns), a certain confidence is to be attributed to the average removal rates presented in Table 2, stating that TSUAR with an organic loading rate (OLR) of 760 for R_1 and 400 g m^{-3} d^{-1} for R_2, and a global HRT of 48 h removed an average of 80% of CODst or 70% of BOD$_5$ (Table 2). However, TSUAR performance in SS removal under configuration I was not satisfactory. The removal rate did not exceed 30% and might be explained by the maximum sludge hold up mode adopted and also by the occurrence of low-density particles for which the settler was ineffective. Indeed, particle distribution analysis of the bulk of SS leaving reactors R_2 was dominated by two types of particles: i) reticulated particles of 100 and 350 μm diameter having an SVI of 20 mg/l; and ii) low-density particles (probably biological material in an advanced stage of digestion) of 60 μm diameter and an SVI of 35 mg/l. The settler was inefficient in trapping the 60 μm particles. They were carried away to the first component of the post-treatment unit, the HRAP, in which they reduce the light penetration through the water column and contributed to the build up of unwanted sediment in the pond (El Hafiane et al., 2002). In configuration II, these troublesome particles were successfully stopped in the GF.

Recorded specific biogas production in the TSUAR was 0.25 m^3 kg^{-1} of COD removed. Methane represented 77%; nitrogen, 14%; carbon dioxide, 2% and H$_2$S was only found in traces. The noticeable N content might be due to an uncommon anaerobic denitrification process, most likely the ANAMMOX process described by Mulder et al. (1995).

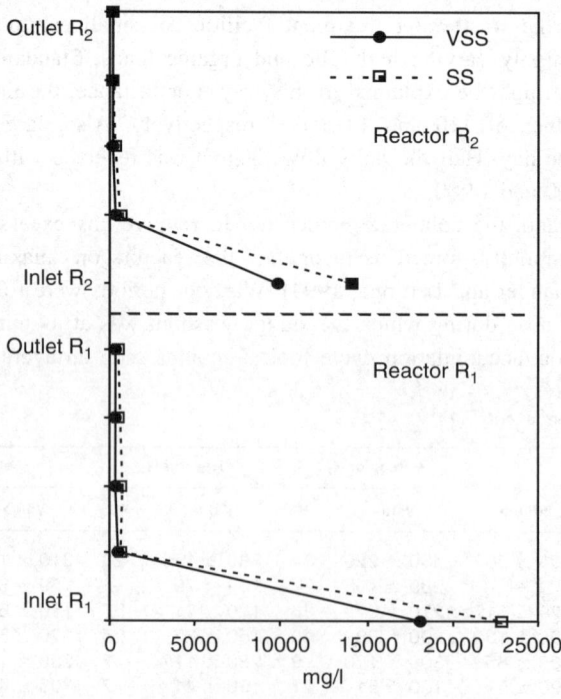

Figure 2 SS and VSS profiles in reactors R_1 and R_2

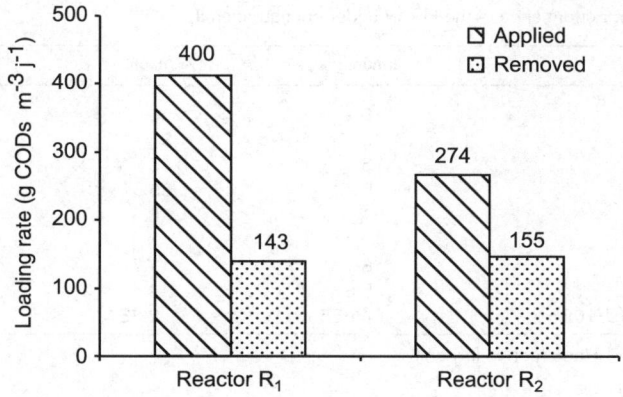

Figure 3 Applied and removed CODs for reactors R_1 and R_2

Table 3 Physicochemical characteristics of importance for anaerobic treatment in the TSUR

Parameter	Influent	Reactor R_1	Reactor R_2	Settler
pH	6.9 ± 0.29	6.6 ± 0.2	6.8 ± 0.12	6.8 ± 0.15
Temperature (°C)	19.5 ± 3.1	20 ± 3.3	21.5 ± 3.6	21 ± 3.3
EC (µS/cm)	1290 ± 260	1400 ± 116	1415 ± 121	1420 ± 108
VFA (mg/l)	120 ± 56	170 ± 44	70 ± 35	–
Alkalinity (mg CaCO$_3$/l)	120 ± 42	164 ± 63	204 ± 42	–

VFA: volatile fatty acids

Figure 3 shows that R_2 achieved similar CODs removals to R_1. However, R_2 load was 2/3 of that of R_1, indicating that CODs removal rate in R_2 was 1.5 times higher than in R_1. We concluded from this and from pH and volatile fatty acids concentrations analysis (Table 3) that reactor R_1 might function as a trap for particulate COD (CODp) and as a digester; in which the acidogenesis process and the hydrogenotrophic methanogenic bacteria could dominante. On the other hand, reactor R_2 might function as a digester with a domination of acetotrophic methanogenic bacteria.

Post-treatment, high-rate algal pond unit

Under configuration II, organic matter removal rate was almost nil in the HRAP while those of N, P and pathogens were improved (Table 4). Nitrogen was removed at 86% among which 39% was removed by algae uptake and 46% was lost by ammonia stripping (Figure 4). Phosphorus was removed at 66% for which algae uptake and P precipitation under the effect of high pH values each accounted for 50%. Residual concentrations of N and P were 8.3 and 2.7 mg/l respectively. These figures are to be compared with those previously reported by El Hafiane et al. (2003) for configuration I.

First-order reaction rate constant, $k_{20°C}$ for CODt removal, calculated following the method published earlier (El Ouarghi et al., 2000; El Hamouri et al., 2003) decreased from $+0.038$ in configuration I to the negative value of $-0.250\,d^{-1}$ in configuration II. This indicates that no more organic matter was degraded in the HRAP under configuration II. At the same time, $k_{20°C}$ for N and P removals were multiplied by 2.3 and 1.6 respectively (Table 5). Faecal coliforms removal rate also improved under configuration II; an average residual concentration of 2.4×10^3 unit/100 ml was recorded on MP1 (result not shown) leading to an overall removal of 3.92 log unit.

The significance of the negative $k_{20°C}$ CODt, found under configuration II, is that the HRAP did add organic material instead of removing it from the effluent. We believe the

Table 4 Treatment performance of the HRAP under configuration II

	Influent	Effluent	Removal rate (%)
CODt (mg O_2/l)	110	250	–
BOD_5 (mg O_2/l)	45	35	22
SS (mg/l)	15	115	–
VSS (mg/l)	5	85	–
TKN* (mg/l)	61	8.3	86
$N-NH_4^+$ (mg/l)	49	7	86
Pt* (mg/l)	8	2.7	66
$P-PO_4^{3-}$ (mg/l)	5.8	2.4	59
Faecal coliforms (U/100 ml)	4.6E5	2.7E4	1.23**

*Filtered effluent; ** Reduction in log unit

Table 5 First-order reaction rate constants for CODt, N and P removals in the HRAP

	Configuration I	Configuration II
$k_{20°C}$ N (d^{-1})	0.282	0.653
$k_{20°C}$ P (d^{-1})	0.153	0.249
$k_{20°C}$ DCOt (d^{-1})	0.038	−0.245

HRAP was forced to do so because the carbon concentration left by the TSUAR system in the effluent was not enough to support the algae production that could be supported by the available N and P concentrations in the HRAP. Carbon was then imported from the atmosphere and used for that purpose. This means that the HRAP shifted from a combined secondary/tertiary treatment unit, observed under configuration I (see also El Ouarghi et al., 2000), to a strictly tertiary unit under configuration II. With this configuration the HRAP operated in a way similar to that used in Chlorella farms, where CO_2 and nutrients are supplied to produce algae biomass on clean waters (Oswald, 1988).

One of the main consequences of such a fundamental change in the way to operate the HRAP was that the algae concentration in the HRAP was more stable and that the treatment process (N, P and pathogen removal) was sustainable. Algae cell concentrations were kept within optimal limits values (Table 6) when compared with the succession of high and low concentrations observed in configuration I, where the HRAP was operated as a secondary/tertiary unit. Average DO concentrations recorded during 7 days in the HRAP during the coldest period of the year 2004 is shown on Figure 5. The typical diurnal DO profile shown is different from that reported earlier on the same HRAP operated under configuration I (El Ouarghi et al., 2000). The anoxic period recorded in the night, which extended for approximately 4 h was absent under configuration II.

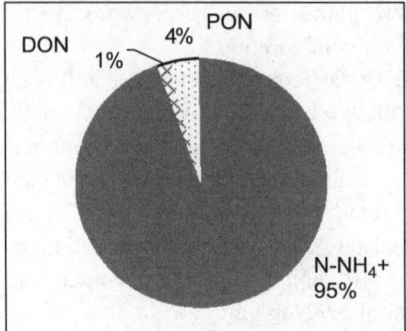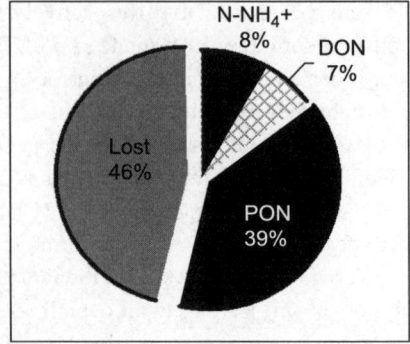

Figure 4 Nitrogen mass balance in the HRAP in configuration II

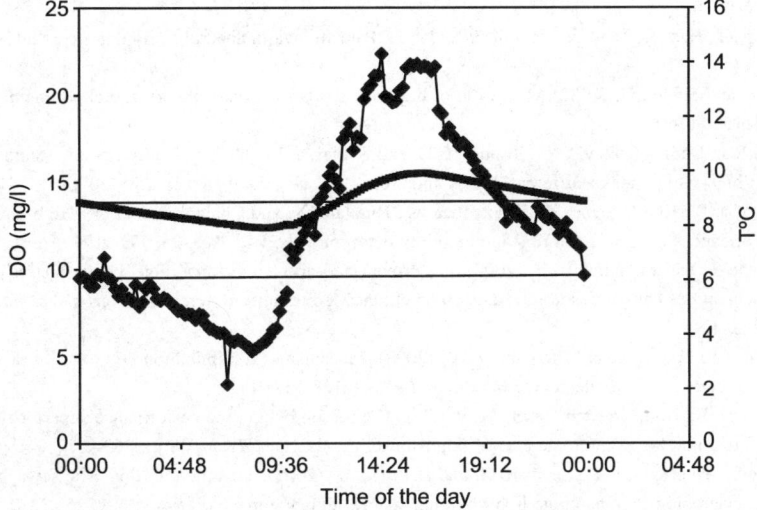

Figure 5 One-week continuous recording of diurnal water temperature and dissolved oxygen in the HRAP during the coldest days of the year

Table 6 Changes in HRAP operation parameters upon the switch from configuration I to II

	Configuration I	Configuration II
Organic loading rate (kg CODt ha^{-1} d^{-1})	233	83
HRT (d)	5.2	3
Water depth (m)	0.35	0.35
Chlorophyll-a (mg/l)	2.0	0.6
Algae cell counts (10^6/ml)	3.0	0.8

Conclusions

The two-step up flow anaerobic reactor (TSUAR) presented in this paper showed an excellent behaviour and operation simplicity. Not only were the performances high when compared to open anaerobic ponds, but the troublesome task of removing sludge from the pond every four to five years is no longer necessary. Also, the absence of any contact between wastewater being treated inside the reactors and the atmosphere prevents offensive odour release, which is a significant drawback of open anaerobic ponds. On another hand, the paper demonstrates that the combination of a TSUAR and an HRAP, as a post-treatment unit, is attractive. The organic matter degradation occurs in the TSUAR, while the HRAP, operated as a tertiary treatment unit, removes N and P, and helps to reduce FC survival. The advantages of such a combination are numerous: the effluent is of good quality, the land area requirements are low (1.0 m^2/capita for a complete treatment following configuration II conditions), odour problems are tackled and sludge management is simplified.

Acknowledgements

The authors wish to thank the Belgian cooperation, the Commission of the European Union and the Administration du Génie Rural, Ministère de l'Agriculture et du Développement Rural for their support. They also would like to thank Mohamed Marghich, Lahoucine Berraoui, Mohamed Zakour and Fatiha Boulainine for their help.

References

Arther, R.G., Fox, J.C. and Fitzgerald, P.R. (1981). Parasite ova in anaerobically digested sludge. *JWPCF*, **53**, 1334–1338.

Azov, Y. and Shelef, G. (1982). Operation of high rate oxidation ponds: theory and experiment. *Wat. Res.*, **16**, 1153–1160.

Craggs, R.J., Davies-Colley, C.C., Tanner, C.C. and Sukias, J.P. (2003). Advanced pond system: performance with high-rate ponds of different depths and areas. *Wat. Sci. Tech.*, **48**(2), 259–267.

El Hamouri, B., Khallayoune, K., Bouzoubaa, K., Rhallabi, N. and Chalabi, K. (1994). High-rate algal pond performances in fecal coliforms and helminth egg removals. *Wat. Res*, **28**, 171–174.

El Hafiane, F. and El Hamouri, B. (2002). Performances d'un système anaérobie à deux phases dans le traitement des eaux usées domestiques sous climat Méditerranéen. *Actes Inst. Agron. Vet (Maroc)*, **22**(3), 133–141.

El Hafiane, F., Rami, A. and El Hamouri, B. (2003). Mécanismes d'élimination de l'azote et du phosphore un chenal algal à haut rendement. *Sciences de l'eau*, **16**(2), 157–172.

El Hamouri, B., Rami, A. and Vasel, J.L. (2003). The reason behind the performance superiority of a high rate algal pond over three facultative ponds in series. *Wat. Sci. Tech.*, **48**(2), 269–276.

Elmitwalli, A.T., Kim, L.T., Zeeman, G. and Lettinga, G. (2002). Treatment of domestic sewage in a two-step anaerobic filter/anaerobic hybrid system at low temperature. *Wat. Res.*, **36**, 2225–2232.

El Ouarghi, H., Boumansour, B.E., Dufayt, O., El Hamouri, B. and Vasel, J.L. (2000). Hydrodynamics and oxygen balance in a high rate algal pond. *Wat. Sci. Tech.*, **42**(10–11), 349–356.

Evans, R.A., Followfield, H.J., N, J. and Cromar, N.J. (2003). Characterisation of oxygen dynamics within a high-rate algal pond system used to treat abattoir wastewater. *Wat. Sci. Tech.*, **48**(2), 61–68.

Green, F.B. and Oswald, W.J (1993). Engineering strategy to enhance microalgal use in wastewaer treatment *Proceedings of the 2d IAWQ International Specialist Conference*, Oklnd, California, 1993

Hulshoff Pol, L., Euler, H., Eitner, A. and Grohganz, T.B.W. (1997). State of the art sector review. *Anaerobic Trends. WQI*. July/August, 31–33.

Laguna, A., Outtara, A., Gonzalez, R.O., Baron, O., Famá,, El Mamouni, R., Guiot, S., Monroy, O. and Macarie, H. (1999). A simple and low cost technique for determining the granulometry of upflow anaerobic sludge blanket reactor sludge. *Wat. Sci. Tech.*, **40**(8), 1–8.

Lettinga, G., Van Velsen, A.F.M., Hobma, S.W., de Zeeuw, W.J. and Klapwijk, A. (1980). Use of the upflow sludge blanket (USB) concept for biological wastewater treatment, especially anaerobic treatment. *Biotechnol. bioeng.*, **22**, 699–734.

Malina, J.F.J. (1962). Variable affecting anaerobic digestion. *Public Works*, **93**(9), 113–116.

McCarty, P.L. (1964). Anaerobic wastewater fundamentals. *Public Works*, **95**(10), 123–126.

Mulder, A., Vandergraaf, A.A., Robertson, L.A. and Kuenen, J.G. (1995). Anaerobic ammonium oxidation discovered in a denitrifying fluidized bed reactor. *FEMS Microbiol. Ecol.*, **16**(3), 177–183.

Oswald, W.J. and Glueke, C.G. (1959). Biological transformation of solar energy. Symposium on engineering advances in fermentation practice. Division of Agricultural and food chemistry. 136th Meeting, American Chemical Society, Atlantic City, New Jersey.

Oswald, W.J. (1988). Large-scale algal culture systems (engineering aspects). In *Microalgal Biotechnology*, Borowitzka, M.A. and Borowitzka, L.J. (eds), Cambridge University Press, U.K..

Pearson, H.W., Mara, D.D. and Mills, S.W. (1987). Physicochemical parameters influencing fecal bacterial survival in waste stabilization ponds. *Wat. Sci. Tech.*, **19**(12), 145–152.

Picot, B., Bahlaoui, A., Moersidik, S., Baleux, B. and Bontoux, J. (1992). Comparison of purifying efficiency of high rate algal pond with stabilization pond. *Wat. Sci. Tech.*, **25**(12), 197–206.

Standard Methods for the Examination of Water and Wastewater (1989). 17th edn, American Public Health Association, Washington DC, USA.

Van Haandel, A. and Lettinga, G. (1994). Anaerobic sewage treatment. In *A Practical Guide for Regions with a Hot Climate*, Wiley & Sons, Chichester, UK.

Van Lier, J.B., Sanz Martin, J.L. and Lettinga, G. (1996). Effect of temperature on the anaerobic thermophilic conversion of volatile fatty acids by dispersed and granular sludge. *Wat. Res.*, **30**, 199–207.

Wang, K. (1994). *Integrated anaerobic and aerobic treatment of sewage*. Phd Thesis, Agricultural University, Wageningen, The Netherlands.

Zeeman, G. and Lettinga, G. (1999). The role of anaerobic digestion of domestic sewage in closing the water and nutrient cycle at community level. *Wat. Sci. Tech.*, **39**(5), 187–194.

Improving nitrogen reduction in waste stabilisation ponds

H.E. Archer and B.M. O'Brien

Beca Infrastructure Ltd, Level 3, PWC Centre, 119 Armagh Street, Christchurch, New Zealand
(E-mail: *harcher@beca.co.nz; bobrien@beca.co.nz*)

Abstract This paper reviews the performance of two waste stablisation ponds (WSP) systems in the South Island of New Zealand that have been upgraded to multiple ponds-in-series to improve effluent quality. Results of monitoring are provided which show that it is possible to achieve relatively low ammonia (approximately 1 g/m^3) and total nitrogen (approximately 10 g/m^3) effluent concentrations through the use of nitrification filter beds (rock trickling filters) and sand filters. Evidence suggests that the nitrification and denitrification processes in the extra biofilm surface area provided by the rock filters or rock bank protection is primarily responsible for the improved effluent quality. The paper also compares the WSP results with effluent quality predicted by published formulae. It is concluded that these formulae do not reliably predict the performance of WSP systems and the development of universally applicable design guidelines would be useful.
Keywords Waste stabilisation ponds; nitrogen reduction; nitrification; ammonia reduction

Introduction

There are about 200 waste stabilisation pond (WSP) systems in New Zealand. Most are single or two ponds-in-series systems designed to the Guideline for the Design, Construction and Operation of Oxidation Ponds (Ministry of Works and Development, 1974). Over the past five years, about 15 of these WSP systems have been upgraded to multiple ponds-in-series, primarily for improved pathogen reduction, but with other contaminant reductions targeted as well. Along with Rangiora and Kaiapoi, the Christchurch and Blenheim WSP systems are assessed for their ammonia removal performance, compared with that predicted by two formulae.

Case studies description
Rangiora Sewage Treatment Plant

The Rangiora Sewage Treatment Plant (STP) serves a population of 11,000 and has an average flow of 4,000 m^3/d. Prior to upgrading in January 2003, the Rangiora STP comprised two aerator-assisted ponds-in-series plus UV disinfection. Ammonia reductions were variable and higher solids in the effluent reduced the UV disinfection channel performance. The upgrade involved bringing two disused ponds back into service, with the final layout comprising two parallel primary facultative ponds, followed by five maturation ponds-in-series, with an overall retention time of 45 days. Horizontal flow rock filters were constructed after Ponds 3, 4 and 5. The first two rock filters were fitted with perforated pipes at the base to provide aeration to the submerged biofilm. A pump and surface sprays were also installed to recirculate Pond 5 liquor over the non-submerged part of the rock filters, to mimic the trickling filter process. These enhancements were designed to reduce ammonia by nitrification.

Results of effluent quality monitoring before and after the upgrade are shown in Table 1. The results to date show consistent performance in summer but lesser removal of ammonia in winter (Figure 1). The strong linkage of ammonia removal with

Table 1 Rangiora STP effluent quality data

	Before upgrade[1]			After upgrade[2]		
	Median	90%ile	Max.	Median	90%ile	Max.
NH_3-N (g/m^3)	18	20	21	7	18	21
NO_x-N (g/m^3)[3]	0.2	0.45	0.60	2.2	6.2	8.2
BOD_5 (g/m^3)	31	49	160	24	47	73
TSS (g/m^3)	88	174	210	66	98	140
Faecal coliforms (cfu/100 mL)	1500	3.8×10^4	1.4×10^5	400	1320	2000

[1]Data collected between July 2001 and January 2003.
[2]Data collected between January 2003 and April 2004.
[3]NO_x-N is the sum of nitrate nitrogen and nitrite nitrogen.

temperature suggests that nitrification is the dominant mechanism, rather than volatilisation (see later discussion). Furthermore, when there was almost complete ammonia reduction, the concentration of nitrate increased (nitrite concentrations were normally negligible).

Kaiapoi STP

The Kaiapoi STP plant serves a population of 10,000 and has an average flow 3,400 m^3/d. Until mid 2003 it comprised an aeration pond, two parallel facultative oxidation ponds, a large 32 ha infiltration wetland, and a sand filter and UV disinfection channel designed for a flow rate of 1,100 m^3/d. Infiltration to groundwater from the wetland area was less than anticipated and discharge rates of more than 4,000 m^3/d occurred, which hydraulically overloaded the sand filter and UV channel. When the sand filter was operating at its design capacity, the median parameter values achieved were very good, indicating the effectiveness of sand filtration and UV after ponds or wetlands. The STP was upgraded by constructing bunds in the wetlands to eliminate short-circuit paths thus creating three cells-in-series, constructing two additional sand filters and increasing the UV disinfection hydraulic capacity. The "wetlands" have scattered planting (less than 2% coverage) and function more as shallow (0.2 m deep) polishing ponds.

Effluent quality data before and after the upgrade is shown in Table 2. There are only five results since upgrading, and one result with high concentrations (possibly due to commissioning bypassing) has distorted the statistical performance.

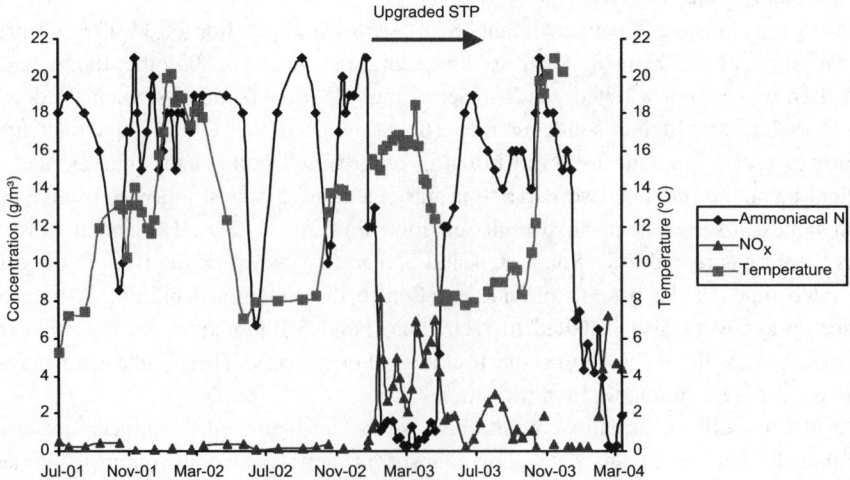

Figure 1 Rangiora STP seasonal trends in nitrogen forms

Table 2 Kaiapoi STP effluent quality data

	Before upgrade[1]			After upgrade[2]		
	Median	90%ile	Max.	Median	90%ile	Max.
NH_3-N (g/m^3)	1.5	12	25	4	13	19
NO_x-N (g/m^3)	0.5	4.6	16	0.4	8	19
BOD_5 (g/m^3)	10	30	45	8	11	11
TSS (g/m^3)	16	54	100	14	36	48
Faecal coliforms (cfu/100 mL)	30	240	8000	105	380	660

[1]Data collected between April 1997 and July 2003.
[2]Data collected between August 2003 and April 2004.

Sampling was carried out in August 2003 (winter) to determine the reductions of contaminants across the three wetland cells as the retention time increased. Historically, the average flow for August is 2,900 m^3/d. The averages of these results are shown in Table 3 and show a good reduction of ammonia, even though the wetland temperature was only about 8°C, when nitrification would be diminished. The longer retention times could have resulted in volatilisation being significant.

Christchurch Wastewater Treatment Plant

The much larger Christchurch Wastewater Treatment Plant (CWTP) serves a population equivalent of about 380,000 and has an average flow of 160,000 m^3/d. It consists of primary sedimentation, trickling filter/solids contact process and secondary clarification, followed by six polishing ponds with a combined area of 220 ha. Historically, these ponds operated as two parallel trains of three ponds-in-series, but were reconfigured in early 2004 to a single train of seven ponds-in-series. The average overall hydraulic retention time is about 20 days.

The influent ammonia nitrogen concentration is typically around 23 g/m^3, with a maximum of 28 g/m^3. The influent TKN concentration is typically 34 g/m^3. Analysis of the final effluent monitoring data collected between October 2003 and March 2004 gives a median pH of 8.3, median temperature of 18°C and a median ammonia concentration of 28 g/m^3. The pH is quite high for ponds, at around 8.3, which would be expected to encourage volatilisation to occur. However, there is virtually no ammonia removal through the ponds.

Discussion

The most significant mechanisms for nitrogen removal in pond systems are generally considered to be nitrification and denitrification, ammonia volatilisation and biomass assimilation, although there is disagreement about which is most important (Strang, 2001). To illustrate, Hurst and Connor (1997) concluded that nitrification was the major mechanism for ammonia removal based on extensive studies of the large Western Treatment Plant ponds at Werribee near Melbourne Australia. In describing the Mèze stabilisation

Table 3 Kaiapoi STP wetland cell average water quality data

	Oxidation pond	Wetland cell 1	Wetland cell 2	Wetland cell 3
NH_3-N (g/m^3)	23	12	3	1.5
BOD_5 (g/m^3)	38	49	41	32
TSS (g/m^3)	68	134	75	44
Faecal coliforms (cfu/100 mL)	1400	224	13	35
Retention time (d)	30	3	13	6

ponds in France, Picot et al. (2005) also found that nitrification and denitrification appeared to be the main process for nitrogen removal during warm periods. This contradicts the statement by Pano and Middlebrooks (1982) that ammonia stripping is the major process for ammonia removal in facultative stabilization ponds.

The experience at Werribee WSP, and also in New Zealand, is that when nitrification occurs, the nitrate produced is readily denitrified to nitrogen gas. Thus, the absence of nitrate in a pond effluent does not indicate that nitrification has not occurred – a statement made by Pano and Middlebrooks (1982) in justifying their focus on ammonia volatilisation. The major mechanisms are summarised in the following paragraphs and then applications of the various formulae are discussed.

Ammonia volatilisation. Ammonia is present in wastewater in two forms, ionised ammonia (NH_4^+) and unionised ammonia (NH_3). The unionised is relatively volatile and can be lost to the atmosphere. The two most commonly used formulae for predicting nitrogen removal from ponds are those by Pano and Middlebrooks (1982) for ammonia nitrogen, and Reed (1985) for total nitrogen. The Pano and Middlebrooks (1982) formula is empirical, based on relatively small ponds with retention times of 29–92 days. It was modelled on the volatilisation mechanism with only pH, temperature, ammonia concentration and pond surface area being taken into account. For pH 7, it predicts that the ammonia concentration will not drop below $19 \, g/m^3$ for temperatures up to 20°C.

Biomass assimilation. Algae use nitrogen for cell growth. When algae die and decompose, most of the nitrogen is released back into the pond. However, a small proportion of the nitrogen is non-biodegradable and is removed from the pond water by settling out in the sludge. A literature review by Strang (2001) suggests that biomass assimilation may be a significant removal mechanism for heavily loaded primary ponds, but could be much less significant for maturation ponds.

Nitrification and denitrification. In nitrification, ammonia is converted by bacteria to nitrite and nitrate in aerobic conditions. In denitrification, nitrate is converted to nitrogen gas by a different set of bacteria in anoxic conditions, such as is found at the base of ponds. The nitrogen gas is then lost to the atmosphere. Reed et al. (1995) present the following formula for calculating ammoniacal nitrogen removal in a free water surface wetland:

$$C_e = C_o \exp(-K_T t)$$

where C_e is the ammonia nitrogen concentration in the pond effluent ($g \, N/m^3$), C_o is the TKN concentration in the pond influent ($g \, N/m^3$), t is the hydraulic residence time (d) and K_T is a temperature dependent rate constant (d^{-1}). For a temperature of 0°C, K_T is $0 \, d^{-1}$; for temperatures of 1–10°C, K_T is given by $(0.1367)(1.15)^{(T-10)} \, d^{-1}$; for temperatures of greater than 10°C, K_T is given by $(0.2187)(1.408)^{(T-20)} \, d^{-1}$. This formula assumes that ammonia removal is entirely due to nitrification, with no allowance for plant uptake, so is therefore potentially applicable to ponds as well as wetlands.

Reed et al. (1995) also present a nitrification filter bed (NFB) concept for enhancing the removal of ammonia in free water surface wetlands, which involves spraying effluent over mounds of rock or gravel to allow the formation of a nitrifying biofilm. Significantly more biofilm surface area is required at lower temperatures, which aligns with conventional wastewater treatment experience for nitrification. This NFB concept is similar to that installed at the Rangiora STP.

Table 4 Comparison of predicted and actual ammonia effluent concentrations

Sewage treatment plant	Equivalent population	Measured median ammonia effluent concentration (g/m³)	Predicted ammonia concentration (g/m³)	
			Pano and Middlebrooks (1982)	Reed et al. (1995)
Christchurch	400,000	28	7	0.7
Kaiapoi wetland	11,000	1.5	6	2
Rangiora	12,000	1	15	<1
Blenheim	40,000	1.5	16	<1

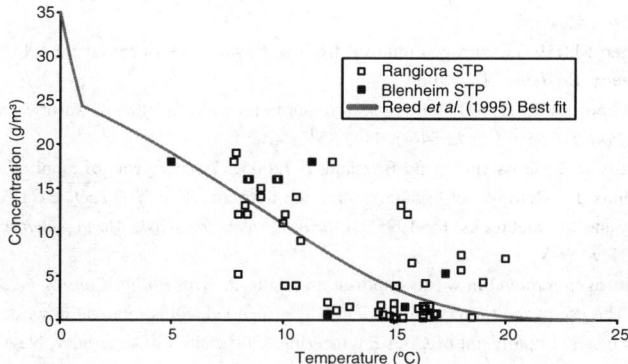

Figure 2 Rangiora and Blenheim STP ammonia effluent concentrations

Predictions from the Pano and Middlebrooks (1982) and Reed et al. (1995) formulae are compared with actual effluent ammonia results for four WSP systems in Table 4. It can be seen that there is no ammonia removal through the Christchurch WSP, in contrast to the formulae predictions. This may be because nitrification and denitrification is a more important mechanism than volatilisation, and the CWTP ponds have a small area of rock protection along the banks for aerobic biofilm to develop, relative to the large surface area of the ponds. The Kaiapoi wetland cells achieved a good level of ammonia reduction (see also Table 3), better than is predicted by the Pano and Middlebrooks (1982) formula, but similar to the Reed et al. (1995) formula. It is possible that both volatilisation and nitrification are occurring in these wetland ponds; volatilisation because of the large surface area, and nitrification because of the shallow water depth allowing oxygen to diffuse throughout the water column to a nitrifying biofilm on the base.

The Reed et al. (1995) style of formula was compared with the effluent ammonia concentration results from the Rangiora and Blenheim STPs (see Figure 2). Both WSP systems have a total retention times of 45 days and rock filters (see Archer and Donaldson (2003) for a full description of the Blenheim STP). A reasonable fit to the data was found when K_T was adjusted using the least squares method. It was found that for temperatures of 1–10°C, K_T is given by $(0.0259)(1.138)^{(T-10)} \, d^{-1}$ and for temperatures of greater than 10°C, K_T is given by $(0.0985)(1.143)^{(T-20)} \, d^{-1}$.

Conclusions

Large surface area ponds do reduce ammonia effectively over a range of temperatures. It is possible to upgrade WSPs to achieve low ammonia and total nitrogen concentrations without using large land areas, by installing NFBs (rock or gravel) as described by Reed et al. (1995), or sand filters. Significantly more biofilm surface area is needed at reduced temperatures, which aligns with conventional wastewater treatment practice to achieve nitrification. There are conflicting statements in the literature as to the relative importance of the ammonia removal mechanisms in WSPs. The published formulae do not reliably predict the performance of the WSP systems studied in this paper. It would be helpful to designers for the mechanisms of ammonia and total nitrogen reduction to be researched further and universally applicable design guides produced.

References

Archer, H. and Donaldson, S. (2003). Waste stabilisation ponds upgrading at Blenheim and Seddon, New Zealand – Case studies. *Wat. Sci. Tech.*, **48**(2), 17–23.

Guideline for the Design, Construction and Operation of Oxidation Ponds (1974). Ministry of Works and Development, New Zealand.

Hurst, T. and Conner, M. (1997). Nitrogen removal from wastewater treatment lagoons. In *Proceedings of BNR 3 Conference, Brisbane, Australia.*

Pano, A. and Middlebrooks, E. (1982). Ammonia nitrogen removal in facultative wastewater stabilization ponds. *J. Wat. Pollut. Control Fed.*, **54**(4), 344–351.

Picot, B., Andrianarison, T., Gosselin, J. and Brissaud, F. (2005). Twenty years of monitoring at Mèze stabilisation ponds. I – Removal of organic matter and nutrients. *Wat. Sci. Tech.* **51**(12), 23–31.

Reed, S., Crites, R. and Middlebrooks, E. (1995). *Natural Systems for Waste Management and Treatment*, McGraw-Hill, New York.

Reed, S. (1985). Nitrogen removal in waste stabilisation ponds. *J. Wat. Pollut. Control Fed.*, **57**(1), 39–45.

Strang, T. (2001). The use of rock filters for nutrient and suspended solids removal in waste stabilisation ponds. Masters thesis, Department of Civil Engineering, University of Canterbury, New Zealand.

Modeling ammonia removal in aerated facultative lagoons

C.D. Houweling*, L. Kharoune*, A. Escalas** and Y. Comeau*

*Department of Civil, Geological and Mining Engineering, Ecole Polytechnique of Montreal, P.O. Box 6079, Station Centre-Ville, Montreal (Quebec) Canada, H3C 3A7 (E-mail: *dwight.houweling@polymtl.ca*; *lynda.kharoune@polymtl.ca*; *yves.comeau@polymtl.ca*)

**Centro de Investigación y Estudios de Posgrado, Facultad de Ingeniería, Universidad Autónoma de San Luis Potosí. Av. Dr. Manuel Nava No.8, Edificio P, Zona Universitaria. C.P. 78290, San Luis Potosí, SLP, México (E-mail: *antoni.escalas@uaslp.mx*)

Abstract A mechanistic model has been developed to model ammonia removal in aerated facultative lagoons. Flow is modeled through the water column by a continuously stirred tank reactor and exchanges between the sludge layer and the water column are simulated by a solids separator. The biological model is based on an activated sludge model with reactions added for anaerobic bacterial growth and degradation of inert organic material. Results show that the model is able to predict seasonal variation in ammonia removal as well as sludge accumulation in the lagoons.
Keywords Aerated facultative lagoons; mathematical model; nitrification; sludge accumulation

Introduction

Over 450 installations of aerated facultative lagoons are operated in Quebec for the removal of biochemical oxygen demand (BOD_5), total suspended solids (TSS) and phosphorus (by the addition of alum) from municipal wastewaters. Despite the proven effectiveness of lagoons for the removal of these pollutants, there is increasing concern about their inability to achieve reliable ammonia removal. Ammonia removal is observed to occur only seasonally as a result of changing lagoon temperatures, which vary from 0.5 °C in winter to 25 °C in summer. A discharge limit governing ammonia effluent concentration from lagoon installations treating a flow greater than 5000 m^3/d is expected from the Canadian government in December 2004 (Canada Gazette, 2003). Studies are therefore underway to evaluate how the current lagoon installations can be optimized or upgraded to provide year-round ammonia removal. A mechanistic model has been developed to help better understand the performance of aerated lagoons and to serve as a tool to evaluate different optimization and upgrade scenarios.

Previous efforts at modeling aerated facultative lagoons have been limited in part because of the complexity of their hydrodynamic behaviour (Dorego and Leduc, 1996; Namèche and Vasel, 1998) as well as a lack of data with which to calibrate models. The phenomenon of sludge accumulation in the lagoons also affects nutrient loads and oxygen demand in the water column (Chabir *et al.*, 2000), which introduces additional uncertainty to the process of model calibration. Numerous processes must therefore be accounted for in order to properly model ammonia removal in aerated facultative lagoons, many of which are not yet properly understood. The research described in this paper represents an ongoing work whose objective is to develop a reliable mechanistic model of an aerated facultative lagoon for diagnosing lagoon performance and for evaluating upgrade and optimization scenarios.

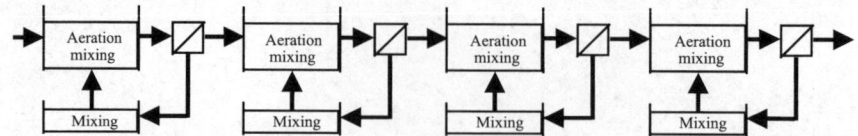

Figure 1 Model layout of four aerated facultative lagoons in series including water column, sludge layer and solids separator

Methods

The municipal lagoons at Drummondville, Quebec, Canada in operation since 1997, were used as the basis for calibrating the model. The plant comprises 8 lagoons, which are divided in 2 parallel series, and have a combined hydraulic retention time (HRT) of close to 50 d. This is higher than the combined HRT of 20–30 d typical of lagoon designs in Quebec. Aeration intensities are insufficient to maintain complete mixing, $0-1.5\,W/m^3$, and significant solids accumulation has been measured in the first two lagoons. Design criteria are presented in Table 1.

The model simulates hydraulic flow in a lagoon using two parallel reactors, the larger, a continuously stirred tank reactor (CSTR) which represents the aerated water column, and a smaller CSTR, which represents the non-aerated sludge layer (Figure 1). Liquid and solids fluxes between the two reactors are controlled by a solids separator which also controls the amount of solids in the lagoon effluent. These fluxes are meant to represent the exchanges of soluble and particulate substances between the sludge layer and the water column.

The biological model has been based on ASM2d (Henze *et al.*, 1999) without the biological phosphorus removal processes. Nitrogen removal mechanisms include assimilation for biomass growth, nitrification and denitrification. Equations have been added to account for anaerobic growth of methanogenic bacteria in the sludge and the slow degradation of non-biodegradable COD into BOD: $nbCOD \rightarrow BOD$.

Results and discussion

The model was simulated over the first 5 years of operation of the lagoons using average values for influent flow, composition and aeration flow. The only dynamic input to the model was the lagoon temperature. Results show that simulating the model in this way is sufficient to represent seasonal variation in ammonia removal from the lagoons (Figure 2). A direct relationship between nitrifying biomass concentration in the lagoons and ammonia removal is observed in the model indicating that nitrification is the primary mechanism for ammonia removal. Measured effluent nitrite–nitrate data shown in Figure 3 validate this claim.

Table 1 Design criteria for the lagoons of Drummondville

Parameter	Units	Lagoon No.				Total
		1A, 1B	2A, 2B	3A, 3B	4A, 4B	
V	m^3	384 000	384 000	384 000	384 000	3 070 000
HRT	D	11.6	11.6	11.6	11.6	46.4
Surface area	m^2	68 800	68 800	49 700	49 700	474 000
Water depth,	m	6.5	6.5	6.5	6.5	–
Aeration intensity	W/m^3	0.47–1.23	0.14–0.36	0.08–0.20	0.06–0.15	–

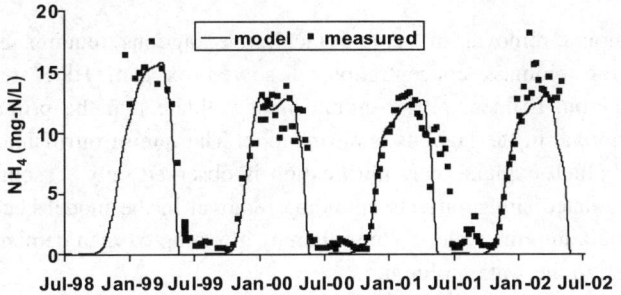

Figure 2 Ammonia effluent concentrations in the final lagoon at Drummondville

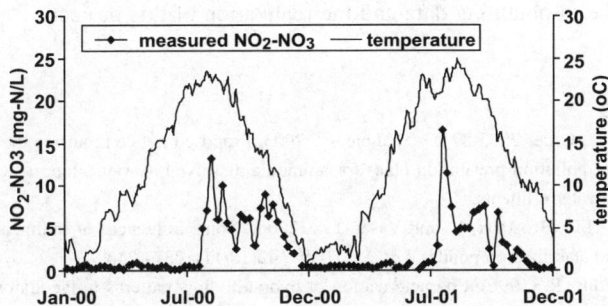

Figure 3 Relationship between NO_2-NO_3 effluent concentrations and temperature in the first lagoon

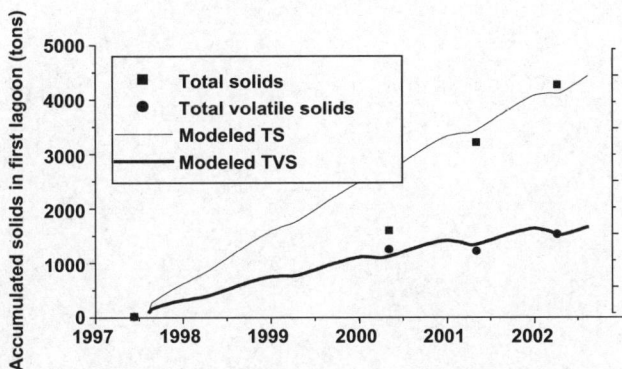

Figure 4 Solids accumulation in the first lagoon (V = 384 000 m^3) over five years

Volatile solids accumulation data (Figure 4), measured annually since the third year of operation of the lagoons, enabled calibration of the hydrolysis reaction: $nbCOD \rightarrow BOD$. Solids and liquid fluxes between the sludge layer and the water column cause both soluble and particulate BOD to re-enter the water column and exert an oxygen demand. This has an important effect on the modeled dissolved oxygen concentration of the lagoon.

Conclusions

Predicting ammonia removal in aerated facultative lagoons requires a model which predicts nitrifying biomass concentration, dissolved oxygen, HRT, temperature and sludge accumulation. Effluent nitrite–nitrate data validate that the primary mechanism for ammonia removal in the lagoons is nitrification. The rate of nitrification is dependent on temperature which explains why nitrification is observed only seasonally in northern climates. Accumulated sludge affects ammonia removal in the model because it releases BOD and ammonia into the water column thereby affecting oxygen demand and available dissolved oxygen in the water column.

Acknowledgements

The authors wish to thank Marc-André Desjardins, Gino Bélanger and Alain Rousseau of AXOR Experts-Conseils, Janick Lemay of the Ministry of Municipal Affairs, Sports and Recreation and François Chabot of the municipality of Drummondville, Quebec, Canada, for their assistance in obtaining data and the realization of this project.

References

Canada Gazette, Part I, Order 2003-87-03-02, June 7th, 2003, Proposed notice requiring the preparation and implementation of pollution prevention plans for ammonia dissolved in water, Inorganic chloramines and chlorinated wastewater effluents.

Chabir, D., Ouarghi, H.E., Brostaux, Y. and Vasel, J.L. (2000). Some influences of sediments in aerated lagoons and waste stabilization ponds. *Wat. Sci. Tech.*, **42**(10/11), 237–246.

Dorego, N.C. and Leduc, R. (1996). Characterization of hydraulic flow patterns in facultative aerated lagoons. *Wat. Sci. Tech.*, **34**(11), 99–106.

Henze, M., Grady, C.P.L., Gujer, W., Marais, G.v.R. and Matsuo, T. (1999). Activated sludge model no. 2D ASM2D. *Wat. Sci. Tech.*, **39**(1), 165–182.

Namèche, T. and Vasel, J.L. (1998). Hydrodynamic studies and modelization for aerated lagoons and waste stabilization ponds. *Wat. Res.*, **32**(10), 3039–3045.

Light attenuation parameters for waste stabilisation ponds

S. Heaven*, C.J. Banks* and E.A. Zotova**

*School of Civil Engineering and the Environment, University of Southampton, Southampton SO17 1BJ, UK
(E-mail: *sh7@soton.ac.uk; cjb@soton.ac.uk*)
**BG Chair of Environmental Technology, AIPET, 126 Baytursynov Street, Almaty 480013, Kazakhstan
(E-mail: *bgchair@aipet.kz*)

Abstract Effective modelling of shallow water ecosystems, including waste stabilisation ponds, is strongly dependent on the availability of good estimates of the light attenuation coefficient k (m^{-1}). Experimental data is presented on its determination using purpose-built laboratory apparatus with a near-parallel halogen light source and an array of photodiodes allowing measurements of irradiance at different depths. The equipment was used to compare k values from 4 different pure cultures, and mixed cultures of algae taken from a pilot-scale WSP. Laboratory values were compared with in situ measurements in the pond. At concentrations above 50 mg l^{-1} the relationship between k and suspended solids is non-linear; k also varied with depth. This could be modelled by a single equation, suggesting similarity of response in different cultures. At shallow depths and low suspended solids concentrations k values are variable and hard to measure reliably. The results highlight the need to standardise on a method for the measurement and reporting of k values if these are to be widely applicable in the development of pond models.
Keywords Extinction coefficient; light attenuation; waste stabilisation ponds

Introduction

Light plays a vital role in the functioning of a waste stabilisation pond (WSP), providing the energy source for photosynthesis and thus oxygen production. Knowledge of the parameters affecting light intensity or irradiance within the pond is thus crucial to modelling and prediction of WSP behaviour. There is an extensive scientific literature, both theoretical and experimental, on light attenuation in oceans and freshwater bodies, and also in photobioreactors. WSPs fall between these two applications, however, and relevant parameter values can be more difficult to find.

Attenuation of light follows an exponential relationship of the form

$$I_z = I_0 e^{-kz} \tag{1}$$

where I_0 is the subsurface irradiance, I_z the irradiance at depth z and k is a light attenuation coefficient. A huge body of work in the fields of limnology and oceanography concerns values and expressions for k (see Kirk, 1994). For many purposes k is assumed to be a linear function of one or more components such as suspended solids (SS), dissolved solids or chlorophyll. Numerous expressions have been proposed for conditions similar to those in WSPs, such as eutrophic lakes and estuaries (e.g. Tsirtsis, 1995; Lonin and Tuchkovenko, 2001). Values of k in excess of 10 m^{-1} are quoted by Wetzel (2001) for stained and eutrophic lakes. Brawley *et al.* (2003) note that effective modelling of shallow water ecosystems is strongly dependent on the availability of good estimates for k.

Photobioreactors are designed to operate at concentrations of algal biomass far higher than those usually found in WSPs. In these conditions the assumption of linear dependence of k on biomass concentration is known to be invalid. Yun and Park (2001) tried a theoretically based approach to modelling k for *Chlorella vulgaris* in the biomass concentration range 0–2000 mg l^{-1}, using a linear approximation, an equation proposed by Cornet et al. (1992), and a hyperbolic model. While Cornet's equation was more satisfactory in the sense of having a physical basis, the hyperbolic model was found to give the best fit. Fernandez et al. (1997) also found a hyperbolic model gave the best results for data from *Phaeodactylum tricornutum* at concentrations of up to 3000 mg l^{-1}. At low biomass concentrations the effect of scattering was significant: for the range observed the highest value of k was found at 24 mg l^{-1}, while 40–294 mg l^{-1} the relationship between k and biomass concentration appeared linear. Privoznik and Incropera (1978) also noted the relative importance of scattering in cultures of *Chlorella pyrenoidosa* at low concentrations. Ogbunna et al. (1995) calculated a specific k value per kg m^{-3} of 200 m^2 kg^{-1} for *C. pyrenoidosa*, and applied this to modelling growth curves the range 17–3000 mg l^{-1}.

A definitive study of light penetration in WSPs, looking at both photosynthetically active radiation (PAR) and monochromatic light, was carried out by Curtis et al. (1994). Absorbance played a far more important role than scattering for all ponds in the study, pond-to-pond variation was mainly attributable to differences in algal biomass, and variations in attenuation were observed at different wavelengths and depths. Despite this, in practice many models are based on simple empirical linear relationships, and measured absolute values for k are hard to find. Bartsch (1961) found k values of 6–11 m^{-1} in Dakota WSPs in summer, while Thomann and Mueller (1987) gives a value of 23 m^{-1}. Mesplé et al. (1994) obtained a specific k value for total SS minus phytoplankton of 0.05 m^{-1} per mg dry weight l^{-1}, for high rate algal ponds. Juanico et al. (2003) used a specific k value of 0.29 m^{-1} per mg l^{-1} of carbon, based on literature values, and found that a change in light absorption and self-shading factors produced a two-fold change in average irradiance in spring, but was not significant in summer due to higher concentrations of algae and organic matter. A survey of WSPs in New Zealand found a median euphotic depth of 0.35 m corresponding to a k value of 13 m^{-1} (Davies-Colley et al., 1995).

In the current work, k values for mixed and pure cultures were measured in purpose-built laboratory apparatus, and also in three pilot-scale pond systems. As photosynthetic sulphate-reducing bacteria can also be present in pond systems in certain conditions, imparting a red-purple colour to the water, a mixed culture rich in these was also grown and tested.

Materials and methods

Light sensors. Light intensity was measured using type BPW 21 photodiodes (RS components, UK), used in similar applications elsewhere for measurement of PAR (Ensminger et al., 2001). Photodiodes for laboratory use were calibrated against a LI-210SA photometric sensor (LiCor, USA), while those used for external measurements were calibrated against a RC/0308 standard photovoltaic cell (PV Systems, UK). Where several photodiodes were to be used in one set of measurements, the output from each was checked against the average output from all under different conditions of illumination. A strong linear relationship ($R^2 > 0.998$) was found, allowing calculation of a normalising factor. Photodiode outputs were continuously sampled using a datalogger (DataTaker D500 and expansion unit). Readings were averaged over a 30 second period and then over longer periods as required. Output was measured in milliamps unless noted.

Waste stabilisation ponds. Measurements were made in three sets of pilot-scale ponds, two located in Almaty, Kazakhstan and one in Southampton, UK. The first set of Almaty ponds A(I) consisted of four circular concrete tanks 2 m in diameter with a water column depth of 1.5 m. The ponds were fed on screened wastewater from Almaty sewage works, with a typical 5-day biochemical oxygen demand (BOD$_5$) of 200 mg l^{-1}. They were batch fed over a one-hour period each day, to give hydraulic retention times of 7.5, 15, 22.5 and 30 days. An array of photodiodes constructed to read at the surface and at depths of 0.3, 0.6, 0.9 and 1.2 m was moved between the ponds on a 3-day cycle. Readings taken in millivolts were averaged over a 15-minute period. Construction and operation of the second set of Almaty ponds A(II) is described elsewhere (Banks et al., 2002). These ponds were instrumented each with a single photodiode at a depth of 0.25 m, with readings averaged over a 15-minute period. The work in Southampton was carried out on two ponds (SP1 and SP2) each with a surface area of 0.9 m^2 and a water column depth of 0.6 m. The ponds consisted of semi-translucent polypropylene tanks externally insulated with 50 mm of polystyrene foam, preventing any light entering through the tank walls. They were housed in a south-facing greenhouse, and received supplemental lighting from an array of halogen floodlights capable of providing a surface illumination of 300 W m^{-2}. They were batch fed on a synthetic wastewater of the type used for Almaty A(II) ponds, at a hydraulic retention time of 22 days. The ponds were instrumented with photodiodes at depths of 0, 0.15, 0.33 and 0.53 m and in normal operation readings were averaged over 10-minute intervals. For detailed comparison with laboratory measurements an array of eight photodiodes at depths of 0 (two diodes), 0.09, 0.18, 0.27, 0.36, 0.45, and 0.54 m was used, with measurements averaged over 1-minute intervals. In all cases readings were taken immediately after cleaning of the photodiode surface.

Microbial suspensions. Mixed cultures were taken from the Southampton ponds at different seasons; and a culture dominated by purple sulphur bacteria was grown in shallow pond water covering a layer of sulphur-rich pond sediment. Cultures of *Scenedesmus subspicatus* (CCAP 276/20), *Chlorella vulgaris* (CCAP 276/20), *Chlamydomonas reinhardtii* (CCAP 11/32b), and *Microcystis aeruginosa* (CCAP 1450/16) were obtained from the Culture Collection of Algae and Protozoa, Dunstaffnage Marine Laboratory, UK. Cultures were grown on Jaworski's medium (CCAP JM recipe), modified for *M. aeruginosa* by the addition of 1 ml l^{-1} of trace element solution (Pfennig et al., 1981). Cultures were activated by inoculation into 250 ml flasks containing 100 ml of medium, and incubating for 4–7 days at 20 °C on an illuminated orbital shaker (Gallenkamp, UK). The contents of each 250 ml flask were then transferred to a 2 litre flask containing 1 litre of medium, and incubated under the same conditions for a further 4–7 days. Two 2 litre flasks were then used to inoculate a 20 litre glass container aerated by a filtered air supply and illuminated by an array of eight 35 W white fluorescent tubes at 18–22 °C for a further 4–7 days.

Light apparatus. Light attenuation measurements were made in a purpose-built column apparatus consisting of a dark grey non-reflective PVC tube 150 mm in diameter and 1.5 m deep, fitted with a horizontal array of six photodiodes located centrally to minimise wall effects. The array could be moved vertically through the column of water and positioned at any depth. Illumination of the water column could be achieved using different light sources, but the source used in the current work was a PAR 36 light with a sealed beam 30 W halogen lamp (General Electric 4515). This allowed near-parallel light to be directed at the water surface. A mathematical correction for divergence is possible (Privoznik and Incropera, 1978) but as each bulb had slightly different characteristics in practice correction was made by measuring the light intensity in air, and deducting the resultant value or slope of the line from values measured in the algal cultures (Fernandez et al., 1997). Sedimentation of algae was prevented by recirculation at a pumping rate in excess of algal settling rate (Stutz-McDonald and Williamson, 1979).

Sampling and analysis. Suspended solids were measured by filtration of an appropriate volume through a pre-dried and weighed GFC filter (Whatman, UK), in accordance with the procedures in *Standard Methods for the Examination of Water and Wastewater (1998)*. Chlorophyll was determined by filtering through a GFC filter (Whatman, UK) previously dosed with 1 ml of a saturated solution of $MgSO_4$. Extraction was by grinding followed by treatment with acetone and centrifugation for 15 minutes at 3000 rpm. The resultant colour was measured at 664 and 665 nm using a Cecil Instruments spectrophotometer (3000 series). Absorbance was measured at 664 nm in a colorimeter (Camlab DREL/5). Samples were analysed in triplicate.

Data handling. At low to medium suspended solids concentrations, where the relationship appeared linear, k values were obtained by plotting $\ln(I_z)$ against depth z, with k as the gradient of a line fitted by the method of least squares. At higher concentrations where the non-linearity of the relationship between $\ln(I_z)$ and z is apparent, local k_z values were calculated for a given depth using the formula $k_z = \ln(I_z/I_o)/z$. For calculation of daily in-pond k values, measurements were rejected if the correlation coefficient for z and $\ln(I_z)$ was $R^2 < 0.98$. Measurements taken in Kazakhstan were also discarded if the irradiance readings indicated passage of clouds during the measurement period. All data processing was carried out using Excel spreadsheet software (Microsoft Excel).

Results and discussion
Light column experiments

The results of more than 60 sets of light attenuation measurements in the column apparatus described above carried out with pure and mixed cultures at different dilutions indicated that repeatability and reliability of measurements was good. The difference between values of k obtained from separate runs for the same culture and dilution was usually less than $0.1\,m^{-1}$. At low to medium concentrations of SS, where the relationship between depth and the natural logarithm of irradiance can be considered as linear, correlation coefficients between z and $\ln(I_z)$ were generally in excess of 0.995 and often of 0.999. The relationship between k values and SS concentrations obtained by dilution of a given culture was also strongly linear at low concentrations, with $R^2 > 0.995$. Gradients for k versus SS were generally in the range of $0.12\text{--}0.2\,m^{-1}$ per mg l^{-1} of suspended solids. These values agree well with Bowen (2004), who suggested an average value for algal biomass of 0.17 and a range of $0.06\text{--}0.34\,m^{-1}\,mg^{-1}l$, based on a value of $19\,m^{-1}\,mg^{-1}l$ for chlorophyll.

At higher concentrations the assumption of linearity no longer holds true, initially for the relationship between k and SS and then for depth and $\ln(I_z)$. Once the latter relationship is non-linear, k values can no longer be obtained from the gradient of z versus $\ln(I_z)$ and local values for k_z must be calculated. Figures 1 and 2 show the variation of k_z with suspended solids and with depth for a culture of *S. subspicatus*. At concentrations below $50\,mg\,l^{-1}$, values of k_z are similar at all depths and the relationship between k_z and SS is close to linear (Figure 1). Above this concentration, linearity is lost and the overall shape of the curve can be approximated by a hyperbola, in accordance with the findings of other researchers (Fernandez *et al.*, 1997; Yun and Park, 2001). Values for k_z at different depths also show increasing divergence, as described by Kirk (1994). This can be seen more clearly in Figure 2 where for SS concentrations of 10 and $20\,mg\,l^{-1}$ values of k_z are almost constant. Above this value the line shows increasing curvature, resulting in a steep decline of k value versus depth at the highest measured concentration of 119 mg l^{-1}. This degree of variation supports the decision to investigate actual values of k and k_z on a scale appropriate to typical WSP design.

The results for *S. subspicatus, C. reinhardtii, C. vulgaris* and *M. aeruginosa* showed a similar pattern of variation of k_z with both depth and SS. Figure 3 shows the variation of k_z for *C. vulgaris* plotted as a 3D surface. The response surface for *S. subspicatus* can be modelled by an equation of the form

$$k_z = 0.4 \times (\text{depth in metres}) + 0.0001 \times (\text{SS in mgl}^{-1})^2 - 0.2(\text{SS in mgl}^{-1}) \qquad (2)$$

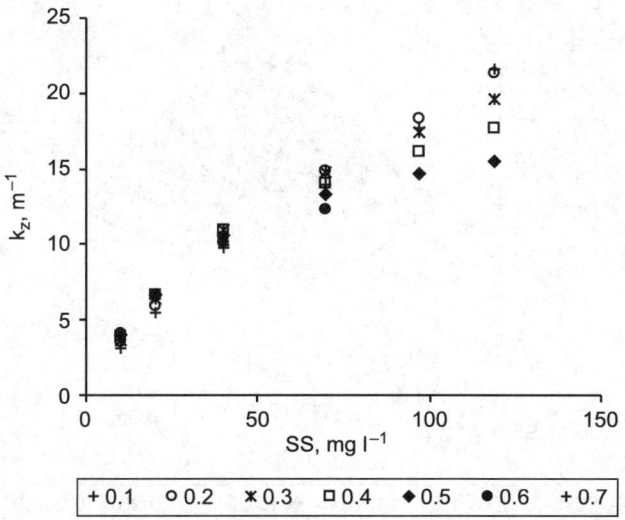

Figure 1 Variation of k_z with SS at different depths (metres) for a culture of *S. subspicatus*

giving a correlation of $R^2 = 0.94$ with the experimental data. The same equation applied to results for *C. reinhardtii*, *C. vulgaris* and *M. aeruginosa* gives $R^2 = 0.88$, 0.95 and 0.84 respectively, indicating good similarity. The lower value for *M. aeruginosa* may be due in part to the lower SS concentration in the experiment, and the low chlorophyll content of 0.04 mg l^{-1} or 0.001 mg mg^{-1}. While equation (2) has no physical basis it does indicate that at depths typical for a WSP a linear correction for k may be satisfactory, but variations with respect to SS are significant in the range of concentrations likely to be encountered. Figure 4 shows k_z values for a mixed culture dominated by purple sulphur bacteria, with a high suspended solids content. The correlation coefficient for experimental and modelled data with the above expression was $R^2 = 0.94$.

Values of k and k_z were determined for mixed cultures from SP1 and SP2 in the concentration range 5–58 mg l^{-1} SS. In the region of assumed linearity, k values ranged from 5.2 to 14.7 m^{-1}. The results were compared with values measured in situ in SP1 and SP2 using an array of eight sensors. At SS concentrations above 25 mg l^{-1}, values

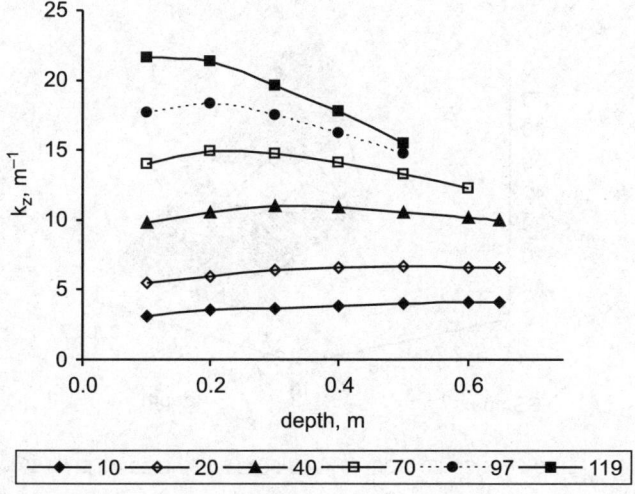

Figure 2 Variation of k_z with depth at different SS (mg l^{-1}) for a culture of *S. subspicatus*

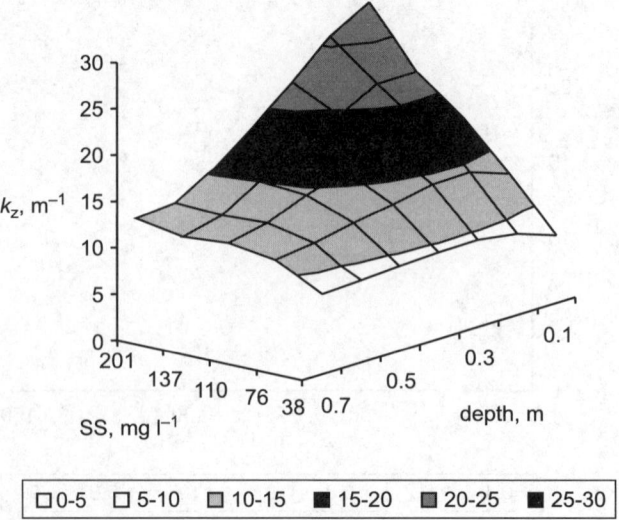

Figure 3 Variation of k_z with depth and SS for *C. vulgaris*

measured by the two methods showed reasonably good agreement. The main difficulty was in obtaining reliable and reproducible k values from measurements in the pond: accuracy requires bright sunlight, high solar elevation, a cloud-free sky, the absence of local shading, and minimal depth variations, as noted by Curtis *et al.* (1994). Results vary depending on whether the light is bright or diffused Kirk (1994) note that k may decrease with depth in diffuse light. While the photodiode array proved highly effective, optimum measurement conditions are often difficult to achieve in practice, providing an argument in support of laboratory-based methods. Values for k and k_z at SS concentrations below 20 mg l^{-1} were more difficult to measure in either the ponds or the column apparatus, especially at shallow depths (up to 0.2 m). In the column apparatus, this could be due in part to limitations of the equipment: at very low concentrations of suspended solids little absorption takes place, scattering plays a greater role and edge effects from the column walls may become apparent. At shallow depths small errors in depth measurement also have a greater effect. On dilution of a given culture from 25 mg l^{-1} to 10 and then 5 mg l^{-1}, however, the

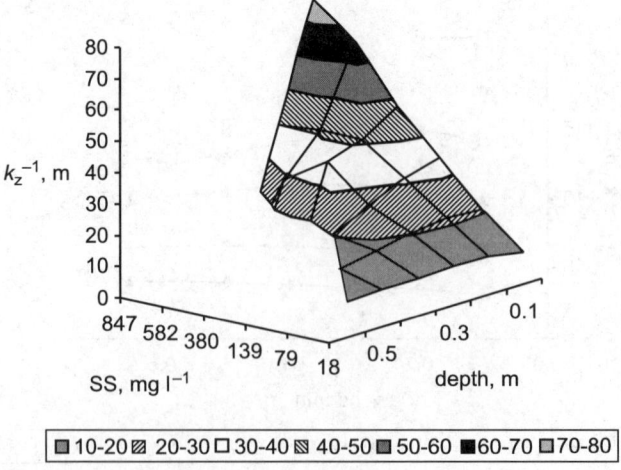

Figure 4 Variation of k_z with depth and SS for mixed culture with purple sulphur bacteria

non-linearity of the relationship between k and SS was clearly seen. Further support for the degree of variation in k values at shallow depths can be found from several sources. Curtis *et al.* (1994) noted that rates of attenuation were sometimes lower near the surface. The results in Figure 2 suggest that k values at shallow depth may be lower even at relatively high SS concentrations. In-pond measurements at shallow depth where wall effects are absent also showed a similar pattern of variation. This type of near-surface variation may cause problems for WSP modelling since under normal conditions the majority of photosynthetic activity occurs in the upper 0.2–0.3 metres, and values of k or k_z obtained over a greater depth may therefore not be applicable. The difficulty is compounded at low SS concentrations. This is not generally an issue in operational WSPs with steady-state SS concentrations of 30–100 mg l^{-1}, but may present problems in modelling pond start-up or for non-steady-state conditions such as those encountered in extreme climates.

There is clearly a need for standardisation in reporting of k values in ponds. In photobioreactors operating at high SS concentrations attenuation constants are frequently reported as specific values in m^2 kg^{-1} and may be based on a particular light path length, giving a value of k_z; while limnologists tend to measure over a long path length and assume linearity of k with SS. Algal WSPs fall between these two applications, but in a range where linearity cannot be assumed and it may be necessary to specify depth, SS concentration and possibly even surface irradiance when reporting k values.

Pond measurements

While the above discussion indicates some of the difficulties in measuring and calculating attenuation coefficients in WSPs, it was considered useful to have an idea of the range of values that might be expected under varying operating conditions. For this reason measurements were made in pilot-scale ponds in three locations as described above.

Values of k for the Southampton ponds were in the range 4.8–13.7 m^{-1} throughout a one-year period of observation, while suspended solids were in the range 42–172 mg l^{-1}. As the ponds were kept under semi-controlled conditions of light and temperature, and were subject to different experimental conditions in different periods, no clear seasonal trends were noted. There was a reasonable correlation between k values and suspended solids concentrations in both ponds throughout the year ($R^2 = 0.74$) (see Figure 5). The gradients of graphs of k in m^{-1} versus suspended solids in mg l^{-1} were 0.066 and 0.056 respectively. This is low in comparison with values for algal cultures and probably reflects the presence of non-algal SS. The relationship between absorbance and k in the period for which both parameters were measured was weaker ($R^2 = 0.53$ and 0.65 for

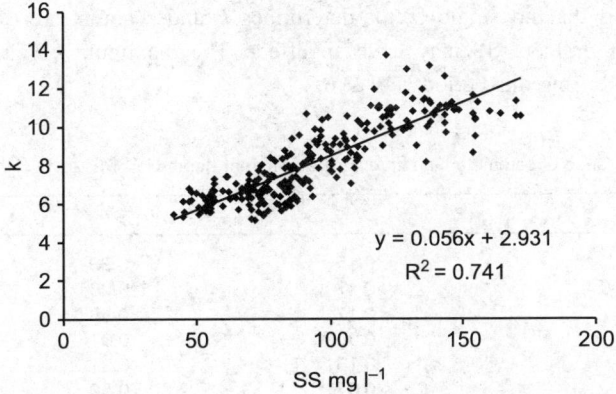

Figure 5 SS and k values for SP2

SP1 and SP2 respectively), while chlorophyll and k were effectively unrelated ($R^2 <$ 0.05 for both ponds).

The absolute value of k on any particular day is of uncertain reliability, due to the method of measurement and in particular to variations in surface irradiance. On average surface irradiance will be less than the clear-sky maximum, although as noted above k values with low correlation or low irradiance were discarded. The results, however, provide an interesting view of the range of k values likely to be encountered for a corresponding range of SS and chlorophyll concentrations.

In addition to the calculation of k values, the relationship between irradiance at a given depth and a number of other parameters was investigated by means of regression equations. Results for SP2 are summarised in Table 1. As might be expected, surface irradiance is the most influential single factor. Suspended solids appear to be better predictor of irradiance at 0.33 m than 0.15 m, once again confirming the influence of other factors and the problems of measurement in the upper layers. Surface irradiance and SS taken together are the strongest predictors, accounting for over 50% of variation in the top two layers of SP2. Chlorophyll by itself was a poor indicator of irradiance at depth but improved when considered in conjunction with surface irradiance.

For comparative purposes a similar study was carried out from late March to late May in Almaty, Kazakhstan where seasonal and climatic factors have a strong influence: in winter algal biomass concentrations in the water column fall sharply when the ponds freeze, followed by revival in spring. Values of k based on measurements in ponds A(II)1–3 rose rapidly from around $3-5\,\mathrm{m}^{-1}$ immediately after thawing in late March to $13-19\,\mathrm{m}^{-1}$ by early May, while suspended solids rose from $3-6$ to $60-70\,\mathrm{mg\,l}^{-1}$. Pond A(II)3 showed the strongest correlation between SS and k values for the whole period of observation, with $R^2 = 0.63$ and a gradient of $0.14\,\mathrm{m}^{-1}\,\mathrm{mg}^{-1}\mathrm{l}$, showing good agreement with Bowen (2004) and the above results. During this period the pond was not fed, and the suspended solids were therefore mainly of algal origin, arising from nutrients remaining in the pond over winter or released from the bottom sediments in spring. Ponds A(II)1 and 2 showed a weaker correlation between SS and k values in the same period. During this time these ponds were fed on a synthetic wastewater containing suspended solids. The correlation between SS and k was stronger for Pond A(II)2, which was fed on half-strength wastewater, than for Pond A(II)1 fed on full-strength wastewater containing correspondingly more SS ($R^2 = 0.45$ and 0.29 respectively). After peaking in late April, k values for these ponds fell until late May, although SS concentrations remained steady or rose slightly. The correlation of k with chlorophyll concentration in A(II)1 and 2 appeared promising but data were too few for reliability. In general the results suggested that no single factor determines k under non-steady-state conditions, especially where influent SS may have an effect. The maximum k value measured in Ponds A(II)1–3 during this period was $23\,\mathrm{m}^{-1}$.

Table 1 R^2 for influence of parameters on irradiance at different depths in SP2

Parameters			
Depth (m)	0.15	0.33	0.53
Surface irradiance (SI)	0.47	0.28	0.14
SS	0.17	0.38	0.37
Chlorophyll	0.00	0.01	0.01
Absorbance	0.16	0.31	0.16
SI and SS	0.57	0.50	0.15
SI and chlorophyll	0.45	0.29	0.10

No detailed analysis of relationships between k and other parameters was carried out for the Almaty A(I) ponds due to insufficient data on SS and chlorophyll, but results for measurements of k values support those from the A(II) ponds, showing a steady rise from the end of April into May. Subsequent falls appear to have been linked with the appearance of large numbers of grazing organisms. The maximum value of k measured during this period was $25\,m^{-1}$. Values of k measured by this method in Almaty in this period are likely to be relatively reliable due to the high solar elevation and long periods of cloudless weather. None of the results from the A(I), A(II) and Southampton ponds showed a good fit with empirical equations devised for other locations (eg. Xu et al., 2002), indicating that these depend on other parameters.

Conclusions

The light attenuation coefficient k was found to be significantly affected by depth and SS concentrations in the range of values typically found in WSPs. For practical purposes it may often be sufficient to consider k values as constant, but in certain conditions a more sophisticated approach may be needed taking local variation into account. Examples include modelling of WSP start up, or of the annual spring revival in strongly seasonal climates. Typical values for k in ponds appear to lie in the range $5-25\,m^{-1}$. The use of photodiodes to measure local irradiance proved highly successful. In-pond measurement presents many practical difficulties, however, and the column apparatus may offer a reliable means of measurement under standard conditions. It is recommended that a standardised approach is adopted to the measurement and reporting of k values.

Acknowledgements

The authors wish to acknowledge the support of INTAS Project KZ 96-1864, EU INCO-Copernicus Project CT98-0144, and of the BG Foundation, which enabled them to carry out this work.

References

Banks, C.J., Pak, L.N., and Rspaev, M.K. (2002). Springtime acclimatisation of a winter ice-covered waste stabilisation pond: operational data from 4 experimental units. In: *5th Int. IWA Conf. on Waste Stabilisation Ponds*. Auckland, New Zealand, 671-678.

Bowen, J.E. (1997). *EGET 3000 - Water Quality Modeling* http://www.coe.uncc.edu/~jdbowen/eget3000/1997/con_kin.html#EXTINCTION_COEFFICIENTS (Accessed April 2004).

Bartsch, A. (1961). Algae as a source of oxygen in waste treatment. *J. Wat. Poll. Contr. Fed.*, **33**(3), 239-249.

Brawley, J.W., Brush, M.J., Kremer, J.N. and Nixon, S.W. (2003). Potential applications of an empirical phytoplankton production model to shallow water ecosystems. *Ecol. Model.*, **160**(1-2), 55-61.

Cornet, J.F., Dussap, C.G. and Dubertret, G. (1992). A structured model for simulation of cultures of the cyanobacterium *Spirulina platensis* in Photobioreactors. I. Coupling between light transfer and growth kinetics. *Biotechnol. Bioeng.*, **40**, 817-825.

Curtis, T.P., Mara, D.D., Dixo, N.G.H. and Silva, S.A. (1994). Light penetration in waste stabilization ponds. *Wat. Res.*, **28**(5), 1031-1038.

Davies-Colley, R.J., Hickey, C.W. and Quinn, J.M. (1995). Organic-matter, nutrients, and optical characteristics of sewage lagoon effluents. *New Zeal. J. Mar. Fresh.*, **29**(2), 235-250.

Ensminger, I., Xylander, M., Hagen, C. and Braune, W. (2001). Strategies providing success in a variable habitat: III. Dynamic control of photosynthesis in Cladophora glomerata. *Plant Cell Environ.*, **24**, 769-779.

Fernandez, F.G.A., Camacho, F.G., Perez, J.A.S., Sevilla, J.M.F. and Grima, E.M. (1997). A model for light distribution and average solar irradiance inside outdoor tubular photobioreactors for the microalgal mass culture. *Biotechnol. Bioeng.*, **55**, 701-714.

Kirk, J.T.O. (1994). *Light and photosynthesis in aquatic ecosystems*, Cambridge University Press, Cambridge, UK.

Lonin, S.A. and Tuchkovenko, Y.S. (2001). Water quality modelling for the ecosystem of the Ciénaga de Tesca coastal lagoon. *Ecol. Model.*, **144**(2-3), 279–293.

Mesplé, F., Casellas, C., Troussellier, M. and Bontoux, J. (1995). Some difficulties in modelling chlorophyll a evolution in a high rate algal pond system. *Ecol. Model.*, **78**, 25–36.

Ogbunna, J.C., Hirokazu, Y. and Tanaka, H. (1995). Kinetic study on light-limited batch cultivation of photosynthetic cells. *J. Ferment. Bioeng.*, **80**(3), 259–264.

Pfennig, N., Widdel, F. and Truper, H.G. (1981). The dissimilatory sulphate-reducing bacteria. In: *The Prokaryotes*, in Starr, M.P., Stolp, H., Truper, H.G., Balows, A. and Schlegel, H.G. (eds.), Springer-Verlag, New York, pp. 926–940.

Privoznik, K.G. and Incropera, F.P. (1978). Absorption, extinction and phase function measurements for algal suspensions of *Chlorella pyrenoidosa*. *J. Quant. Spectrosc. Ra.*, **20**, 345–352.

Standard Methods for the Examination of Water and Wastewater (1998). American Public Health Association/American Waterworks Association/Water Environment Federation, 20th edn., Washington DC, USA.

Stutz-McDonald, S.E. and Williamson, K.J. (1979). Settling rates of algae from wastewater lagoons. *J. Env. Eng. Div. ASCE*, **105**(2), 273–282.

Thomann, R.V. and Mueller, J.A. (1987). *Principles of Surface Water Quality Modeling and Control*. Harper & Row, New York.

Tsirtsis, G.E. (1995). A simulation model for the description of a eutrophic system with emphasis on the microbial processes. *Wat. Sci. Tech.*, **32**(9-10), 189–196.

Wetzel, R.G. (2001). *Limnology: Lake and River Ecosystems*. Academic Press, London.

Xu, P., Brissaud, F. and Fazio, A. (2002). Non-steady-state modelling of faecal coliform removal in deep tertiary lagoons. *Wat. Res.*, **36**, 3074–3082.

Yun, Y.S. and Park, J.M. (2001). Attenuation of monochromatic and polychromatic lights in *Chlorella vulgaris* suspensions. *Appl. Microbiol. Biotech.*, **55**, 765–770.

Optical characteristics of waste stabilization ponds: recommendations for monitoring

R.J. Davies-Colley, R.J. Craggs, J. Park and J.W. Nagels

National Institute of Water and Atmsopheric Research Ltd, P.O. Box 11-115, Hamilton, New Zealand
(E-mail: r.davies-colley@niwa.co.nz)

Abstract The optical character of waste stabilization ponds (WSPs) is of concern for several reasons. Algal photosynthesis, which produces oxygen for waste oxidation in WSPs, is influenced by attenuation of sunlight in ponds. Disinfection in WSPs is influenced by optical characteristics because solar UV exposure usually dominates inactivation. The optical nature of WSPs effluent also affects assimilation by receiving waters. Despite the importance of light behaviour in WSPs, few studies have been made of their optical characteristics. We discuss simple optical measures suitable for routine monitoring of WSPs (including at sites remote from laboratories): optical density of filtrates – an index of dissolved coloured organic (humic) matter, visual clarity – to provide an estimate of the beam attenuation coefficient (a fundamental quantity needed for optical modelling) colour (hue) – as an indicator of general WSP 'condition' and irradiance attenuation quantifying depth of light penetration. The value of optical characterisation of WSPs is illustrated with reference to optical data for WSPs in NZ (including high-rate algal ponds) treating dairy cattle wastewater versus domestic sewage. We encourage increased research on optical characteristics of WSPs and the incorporation of optical measures in monitoring and modelling of WSP performance

Keywords WSPs; disinfection; algal growth; solar radiation; photo-oxidation; UV radiation; optics

Introduction

Bacteria and algae in waste stabilization ponds (WSPs) may be viewed as symbiotically related. The bacteria oxidize organic matter in organic wastes, resulting in mobilization of nutrient elements (notably carbon, nitrogen and phosphorus), while algae assimilate mineral nutrients and provide oxygen for respiration (e.g., Oswald, 1992). Thus aerobic treatment in WSPs is powered by sunlight via algal photosynthesis. Clearly then, optical characteristics influencing sunlight exposure, are fundamental to (aerobic) waste stabilization in WSPs.

Sunlight penetration into WSP water strongly influences several other of their important treatment functions, notably disinfection – the removal of harmful ('pathogenic') micro-organisms from organic wastes. Davies-Colley et al. (2000) reviewed several lines of evidence suggesting that sunlight exposure is the most important driver of WSP disinfection, although sometimes interacting with other factors (notably DO and pH, Davies-Colley et al., 1999).

The optical character of WSPs may also be of concern where WSPs discharge to natural waters. WSP effluent is typically highly light-attenuating due, particularly, to a high content of algal solids that degrade the optical water quality of receiving waters. In New Zealand (NZ), a large number of 2-stage WSPs have generally performed well, but there has been a trend over the past decade towards alternative treatment, driven by legal standards (e.g., NZ Resource Management Act, 1991) protecting the optical quality of natural waters (Davies-Colley et al., 1993).

Despite its importance, the optical character of WSPs and consequent behaviour of sunlight has received little scientific attention. A notable exception is a paper by Curtis et al. (1994) reporting light penetration and spectral light absorption properties of WSPs in Brazil. Davies-Colley et al. (1995) and Sukias et al. (2001) reported optical characteristics of WSPs in NZ. A few other papers have considered the attenuation of sunlight in WSPs as part of studies of particular aspects of treatment. For example, Calkins et al. (1976) measured penetration into WSP waters of skin-burning UV that they considered responsible for inactivation of faecal indicator bacteria.

The goal of this paper is to increase awareness of sunlight behaviour in WSPs. We review the optics of waters with particular reference to WSPs. Simple optical methods suitable for monitoring of WSPs are described, and the utility of these measures is illustrated with summary statistics for different kinds of WSPs treating sewage *versus* dairy cow wastes in NZ. We recommend incorporating optical measures in WSP monitoring, and greater research effort on the optical characteristics of WSPs, particularly in models of treatment performance.

Optics of water applied to WSPs
Components of sunlight

Sunlight comprises a wide range of wavelengths from ultra-violet (UV) at about 300 nm to near-infra-red at about 3000 nm (Iqbal, 1983). Nearly half of sunlight is in the visible range (400–700 nm), which is also the radiation (photosynthetically available radiation, PAR) that drives photosynthesis of plants including the planktonic micro-algae dominating WSPs. Near-infra-red radiation (700–3000 nm, ~ 50% of total solar power) penetrates water bodies minimally owing to strong absorption by molecular water, and is of interest primarily for heating surface waters of WSPs and promoting diurnal thermal stratification. A small fraction of sunlight in the UV range (300–400 nm, comprising about 5% of total solar irradiance near noon) is of disproportionate interest because of its photo-oxidation and disinfection effects in WSPs. The solar UV spectrum is conveniently divided into UV-A (320–400 nm), which is the main driver of photo-oxidation processes; and UV-B (skin-burning UV, 300–320 nm), a variable, but tiny, contribution (<0.2% of solar power), which is strongly absorbed by DNA causing damage to the genome.

Light behaviour in waters: inherent optical properties

Two main optical processes govern the behaviour of light traversing water: *absorption* and *scattering* (Davies-Colley et al., 1993; Kirk, 1994). Light absorption involves transfer of photon energy to another form (heat or chemical energy – as in photosynthesis) with the consequent disappearance of the photon. Absorption is quantified by an *absorption coefficient*, a, which may be conceptualized as the probability of absorption of a photon per unit length of light path through water. Scattering, similarly quantified by a *scattering coefficient*, b, involves change in direction of photons (with no loss of quantum energy) owing to reflection, refraction, diffraction or a combination of these processes. Water molecules and dissolved materials cause weak scattering (equally in all directions), whilst suspended particles, notably algal cells in WSPs, scatter light much more strongly and primarily at small angles to the incident photon direction.

The sum of absorption and scattering coefficients is referred to as the *beam attenuation coefficient*: $c = a + b$ (Kirk, 1994). These three optical coefficients, all with units of 1/m, are termed 'inherent' optical properties (IOPs) because they depend only on the composition of the water.

Absorption coefficients are strong functions of wavelength, and are rather difficult to measure, although special spectrophotometric methods (e.g., Davies-Colley et al., 1993) are satisfactory for strongly light-attenuating waters like WSPs. Figure 1 shows the spectral absorption of light in high rate algal ponds (HRAPs) treating contrasting wastewaters. Total absorption is the sum of contributions by solid particles (a_p), dissolved humic

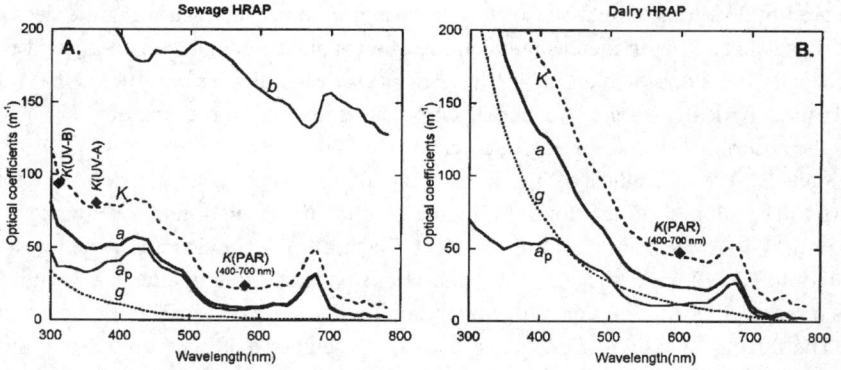

Figure 1 Spectral optical characterization of HRAPs. A. Experimental HRAP treating sewage at Ruakura, near Hamilton, NZ (23 February, 2004), B. Full-scale HRAP treating dairy cattle wastewater from the Dexcel farm near Hamilton, NZ (8 April 2004). Measurements were made using a Shimadzu UV-2501 spectrophotometer with techniques described in Davies-Colley et al. (1993). Absorption spectra are given for dissolved humic matter (g) and particulate constituents (a_p). Total absorption (a) is the sum of these contributions plus that of pure water (a_w, not shown). Also shown is the scattering spectrum (b) (Panel A; scattering is off-scale in Panel B). The spectral irradiance attenuation coefficient (K) was estimated from the spectral absorption and scattering coefficients using empirical equations given by Kirk (1984). Direct estimates of K (from the depth profiles for different wavebands in Figure 2) are shown for comparison (solid diamonds)

matter (or gilvin, g), and water itself (a_w) at each wavelength in the spectrum: $a = a_p + g + a_w$. The solids in WSPs are typically dominated by algae and have absorption peaks at 440 nm in the blue region of the spectrum, and at 676 nm in the red region (Figure 1). These features are the *in vivo* absorption peaks of the main photosynthetic pigment, chlorophyll-a (Chl$a$). Other features in the particulate absorption spectrum are characteristic of various 'accessory' pigments of the dominant green algae (mainly *Scenedesmus*, *Microactinium* and *Ankistrodesmus* in HRAPs), including carotenoids that are responsible for the 450–490 nm absorption 'shoulder' (Davies-Colley et al., 1986). Absorption by gilvin (g) increases with decreasing wavelength in a characteristic exponential pattern. Water contributes almost negligibly to total absorption, consequently a_w is not shown in Figure 1.

Scattering coefficients are very difficult to measure and are seldom directly quantified. In contrast, the beam attenuation coefficient is comparatively easy to measure (by beam transmissometry or simple visibility measurement – as discussed below), and the scattering coefficient may then be estimated by subtracting absorption from total beam attenuation: $b = c - a$ (e.g., Davies-Colley et al., 1993). Figure 1A shows the scattering spectrum for a HRAP treating sewage, estimated from a visibility measurement combined with absorption estimated from spectrophotometric scans.

Apparent optical properties

Behaviour of sunlight in water is described by 'apparent' optical properties (AOPs) that depend mainly on the composition of waters, but also depend weakly on the characteristics of the incident sunlight – which changes with weather conditions and solar altitude. AOPs are more intuitive than the IOPs, but the latter are preferable for optical characterization and modeling because they are rigorously additive and independent of incident lighting (Kirk, 1994).

Penetration of sunlight into waters is quantified by the *irradiance attenuation coefficient*, K, which is the probability of disappearance of a photon per unit depth in the water column. K, like the IOPs, has the units of 1/m and varies with wavelength, and is measured by lowering a light sensor into the water column and measuring the decreasing light (irradiance, E) with increasing depth, z. The irradiance attenuation coefficient is estimated from the slope of plots of $\ln E$ versus z (Davies-Colley et al., 1993). Kirk (1984; 1994) used optical modeling to derive empirical equations for estimating K (an AOP) from the absorption and scattering coefficients (IOPs).

Figure 2 shows profiles of irradiance (on a log scale) as a function of depth in HRAPs. The slopes of the lines in Figure 2 quantify K in three wavebands: UV-B (311 nm), UV-A (365 nm), and PAR (400–700 nm). PAR evidently penetrates more deeply into HRAP water than UV-A, while UV-B is most strongly attenuated with depth. The corresponding K-values are shown overlain on the K-spectra in Figure 1 for comparison. The directly measured K(PAR) is just slightly higher than the spectral minimum in the PAR range because, once light has penetrated a short distance below the water surface, the PAR is mostly confined to this spectral region.

The (dimensionless) ratio of upwelling to downwelling irradiance is an additional AOP known as the *irradiance reflectance* (symbol R) despite being more closely related to scattering than to the optical phenomenon of reflection. R, together with K, can provide useful optical information about a water body (Kirk, 1994). However, R is much less often measured than K and is mentioned here mainly because it relates to water colour. Curtis et al. (1994) reported rather low R values ($\sim 1\%$) which they attributed to high humic content in Brazilian WSPs.

Optical methods applicable to WSPs

Table 1 lists a number of simple optical measurements that we recommend for monitoring WSPs. These measures, and their relation to the fundamental processes of absorption and scattering, will now be discussed in turn. The references cited below (and in Table 1) should be consulted for more detail on measurement methods. Table 2 lists median optical (and related) quantities that we have measured in various WSP monitoring campaigns (Davies-Colley et al., 1995; Sukias et al., 2001, and authors' unpublished data).

Dissolved humic matter

An important contributor to light absorption in the ultra-violet range is dissolved humic matter (alternatively 'gilvin', Kirk, 1976) – which has a characteristic absorption spectrum well-described by an exponential function of wavelength: $g = g_L \exp\{S(L - \lambda)\}$ where λ is wavelength and L is a reference wavelength (Figure 1 gives example spectra).

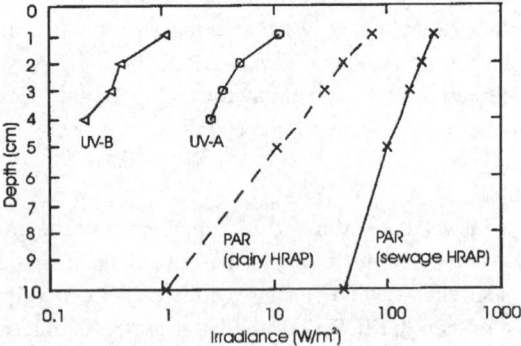

Figure 2 Depth profiles of irradiance in a HRAP. Measurements were made in an experimental HRAP treating sewage at Ruakura, near Hamilton, NZ, (23 February, 2004), using a Macam 3-channel radiometer for irradiance measurements in the UV-B (peak at 311 nm), UV-A (peak at 365 nm) and PAR (400–700 nm). Also shown, for comparison, is the PAR profile (dotted line) in a HRAP treating dairy cattle wastewater from the Dexcel farms near Hamilton, NZ (8 April 2004)

Table 1 Four simple optical variables recommended for monitoring WSPs.

Symbol	Optical variable	Rationale	Measurement notes	References
g_{440} *	Colour of dissolved humic matter	Index of humic matter – coloured organic matter ('gilvin')	Optical density of a membrane filtrate (0.2 μm pore size) of WSP effluent	Kirk (1976)
y_{BD}	Visibility (black disc visual range)	Measure of visual water clarity. Provides an estimate of $c(550)$	Measured *in situ* or (in a trough) on a diluted WSP sample	Davies-Colley (1988); Davies-Colley and Smith (2001)
$K(PAR)(z_{eu})$	Irradiance attenuation coeff. (euphotic depth)	Measures light penetration into WSP	Depth profiling with a submersible radiometer through the WSP water column	Davies-Colley et al., (1993)
–	Colour (hue)	Guide to general optical 'condition' of WSP	Hue of WSP water (viewed *in situ*) matched to colour standards	Davies-Colley et al., (1997)

* g_{440} is a laboratory measurement (other measurements are made in the field)

Table 2 Median values of optical variables (and chlorophyll-*a*) in different WSPs (Chlorophyll-*a* – Chla – is listed because of its strong influence on optical characteristics)

Measure	(units)	Conventional WSPs, treating...		HRAPs, treating...	
		sewage[1]	dairy cow wastes[2]	sewage[3]	dairy cow wastes[4]
Chla	(g m^{-3})	0.48	1.04	2.5	3.3
g_{440}	(m^{-1})	6.5	41	5.4	39
Turbidity	(NTU)	19	140	53	120
y_{BD}	(m)	0.12	0.025*	0.028*	0.022*
z_{eu}(PAR)	(m)	0.35	0.11	0.18	0.07
Colour (hue)	(Munsell hue code)			3GY#	9Y#

In situ black disc range estimated from measurements in a trough on a volumetrically diluted sample
#Colour (hue) values in the Munsell colour system (matched *in situ* to Munsell standards)
[1]Statistics for maturation ponds (second pond) of eleven 2-pond WSP systems treating sewage in NZ (Davies-Colley *et al.*, 1995)
[2]Statistics for facultative ponds (second pond) of six 2-pond WSP systems treating dairy cow wastes in NZ (Sukias *et al.*, 2001)
[3]Authors' unpublished data from an experimental HRAP treating sewage at Ruakura, near Hamilton, NZ
[4]Authors' unpublished data from from a full-scale HRAP treating dairy cattle wastes from the Dexcel farm near Hamilton, NZ (monthly measurements for a year)

The spectral slope parameter, S, is in the range 0.01–0.02 nm^{-1} in a wide variety of natural waters (Kirk, 1994), including WSPs. This predictable spectral absorption means that measurement at just one reference wavelength (usually 440 nm) serves to indicate absorption by dissolved humics.

Dissolved humic matter may be isolated by filtering WSP water through a membrane filter, preferably at 0.2 μm (Kirk, 1976). A double-beam spectrophotometer is ideally used to measure absorbance (D) of the filtrate *versus* pure water in the reference beam, but a simple (single beam) spectrophotometer or filter photometer is satisfactory. At sites remote from laboratory facilities, a visual colour comparator could be used to index dissolved humic matter *versus* stable standards on the Hazen scale (Davies-Colley, 1988; Davies-Colley *et al.*, 1993).

The dissolved humic absorption coefficient is calculated: $g = 2.303D/l$ where D is absorbance and l is cuvette path length. Kirk (1976) advocated reporting g at 440 nm, at the blue absorption peak of chlorophyll-*a*, but we have found measurements at 340 nm (in the UV range) to be particularly useful for studies of sunlight disinfection in WSPs that is caused mainly by solar UV (e.g., Davies-Colley *et al.*, 1999). The median g_{440} value was 6.5 m^{-1} in conventional WSPs treating sewage in NZ (Table 2), which is 'high' considering Kirk's (1994) compilation of g_{440} values for natural freshwaters, and only exceeded by highly humic-stained bog lakes. WSPs treating dairy cow wastes in NZ are very strongly coloured (median g_{440} = 41 m^{-1}).

Visual clarity

The relevance of visual clarity to treatment performance of WSPs is not immediately obvious except where discharge is to visually clear receiving waters sensitive to optical degradation. However, visual clarity is useful in WSPs because it provides an estimate of the beam attenuation coefficient – an IOP needed for optical modelling. Visual clarity is traditionally measured in limnology as the vertical sighting range of a white (or black-and-white) Secchi disc, and this simple tool has sometimes been used in WSPs (e.g., Curtis *et al.*, 1994). However, the use of standard limnological (20 cm diam.) Secchi discs is highly dubious in WSPs with visual ranges ∼ 10 cm, because the presence of the disc distorts the light field (Davies-Colley *et al.*, 1993).

Turbidity has sometimes been measured (by nephelometry) on WSP water samples as an index of visual clarity (e.g., Table 2). However, nephelometric turbidity is only a relative measure (versus an arbitrary standard – formazin). Davies-Colley and Smith (2001) have argued that the visual clarity measurement is preferable for optical characterization of waters, which also goes for WSPs.

A better index of visual clarity is the sighting range of a black disc observed horizontally underwater with a periscope viewer (Davies-Colley, 1988; Davies-Colley et al., 1993). Because a black body (ideally) reflects no light, this observation, unlike the Secchi depth, is independent of incident lighting conditions. Measurement of horizontal visual range in water, y_{BD}, provides a fairly precise (standard error $\sim$ 5%) estimate of the beam attenuation coefficient at the peak sensitivity of the human eye, $c(550) = 4.8/y_{BD}$ (Zanevald and Pegau, 2003). The beam attenuation coefficient can, in turn, be used to quantify scattering by subtracting the absorption coefficient. Visibility is very restricted in the strongly light-attenuating waters of WSPs. For example, Davies-Colley et al. (1995) reported an average $y_{BD} = 12$ cm (20 mm diam. disc) in NZ sewage WSPs (Table 2). Visibilities <10 cm are better measured on a water sample diluted volumetrically (in a trough) rather than directly (Davies-Colley and Smith, 1992), and Table 2 gives median visibilities of a few cm calculated from measurements made on diluted effluents for WSPs treating dairy cow wastes and for HRAPs. Visibility in the sewage HRAP was remarkably strongly correlated to the chlorophyll-a concentration ($R = -0.97$), suggesting the possibility of using visibility (suitably calibrated) as a cheap on-site surrogate for algal biomass in some WSPs.

Light penetration

Light penetration into WSPs is fairly easily measured with standard submersible sensors of light (irradiance), such as PAR sensors used in limnological studies. A pair of (matched) sensors is preferable to a single sensor for measurements under changeable incident light – as is typical under patchy cloud cover. Precise control of depth is needed for obtaining light profiles in WSPs in which light gradients are very steep. A specially constructed profiling frame (to be described elsewhere, Park et al., 2004 in prep.) was used to deploy the 3-channel Macam radiometer (Macam Photometrics Ltd, Livingston, Scotland) for measuring the UV and PAR profiles illustrated in Figure 2. A quantity that is more 'intuitive' than K for quantifying light penetration is the euphotic depth, the depth at which PAR is reduced to 1% of its value at the water surface: $z_{eu} \sim 4.6/K(PAR)$ (Kirk, 1994). For example, Davies-Colley et al. (1995) estimated a median euphotic depth of 0.35 m in sewage WSPs in New Zealand, while Sukias et al. (2001) reported a more restricted euphotic depth (0.11 m) in WSPs treating dairy cattle wastes (Table 2). HRAPs are very light-attenuating with a very shallow euphotic depth (Table 2).

Colour (hue)

Colour of WSP water may be a useful guide to their composition. Colour of waters, in common with colour generally, can be defined with reference to the spectral reflectance (R) of the water body. The most important aspect of water colour is the *hue* which relates to the dominant range of wavelengths in the reflectance spectrum. Two other aspects of water colour, *brightness* and *colour purity*, seem less likely to be useful in WSPs and are not considered further herein. The viewer for black disc visibility measurements may be used to observe water colour *in situ* under water (eliminating surface glare) and hue may be matched to Munsell colour standards (Davies-Colley et al., 1997). WSPs are usually green to yellow in hue because selective absorption of red and blue light by chlorophyll-a and accessory photosynthetic pigments in algae (Figure 1) leaves primarily light in the green-yellow range available to be scattered to the observer's eyes. We have found it useful to record hue of WSP water in routine monitoring (Table 2 summarizes data for HRAPs).

Optical conditions in HRAPs

Recently we have conducted optical monitoring of two HRAPs treating contrasting wastewaters – dairy cow wastes *versus* domestic sewage. These systems display some interesting optical differences between different wastewaters and compared to conventional WSPs. The data in Table 2 serve to illustrate the utility of optical measurements in WSP monitoring. The median chlorophyll-a was very much higher in HRAPs than in conventional ponds, but only slightly higher in the HRAP treating dairy cow wastes than in that treating sewage. The dissolved humic matter (indicated by g_{440}) was much higher in the dairy cow waste than in sewage, irrespective of pond type – apparently reflecting the concentration of coloured organic matter in the source wastewater. HRAPs were appreciably more light-attenuating than conventional WSPs, in terms of both visual clarity and euphotic depth. The colour (hue) of the HRAP treating sewage was usually in the green-yellow (GY) range (averaging Munsell 3GY). In the dairy HRAP, hue was more variable (averaging a greenish-yellow, Munsell 9Y, and ranging from an 'orange' 10YR to 2.5GY).

Light penetration in WSPs, in common with eutrophic lakes, is expected to depend mainly on algal biomass, gilvin, and visual clarity. Monitoring of K, Chl$a$, g_{440}, and y_{BD} (monthly for a year) in the two contrasting HRAPs whose optical quality is summarized in Table 2, provided a suitable dataset for statistical modeling of light penetration. In the HRAP treating sewage, K was fairly strongly related to g_{440} and very strongly related to y_{BD} and Chl$a$ (which themselves are very strongly inter-correlated). In the dairy HRAP, K was most strongly related to g_{440}, and weakly to Chl$a$. The following multiple linear regression equations (variables in SI units) account for about three-quarters of total variance and may be useful for roughly estimating light penetration where PAR sensors are unavailable.

Sewage HRAP:

$$K = -62\, y_{BD} + 4.3\, g_{440} + 28 (R^2 = 87\%), \quad K = 4.4 \text{Chl}a + 2.7\, g_{440} - 0.5 (R^2 = 77\%)$$

Dairy HRAP:

$$K = 1.44\, g_{440} + 5, (R^2 = 69\%), \quad K = 2.1\, g_{440} + 9.6 \text{Chl}a - 38 (R^2 = 79\%).$$

Conclusions

Optical characteristics of WSPs are under-studied and comparatively poorly understood. We believe that optical variables should be more widely studied as part of research on WSPs in order to improve understanding of these systems. Furthermore, certain simple optical variables may provide an indication of WSP 'condition' or 'health', usefully supplementing more familiar (and expensive) variables that are routinely monitored, such as BOD and TSS. Observations of colour (hue) and measurements of visual clarity (y_{BD}, a useful indicator of algal biomass in some WSPs) are recommended. Algal chlorophyll-a (Chl$a$) and dissolved humic matter (g_{440}) can be measured where laboratory facilities are available. (If necessary, humic colour can be indexed on-site using a visual colour comparison method rather than a spectrophotometer). Measurements of light penetration (euphotic depth, z_{eu}) are useful if suitable submersible light sensors are available; alternatively, light penetration may be estimated from g_{440}, y_{BD} and Chl$a$ using empirical equations such as those reported here. We encourage increased research effort on optical aspects of WSPs and incorporation of optical parameters in models of WSP performance.

References

Calkins, J., Buckles, J.D. and Moeller, J.R. (1976). The role of solar ultraviolet radiation in 'natural' water purification. *Photochemistry and Photobiology*, **24**, 49–57.

Curtis, T.P., Mara, D.D., Dixo, N.G.H. and Silva, S.A. (1994). Light penetration in waste stabilisation ponds. *Water Research*, **28**, 1031–1038.

Davies-Colley, R.J. (1988). Measuring water clarity with a black disc. *Limnology and Oceanography*, **33**, 616–623.

Davies-Colley, R.J., Donnison, A.M. and Speed, D.J. (2000). Towards a mechanistic understanding of pond disinfection. *Water Science and Technology*, **42**(10–11), 149–158.

Davies-Colley, R.J., Donnison, A.M., Speed, D.J., Ross, C.M. and Nagels, J.W. (1999). Inactivation of faecal indicator micro-organisms in waste stabilization ponds: interactions of environmental factors with sunlight. *Water Research*, **33**, 1220–1230.

Davies-Colley, R.J., Hickey, C.W. and Quinn, J.M. (1995). Organic matter, nutrients, and optical characteristics of sewage lagoon effluents. *New Zealand Journal of Marine and Freshwater Research*, **29**, 235–250.

Davies-Colley, R.J., Pridmore, R.D. and Hewitt, J.E. (1986). Optical properties of some freshwater phytoplanktonic algae. *Hydrobiologia*, **133**, 165–178.

Davies-Colley, R.J. and Smith, D.G. (1992). Offsite measurement of the visual clarity of waters. *Water Resources Bulletin*, **28**, 1–7.

Davies-Colley, R.J. and Smith, D.G. (2001). Turbidity, suspended sediment, and water clarity: A review. *Journal of the American Water Resources Association*, **37**, 1085–1101.

Davies-Colley, R.J., Smith, D.G., Speed, D.J. and Nagels, J.W. (1997). Matching natural water colors to Munsell standards. *Journal of the American Water Resources Association*, **33**, 1351–1361.

Davies-Colley, R.J., Vant, W.N. and Smith, D.G. (1993). *Colour and Clarity of Natural Waters: Science and Management of Optical Water Quality*, Ellis Horwood, Chichester, UK.

Iqbal, M. (1983). *An Introduction to Solar Radiation*, Academic Press, Toronto, Canada.

Kirk, J.T.O. (1976). Yellow substance (Gelbstoff) and its contribution to the attenuation of photosynthetically active radiation in some inland and coastal south-eastern Australian waters. *Australian Journal of Marine and Freshwater Research*, **27**, 61–71.

Kirk, J.T.O. (1984). Attenuation of solar radiation in scattering-absorbing waters: a simplified procedure for its calculation. *Applied Optics*, **23**, 3737–3739.

Kirk, J.T.O. (1994). *Light and photosynthesis in aquatic ecosystems*, 2nd edn. Cambridge University Press, Cambridge, UK.

Oswald, W.J. (1992). Wastewater treatment with microalgae. *Journal of Phycology*, **28**(3).

Sukias, J.P.S., Tanner, C.C., Davies-Colley, R.J., Nagels, J.W. and Wolters, R. (2001). Algal abundance, organic matter and physico-chemical characteristics of dairy farm facultative ponds. Implications for treatment performance. *New Zealand Journal of Agricultural Research*, **44**, 279–296.

Zanevald, J.R.V. and Pegau, W.S. (2003). Robust underwater visibility parameter. *Optics Express*, **11**, 2997–3009.

Profiling and modelling of thermal changes in a large waste stabilisation pond

D.G. Sweeney*, J.B. Nixon**, N.J. Cromar* and H.J. Fallowfield*

*Department of Environmental Health, Flinders University, GPO Box 2100, Adelaide SA 5001, Australia
(E-mail: *david.sweeney@flinders.edu.au*; *nancy.cromar@flinders.edu.au*; *howard.fallowfield@flinders.edu.au*)
**United Water International, GPO Box 1875, Adelaide SA 5001, Australia (E-mail: *john.nixon@uwi.com.au*)

Abstract A thermal profiling study was undertaken at four depths at each of nine sites, and at the inlets and outlets of a large waste stabilisation pond (WSP). Results were collected simultaneously using a network of 42 thermistors and dataloggers. Profiles at each site were categorised as either "stratified" or "unstratified", and persistence analysis was used to determine the frequency and persistence of stratification events at each of the nine sites. Stratification was found to persist most strongly at the site furthest upwind in the WSP, with respect to prevailing wind during the study, leading to the conclusion that stratification induced short-circuiting will be greatest in this region of the WSP. A computational fluid dynamics (CFD) model was constructed of the WSP, including an energy balance to predict the bulk stratification gradient in the pond. Environmental conditions and WSP inlet temperature during one day in June 2001 were used as boundary conditions. The pond thermal profiles measured during the profiling study, together with outlet temperature during the day, were used to validate the CFD model results. The model predicted mean pond temperature with a high degree of accuracy ($r^2 = 0.92$). However it was evident that even modest winds (≥ 1.5 m/s) partially broke down stratification, leading to poor prediction of the gradient by the CFD model, which did not directly account for the impact of wind shear stress on mixing in the WSP.
Keywords Waste stabilisation pond; thermal stratification; persistence analysis; computational fluid dynamics

Introduction

Previous studies have demonstrated that significant variation in temperature can simultaneously exist throughout all three dimensions of large waste stabilisation ponds (e.g. Weatherell *et al.*, 1999; Sweeney *et al.*, 2002). This has important implications for the study of thermally induced hydraulic behaviour (e.g. stratified short-circuiting), and also for assessing the impact on expected treatment, given the established relationship between biological rate and temperature (Zhao and Zhang, 1991) in waste stabilisation ponds (WSPs).

It is typical in these studies of variability for the thermal sampling through the pond to be undertaken sequentially through a number of depths and sites. However, thermal changes are dynamic, and temporal changes in the thermal gradient of up to 1.9 °C/m hr have been observed at individual sites (Sweeney *et al.*, 2002). As a consequence, the analysis of spatial variability in WSPs is often confounded by the temporal changes associated with moving between sampling sites.

An alternative approach was reported in Sweeney *et al.* (2002), in which a number of sites in a WSP were each profiled over the course of a single day, on sequential days. In this case, the analysis of spatial variability was confounded by the variation in environmental conditions between days at Bolivar Wastewater Treatment Plant (WWTP). While multivariate statistical analysis can be used to test the significance of spatial variability independently of climatic variability, the understanding that this method provides about any systematic trends in variability through a WSP is limited. In order to produce an explicit demonstration of thermal variation, in the first part of this study a simultaneous thermal profiling study was undertaken on a large WSP in 3-D.

Computational fluid dynamics (CFD) modelling is emerging as an accurate and useful method of simulating localised flows in WSPs. This method can be used to compare proposed WSP designs or the impact of physical interventions on WSP hydraulics and ultimately treatment (Shilton and Harrison, 2003). A simplified approach to thermal modelling stratification was used by Salter *et al.* (2000), in which a thermal profile measured at a site was applied to a CFD model and the resulting impact on hydraulics simulated. However, the ability to predict the formation of stratification in a WSP using CFD based on external environmental factors remains largely untested. Such a technique would provide a convenient method of assessing the likely impact of thermal stratification on flow conditions and ultimately treatment efficiency for a specific WSP design at a specific site. The prediction of stratification in a large WSP was undertaken in 3-D using CFD. Results from the physical thermal study provided a basis for validation of the CFD model results.

Method

Profiling simultaneous thermal variation in a WSP

Fieldwork was undertaken on Lagoon 1, a 1.4 m deep, 112 ha pond at Bolivar WWTP, 18 km north of Adelaide, South Australia (34°45′23″S 138°34′15″E). The pond has a nominal residence time of 12 days at typical WWTP flow. A series of 36 thermistors with data loggers were arranged in nine strings of four thermistors, and located at the nine sites shown in Figure 1. A further six thermistors were installed in three of the 22 inlets, and three of the 14 outlets. The thermistors used have an accuracy of ±0.2 °C, and were all calibrated relative to a reference thermistor to minimise relative error. The internal clocks of all loggers were synchronised, and the system programmed to record an entire simultaneous pond profile every 10 minutes. In addition to *in situ* temperature, the wind speed, direction and mean solar radiation at the site in the 10 minutes prior to the sample were also recorded.

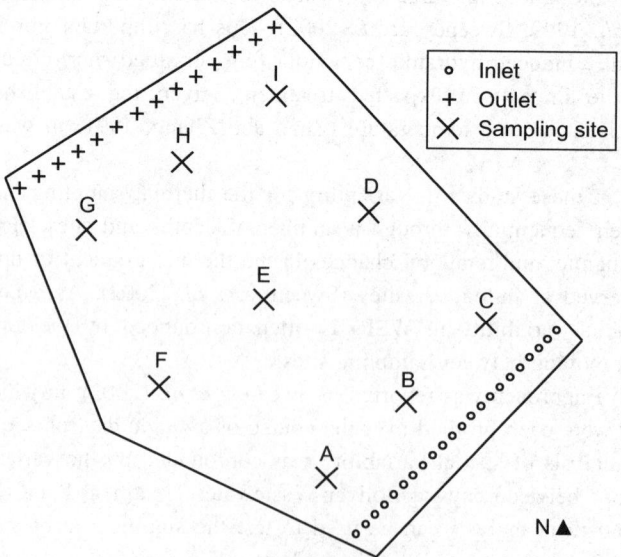

Figure 1 Geometry and sampling sites, Lagoon 1, Bolivar WWTP

In total, over 12,000 thermal profiles were captured covering summer, autumn and winter operation of the WSP. For each profile at each site, the mean temperature through the pond depth (°C) and the normalised stratification gradient (°C/m, as estimated by fitting a cubic spline to the four data points) were calculated. These data were used to construct surface maps of the mean temperature distribution across the WSP, and the distribution in stratification gradient across the pond, every 10 minutes.

The data were further analysed using the technique of persistence analysis described by Kuwashima and Hogben (1986). Each WSP thermal profile was classified as either "unstratified" (0) or "stratified" (1) according to the test parameter described below. The result was a string of binary numbers representing the sequential build-up and breakdown of stratification, which could be analysed to determine the frequency, timing and persistence of stratification events at the site.

$$\text{Test parameter } (i) = \begin{cases} 0 & \text{if mean } \Delta T_i/m < \Delta T_{boundary}/m \\ 1 & \text{if mean } \Delta T_i/m \geq \Delta T_{boundary}/m \end{cases}$$

where: $\Delta T_i/m$ is the temperature difference per metre depth across the whole pond at time i (°C/m); and $\Delta T_{boundary}/m$ is the nominal minimum temperature difference per metre depth for thermal stratification (given to be 0.6 °C/m by Kellner and Pires (2002))

Modelling heat effects in a WSP

Heat balance over a WSP. Over the entire WSP, a heat balance is described as follows. With the exception of the first two terms, all fluxes were assumed to apply specifically to the surface of the WSP.

$$\Delta Q = \begin{bmatrix} \text{solar radiation gain} \\ \pm \text{ advective heat gain and loss through inflow and outflow} \\ \pm \text{ free surface convection to surrounding air} \\ \pm \text{ low temperature radiation exchange between surface and surroundings} \\ - \text{ evaporative heat loss} \end{bmatrix}$$

Heat transfer through free surface convection was estimated using Newton's law of cooling between the pond surface and surrounding air, with a wind speed dependent free convection transfer coefficient (Fritz *et al.*, 1980). Low temperature radiation exchange between the surface and surrounding air was estimated using the Stefan–Boltzmann law for low temperature heat exchange between two grey bodies. Evaporative heat loss was estimated using the method described by Fritz *et al.* (1980) for latent heat loss from small lakes, which describes heat loss rate in terms of wind speed, relative humidity and the saturated vapour pressures of the pond surface and surrounding air.

Absorption of solar radiation through the WSP depth was assumed to decay exponentially, with an attenuation constant determined from fieldwork at the site (33.8 1/m). Radiation absorbed over a depth of pond was calculated from the difference in solar radiation intensity between the top and bottom of the depth slice. Typical absorptivities and emissivities for water from Schenck (1959) were used in the radiation models.

Advective heat gain was specified as a boundary condition, and heat loss was calculated directly by the CFD solver software. A summary of each sub-model used is listed in Table 1.

Modelling heat balance using the FLUENT CFD package. CFD is a computer-based method of simultaneously solving the mass, momentum and energy equations of state in

Table 1 Components of heat balance model

Component	Model	Coefficients
Solar radiation gain (W/m³ in pond)	$\dfrac{Q}{V} = \dfrac{\left(\frac{Q}{A}\right)_{z_1} - \left(\frac{Q}{A}\right)_{z_2}}{z_1 - z_2}$ where $\left(\dfrac{Q}{A}\right)_z = \alpha\left(\dfrac{Q}{A}\right)_{z=0} e^{-K_z z}$	$\left(\dfrac{Q}{A}\right)_z$ is the solar radiation intensity (W/m²) at depth z (m) below the WSP surface. α is the surface absorptivity
Advective heat loss and gain (W/m² through respective surfaces)	Energy conservation	
Free surface convection (W/m² through surface)	$\dfrac{Q}{A} = h(T - T_{surr})$	$h = 1.569\,WS_{av}$ (W/m² K) WS_{av} is the average wind speed (m/s) T, T_{surr} surface and air temperature (K)
Low temperature radiation (W/m² between surface and surroundings)	$\dfrac{Q}{A} = \sigma\varepsilon\left(T^4 - T_{surr}^4\right)$	σ is the Stefan–Boltzmann constant (W/m² K⁴) ε is the surface emissivity
Evaporative heat loss (W/m² from surface) (estimated using method of Fritz et al. (1980))	$\dfrac{Q}{A} = 0.1587\,WS_{av}(e - e_{surr})$	$e = 25.374\,e^{17.62 - \frac{5271}{T}}$ $e_{surr} = 25.374\,R_h\,e^{17.62 - \frac{5271}{T_{surr}}}$ R_h is the relative humidity

discrete sub-volumes throughout a fluid body, to produce a description of flow conditions everywhere therein on a local scale. Modelling of Lagoon 1 at Bolivar WWTP was undertaken using FLUENT, a commercial CFD software package (Fluent Inc., Lebanon, USA).

An existing isothermal CFD model of Lagoon 1 at Bolivar WWTP was expanded to include the components of the heat balance described above. Indigenous convection and low temperature radiation models were available in the software. The latent heat loss and solar radiation heat gain were indirectly specified through the addition of user defined functions (UDFs). The resulting single-phase model of heat exchange in 3-D did not consider the effect of wind on mixing or stratification breakdown in the WSP directly. For the calculation of wind speed dependent parameters in components of the heat balance (evaporation and free convection components), the mean daily wind speed at the site was used.

Validation of the CFD thermal model was undertaken using data which had been collected in Lagoon 1 as part of the thermal profiling study described above. From the numerous profiles, a data set of 24 hours duration was selected, covering influent and air temperature, sunlight levels and the resulting thermal profile in the WSP. The day was selected based on the criteria that mean wind speed was the lowest of any of the days for which data were collected (1.34 m/s, as measured at 3 m elevation), thereby negating the need to directly incorporate wind effects into the CFD thermal model. The data set covered a day of operation during winter (3 June 2001). For the model, the diurnal trends in solar radiation, air temperature and inlet temperature, and mean daily wind speed were used as boundary conditions.

Results
Simultaneous thermal variation measured in the field

A summary of the results measured during the field study is listed in Table 2. Throughout the profiling, the WSP was never isothermal in the length and width dimensions, with a minimum temperature difference between sites of 0.2 °C. It is also evident that the extent of stratification was never consistent between sites, with a minimum variability of 0.2 °C/m. While both values are at the upper level of measurement error, typical variation was, in each case, much higher (mean 1.7 °C and 1.4 °C/m, respectively). Maximum variation in stratification of up to 12.7 °C/m was recorded across the pond—an indication that a range of stratification from extreme (11.8 °C/m) to inverse (−0.9 °C/m) can occur simultaneously at different sites within the WSP.

In total, 137 distinct stratification events were recorded (mean stratification gradient at all nine sites ≥ 0.6 °C/m), ranging in duration from 10 minutes, on a number of occasions, to over 35 hours. Mean event duration was 2.8 hours. Event frequency was 158 events

Table 2 Summary of results from simultaneous thermal profiling study in 3-D of Lagoon 1, Bolivar WWTP

	Mean (± σ)	Maximum (season)	Minimum (season)
Mean pond temperature (°C)	19.2 (± 3.5)	27.4 (summer)	13.1 (winter)
Mean pond stratification gradient (°C/m)	0.4 (± 1.0)	11.7 (summer)	− 0.5 (autumn)
Stratification event duration (hours)	2.8 (± 5.0)	35.5 (summer)	0.2
Variation in depth averaged temperature between sites in pond (°C)	1.7 (± 0.9)	7.8 (summer)	0.2 (autumn)
Variation in stratification gradient between sites in pond (°C/m)	1.4 (± 1.3)	12.7 (autumn)	0.2 (autumn)

per 100 days, or over 1.5 events per day on average. In total, the WSP was stratified for 18% of the study duration.

Figure 2 shows the frequency of stratification events, as categorised by event starting time. As expected, the majority of events began during the warming phase of the solar cycle (09:00–12:00), although 15:00–19:00 also appears to be an important time for stratification build-up. The period between these two times (13:00–15:00) corresponds to the time when wind speed at the site is typically increasing towards the daily maximum. The cumulative total stratified time distribution is shown for all sites in Figure 3. This distribution shows the fraction of total stratified time associated with events of varying durations. It is evident from Figure 3 that stratification was much more persistent at Sites A and G than at any others, with over 30% of the total stratified time at each site associated with events of greater than 50 hours duration. Stratification was also most prevalent at these sites, as shown in Figure 4. By contrast, at Sites B and I 70% of all stratified time was associated with events of less than six hours duration.

Figure 5 shows a wind rose of all wind speed and direction measurements recorded at the site during the thermal profiling study. The wind rose indicates that the most common wind direction at the site during the study was from the southerly direction. In addition, higher wind speeds were more commonly recorded for southerly winds, compared to winds from other directions. Mean wind speed at the site during the study was 3.8 m/s, as measured at an elevation of 3 m.

Thermal mapping results

A comparison of actual and predicted results for the June 2001 data set is shown in Figure 6. Stratification gradient, mean WSP temperature and outlet temperature, as measured and calculated by the model, are compared over a single day. It is evident from these results that the best agreement exists for the mean WSP temperature ($r^2 = 0.92$). Pond outlet temperature predicted by the model follows a similar trend to the measured result, but is approximately 1 °C higher. From the 00:00 starting time, which was initialised to the same starting stratification gradient as the field results (0.1 °C/m), the model accurately predicted the initial period of limited stratification, including the slight increase in inverse stratification recorded during the hours of darkness. From 09:00, the model predicted both the timing and the rate of increase of stratification in the WSP during the major stratification event of the day. However, the model subsequently overpredicted the maximum stratification gradient by approximately 50%, and did not track

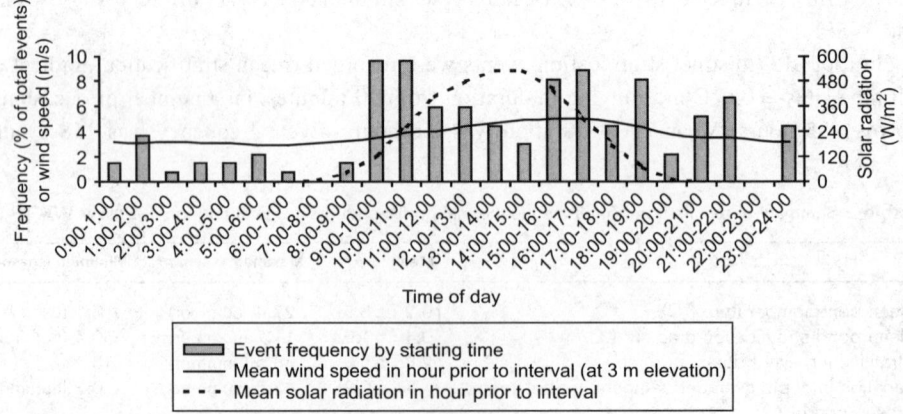

Figure 2 Frequency of total stratification events, by event starting time, for Lagoon 1, Bolivar WWTP. Mean solar radiation and wind speed over day shown for reference

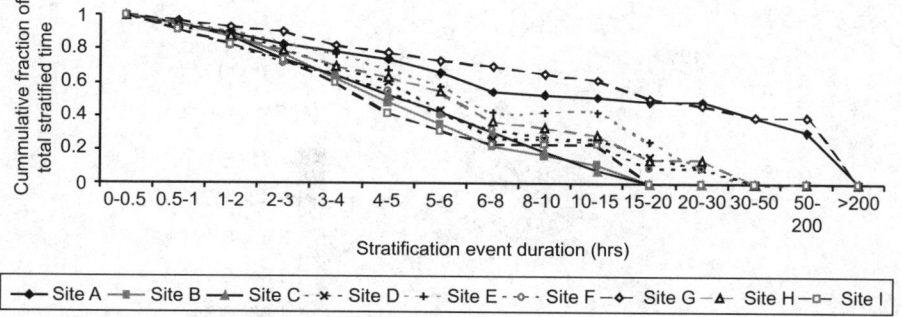

Figure 3 Cumulative stratified time fraction as a function of stratification event duration at each sampling site in Lagoon 1, Bolivar WWTP

the following decrease and increase in stratification gradient which occurred in the WSP during the afternoon and evening. Figure 7 shows the predicted and actual stratification gradient, and corresponding wind speed in the 10 minutes prior to measurement throughout 3 June 2001.

Discussion

It is evident from the results of the simultaneous thermal profiling study that considerable variation occurs in both the frequency and persistence of stratification between sites in Lagoon 1 at Bolivar WWTP. Site A was found to be most often stratified, and stratification events persisted longest at this site compared to any other. As a result, the frequency of distinct stratification events was lowest at this site, as conditions which produced a number of short, separate events at other sites produced a single, concatenated event at Site A. Figures 4 and 5 show that the most often stratified site was furthest upwind with respect to prevailing wind direction.

Compared to the mean residence time at the site, the duration of each stratification event was relatively small. Breakdown of stratification is often accompanied by homogenisation of the WSP contents, at least through the depth dimension on a local scale (Uhlmann, 1979). It is therefore probable that a "minimum stratification duration" exists, below which the impact of stratification induced short-circuiting on overall hydraulics in

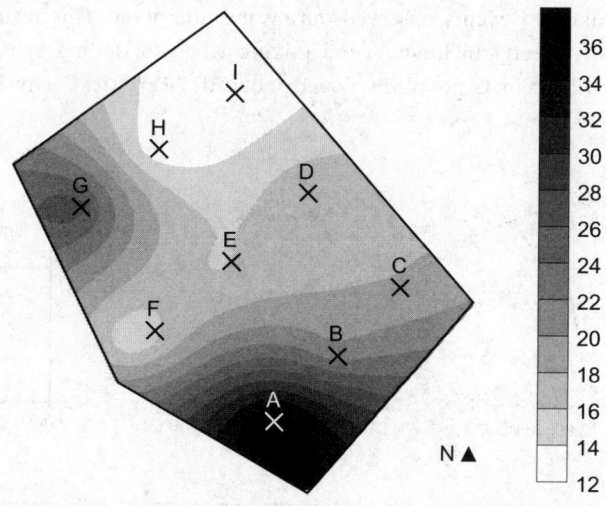

Figure 4 Surface profile of % of total time stratified in Lagoon 1, Bolivar WWTP

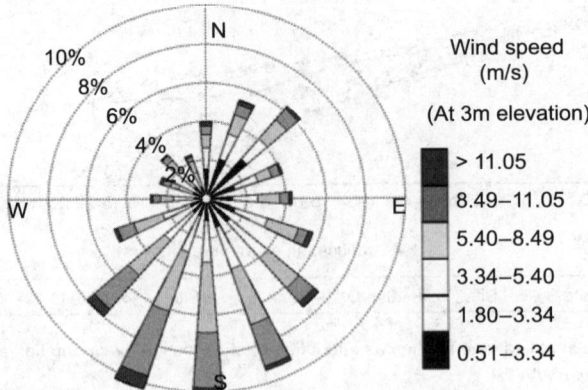

Figure 5 Wind rose of wind speed and direction frequency at site during study

the pond is negligible, given the mixing that subsequently occurs. During a period of stratification, Pedahzur *et al.* (1993) measured that influent short-circuited through to the outlet in approximately 5–10 hours, in a WSP with a mean residence time of 15–20 days. Based on this and other similar findings, it is likely that stratification events of at least the duration required for the influent to short-circuit directly through the WSP will have a serious impact on treatment efficiency. Conversely, it is probable that much shorter events will not, as partial homogenisation of the WSP will occur after each event. The impact of events of interim duration is less certain.

This argument can be extended to include localised short-circuiting at specific locations of the WSP. Due to the high persistence of stratification events at Site A, it is likely that the impact of stratification induced short-circuiting at this site is higher than at other sites in the WSP. However, the impact that this has on the overall residence time distribution of the WSP, and ultimately treatment efficiency, will depend on the variation in stratification throughout the WSP, and the minimum stratification duration required to cause significant short-circuiting at different locations in the WSP.

It is apparent from Figure 7 that, on 3 June 2001, even very low wind (≥ 1.5 m/s) caused a breakdown in mean stratification in Lagoon 1, and hence a divergence between the CFD model prediction (without wind) and actual stratification measured at the site. Wind peaks at 15:00 and 22:00 corresponded (with a slight time lag) to the two partial stratification breakdown events observed during the afternoon. This result demonstrates the importance of directly including wind mixing when modelling energy transfer and stratification behaviour in large WSPs. Wind induced shear stress may be directly esti-

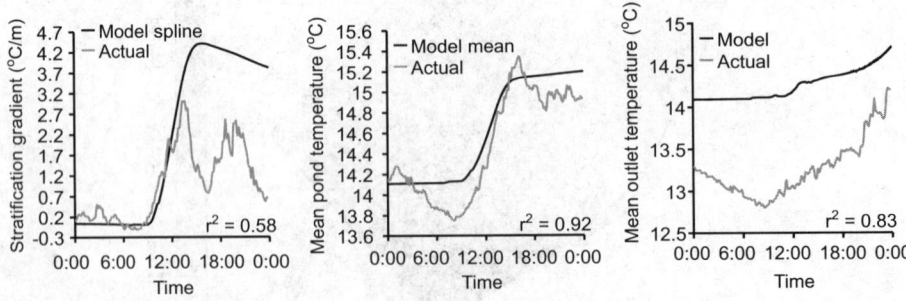

Figure 6 Correlation between modelled and actual results for mean pond stratification gradient (left), mean pond temperature (middle) and mean outlet temperature (right) at Lagoon 1, Bolivar WWTP on 3 June 2001

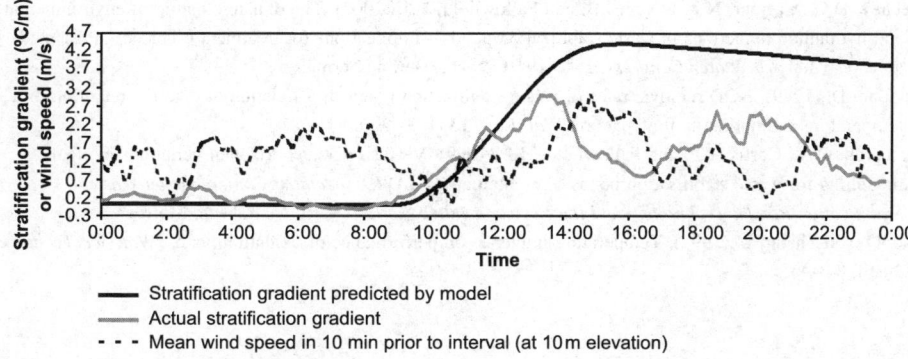

Figure 7 Comparison of modelled and predicted stratification gradient with wind speed (3 m elevation) at Lagoon 1, Bolivar WWTP on 3 June 2001

mated from existing models (e.g. Benqué *et al.*, 1982), however the value of the drag coefficient may require further investigation and calibration in this application.

Conclusions

Based on the results of the thermal profiling study of Lagoon 1 at Bolivar WWTP, considerable variation in the frequency and persistence of stratification events, and consequently the fraction of total time stratified, occurs across a large WSP. Site A was most often stratified (37% of all profiles), while Site I was least often stratified (11% of all profiles).

This will have implications for flow patterns through the WSP, and ultimately treatment efficiency. The majority of stratification induced short-circuiting will occur through regions of the WSP which are most often stratified.

CFD can be used to predict both the mean WSP temperature and the bulk stratification dynamics of Lagoon 1 with some accuracy, based on a heat balance between the WSP and the surrounding environment. However, a comparison of the accuracy of the stratification gradient prediction and wind speed at the site, over the course of a sampling day, suggests that even very low wind speeds cause a partial breakdown in stratification. More accurate prediction may result from the inclusion of wind shear stress in the CFD model.

References

Benqué, J -P., Haguel, A. and Viollet, P -L. (1982). *Engineering Applications of Computational Hydraulics*, Vol. II. Pitman Books, London, UK.

Fritz, J.J., Meredith, D.D. and Middleton, A.C. (1980). Non-steady state bulk temperature determination for stabilization ponds. *Wat. Res.*, **14**, 413–420.

Kellner, E. and Pires, E.C. (2002). The influence of thermal stratification on the hydraulic behaviour of waste stabilization ponds. *Wat. Sci. Tech.*, **45**(1), 41–48.

Kuwashima, S. and Hogben, N. (1986). The estimation of wave height and wind speed persistence statistics from cumulative probability distributions. *Coast. Eng.*, **9**, 563–590.

Pedahzur, R., Nasser, A.M., Dor, I., Fattal, B. and Shuval, H.I. (1993). The effect of baffle installation on the performance of a single-cell stabilization pond. *Wat. Sci. Tech.*, **27**(7–8), 45–52.

Salter, H.E., Ta, C.T., Ouki, S.K. and Williams, S.C. (2000). Three-dimensional computational fluid dynamic modelling of a facultative lagoon. *Wat. Sci. Tech.*, **42**(10–11), 335–342.

Schenck, H. (1959). *Heat transfer engineering*, Prentice-Hall, Englewood Cliffs, NJ, USA.

Shilton, A. and Harrison, J. (2003). *Guidelines for the hydraulic design of waste stabilisation ponds*, Massey University, Palmerston North, New Zealand.

Sweeney, D.G., Cromar, N.J., Nixon, J.B. and Fallowfield, H.J. (2002). The dynamic nature of environmental and hydraulic interactions in waste stabilization ponds—Implications for treatment efficiency. Presented at: 2nd *IWA World Water Congress*, 15th–19th October, Berlin, Germany.

Uhlmann, D. (1979). BOD removal rates of waste stabilization ponds as a function of loading, retention time, temperature and hydraulic flow pattern. *Wat. Res.*, **13**, 193–200.

Weatherell, C.A., Curtis, T.P. and Fallowfield, H.J. (1999). Vertical versus horizontal variations in algal modelling for waste stabilisation ponds. Presented at: 4th *IAWQ Specialist Conference on Waste Stabilization Ponds: Technology and the Environment*, 20th–23rd April, Marrakech, Morroco.

Zhao, Q. and Zhang, Z. (1991). Temperature influence on performance of oxidation ponds. *Wat. Sci. Tech.*, **24**(5), 85–96.

The development and calibration of a physical model to assist in optimising the hydraulic performance and design of maturation ponds

G.J. Aldana*, B.J. Lloyd*, K. Guganesharajah** and N. Bracho*

*Centre for Environmental Health Engineering, University of Surrey, Guildford, Surrey GU2 7XH, UK
(E-mail: B.Lloyd@surrey.ac.uk)
**Mott MacDonald Ltd, Demeter House, Cambridge CB1 2RS, UK

Abstract A physical and a computational fluid dynamic (CFD) model (HYDRO-3D) were developed to simulate the effects of novel maturation pond configurations, and critical environmental factors (wind speed and direction) on the hydraulic efficiency (HE) of full-scale maturation ponds. The aims of the study were to assess the reliability of the physical model and convergence with HYDRO-3D, as tools for assessing and predicting best hydraulic performance of ponds. The physical model of the open ponds was scaled to provide a similar nominal retention time (NRT) of 52 hours. Under natural conditions, with a variable prevailing westerly wind opposite to the inlet, a rhodamine tracer study on the full-scale prototype pond produced a mean hydraulic retention time (MHRT) of 18.5 hours (HE = 35.5%). Simulations of these wind conditions, but with constant wind speed and direction in both the physical model and HYDRO-3D, produced a higher MHRT of 21 hours in both models and an HE of 40.4%. In the absence of wind tracer studies in the open pond physical model revealed incomplete mixing with peak concentrations leaving the model in several hours, but an increase in MHRT to 24.5–28 hours (HE = 50.2–57.1%). Although wind blowing opposite to the inlet flow increases dispersion (mixing), it reduced hydraulic performance by 18–25%. Much higher HE values were achieved by baffles (67–74%) and three channel configurations (69–92%), compared with the original open pond configuration. Good agreement was achieved between the two models where key environmental and flow parameters can be controlled and set, but it is difficult to accurately simulate full-scale works conditions due to the unpredictability of natural hourly and daily fluctuation in these parameters.
Keywords Computational model; dispersion number; hydraulic efficiency; hydraulic performance; hydraulic retention time; pond channels

Introduction

The evaluation of hydrodynamic behaviour in a full-scale waste stabilisation pond system (WSPS) is very complex and difficult to define precisely. It requires knowledge of a number of parameters internal and external to the lagoons. One methodology used to define hydraulic behaviour involves the addition of a dye tracer at the pond inlet and sampling the tracer at the outlet over a defined period. Different techniques are required to determine the pattern of flow in the field such as buoyant objects, whereas chemical salts and dyes have been used as tracers for many years.

The complex hydrodynamic behaviour of ponds has defied precise description using mathematical design formulae and models (Gloyna, 1971; Finney and Middlebrooks, 1980; Agunwamba, 1992). However, numerous authors have asserted that pond design, particularly for the most common facultative systems, is reliably performed using historical organic loading criteria on the basis of surface BOD loading (λ_s kg/ha/d), which is given by the design equation: $\lambda_s = 10\ LiQ/Af$ (Middlebrooks, 1987), where: Li = influent BOD (mg/l); Q = flow (m^3/d) and Af = facultative pond area (m^2). Furthermore, it is currently impossible to reliably predict how various modifications of pond design, such as placement and number of inlets, use of baffles etc, might affect pond performance because these parameters are not included in the design equations.

Early design equations were based on the water volume, number of people contributing waste, flow *per capita* waste contribution and temperature reaction coefficient as in Gloyna's equation (1971). In the 1970s, the retention time was used as a variable for the first time in an equation by Marais (1974). Marais' equation was developed assuming complete mixing and rejecting plug flow conditions. At the end of the 1980s and in the beginning of the 1990s a few researchers like Polprasert and Bhattarai (1985) and Agunwamba (1992), put forward a dispersed design equation which resulted from a combination of plug and completely mixed flow. In those equations, the geometry of the lagoon was considered for the first time together with the retention time. The kinematic viscosity was also considered as a new variable influencing improvement in full-scale lagoon design, but the influence of the shape of WSPs and the wind effect (Wong and Lloyd, 2004) on them was not considered. The majority of these equations are based on false assumptions, such as nominal retention time and complete mixing which is not achieved. Tracer studies carried out in the field, show that the hydraulic retention time is often 50% less than the nominal one (Lloyd *et al.*, 2002). Other calculations for retention time based on nominal capacity and flow (e.g. activated sludge, aerated lagoons), are based on complete mixing being achieved. In these calculations flow is controlled in channels with aeration and recycling (Camp, 1946).

The US Water Pollution Control Federation (WPCF, 1990) described optimal flow as a discharge with a uniform velocity profile. That is, with the water plume moving parallel to the walls and no sideways water movement. Such plug flow conditions were argued to be able to prevent short-circuiting and dead zones, increase hydraulic efficiency, and thus come closer to the nominal (maximum theoretical) retention time. The Federation therefore recommended that ponds be designed to have plug flow. Such a flow is characterised by having a uniform velocity profile, but this does not exist even in ponds with a large length-to-width ratio. Generally the pond water does not move homogeneously, but rather in eddies, waves, with reverse flow and with re-circulation (Persson, 2000, Aldana, 2004). In practice, local velocity profiles are difficult to measure due to low velocity within the lagoon (<1 mm/s). A field investigation using a flow meter resulted in limited information for the evaluation of the hydraulic performance as compared to velocity vectors from running calibrated CFD packages such as HYDRO-3D (Guganesharajah, 2001), and/or tracer experiments using chemical solutions such as dyes, or tritiated water [H^{+3}] (Aldana *et al.*, 1999). These latter methods and techniques are more accurate and allow an understanding of the hydraulic behaviour of lagoons.

The following aspects of hydraulic deficiencies are of particular importance in the design of WSPs: 1) rapid surface flow; 2) arrangement of inlet and outlet (where outlet and inlet are opposite each other); 3) wind effect; and 4) differences between the nominal retention and the mean hydraulic time.

The aim of the present study is to use physical and computational modelling (HYDRO-3D), to analyse how hydraulic performance in full-scale maturation ponds is affected by design parameters, particularly inlet/outlet position, baffle configuration and wind. The overall objective is to identify interventions which will optimise performance by maximising hydraulic efficiency.

Methods

In this study three facilities were used to study the hydraulic performance of WSPs: full-scale maturation ponds, a physical model and a computational model, HYDRO-3D.

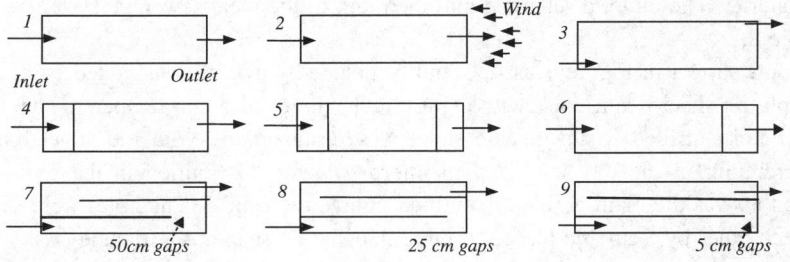

Figure 1 Plan view of nine physical models studied (not to scale)

Full-scale ponds

The full-scale ponds under study were two of three equal sized, parallel, tertiary-stage maturation ponds at a small sewage treatment works serving the village of Lidsey in southern England. The ponds form the final, polishing stage of an otherwise conventional percolating filter treatment works. The central of the three ponds had the open pond configuration shown as number 1 in the physical model schematic Figure 1. Another prototype pond was converted into three channels with the layout shown as 9 in Figure 1. The material used to construct the prototype channel walls was a butyl geo-membrane supported by wood posts. The specifications and hydraulic characteristics from a tracer study of the central and three-channel prototypes are shown in Table 1. The ponds were constructed in the shape of channels to encourage plug flow, but with inlet at the top surface and outlet near the top. The central pond, with a l:w ratio of 8.4:1, had a dispersion number of 0.9, whereas the three-channel configuration (l:w ratio 78:1) had a very low dispersion number of 0.07, much closer to plug flow. However, the low hydraulic efficiency (35%) of both prototypes under natural wind conditions is noteworthy. The prevailing westerly wind conditions at the study site were recorded continuously at the ponds using a wind speed and direction logger throughout the year. The data recorded during field tracer studies were used for the model simulations.

Pond physical model

A physical model, 1/18th of the area of the full-scale ponds, was constructed inside a glass-house at the Centre for Environmental Health Engineering (CEHE), University of Surrey measuring 6.60 m × 0.86 m × 0.37 m (volume = 2,100 litres) to investigate which physical variables could affect the hydraulic efficiency and performance in the full-scale ponds. The physical variables studied as shown in Figure 1, were top inlet/outlet aligned (with (2) and without wind (1)), diagonal inlet/outlet arrangement (3), baffle interventions (4–6) and channels (7–9). The model walls were constructed with concrete blocks. The existing floor and the wall structure were then lined with a white flame-retardant polypropylene sheeting. A storage tank of 500 litres capacity was filled with the mains water supply and pumped into the channel via plastic tubing of 8 mm diameter. At the outlet, a triangular weir was shaped at the top surface using the same

Table 1 Dimensions and hydraulic characteristics of prototype full-scale tertiary ponds at Lidsey

Description	L (m)	W (m)	D (m)	Q (l/s)	Re	d	NRT (d)	MHRT (d)	HE (%)
Central pond	122.4	14.5	1.0	9.0	512	0.9	2.17	0.77	35
Three-channel	122.4 × 3	4.65	1.1	4.5	557	0.07	4.76	1.66	35

Length (L), width (W), depth (D), flow rate (Q), Reynolds Number (Re), dispersion number (d), nominal retention time (NRT), mean hydraulic retention time (MHRT), hydraulic efficiency (HE)

liner material. The outlet discharged into the drain to the local sewerage system as shown in Figure 2.

Layouts shown in Figure 1 as 4, 5 and 6 included baffle(s) constructed from a transparent plastic sheet 8 mm thick with eight circular holes of 5 mm diameter. The arrangement of holes drilled in the plastic sheet was designed to avoid the advection peaks short-circuiting in the top layer and to improve water distribution in the cross-section area, as this was the main problem identified in layouts 1 and 3. The holes were separated from each other by 5 cm and placed in three lines in the shape of a triangle (▲).

To convert the open pond model to the three-channel configuration (Figure 1, 7–9), the baffles were removed and corrugated plastic sheets were used to construct two longitudinal walls in the physical model. They were fixed to a metal frame which was reinforced and held in place with small pieces of the same metal and pieces of wood.

Calibration. A Watson Marlow peristaltic pump was calibrated in the selected flow range to determine linearity at various settings and flow reliability (Aldana, 2004). Flow into and out of the model was routinely checked at regular intervals throughout each experiment. Whenever flow deviated by $> \pm 5\%$ from the set flow (12 ml/s), the pump was reset to the required flow to provide a NRT of 2 days. However, the time required for complete dye washout for each tracer experiment was about 6 days. The basic characteristics of the model are listed in Table 2 and variations in dimensions and flow for individual experiments shown below Figures 3 and 4.

Dye tracer experiments were run without either temperature control or wind effects during the period August 2001 to January 2003. Experiments with wind were carried out in July 2002. The wind was produced over the physical model surface by locating two, 60 W three-speed fans powered by 220 V, 50 Hz, either at the inlet or outlet, depending on whether a following (easterly) or opposite (westerly) wind was required. The wind speed over the surface of the model was calibrated as described by Wong and Lloyd (2004) to produce a velocity of 0.3 m/sec at the mid length of the pond or channels.

Rhodamine WT dye was used as the tracer, having a density of 1.019 kg/m^3 and 20% active volume. All the readings in the experiments were taken on-line using a fluorimeter probe (Chelsea Instruments Ltd, UK) and were recorded every minute by a logger from Marine Instruments (Flexidata 1201). In all 44 tracer studies were conducted of which 10 were replicates. Replication of all tracer experiments could not be undertaken due to time constraints, each experiment taking almost a week. Only tracer experiments with good duplicate agreement were selected for presentation in this paper.

Computational model

A calibrated three dimensional model was required in order to produce simulations which accurately represent the hydraulic conditions of the full-scale and physical model.

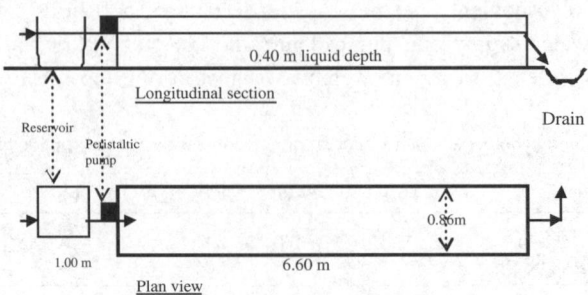

Figure 2 Basic layout of the CEHE pond physical model (not to scale)

Table 2 Basic characteristics of the CEHE physical model

Length	6.60 m
Width	0.86 m
Liquid depth	0.37 m
Flow rate	0.012 l/s
Volume	2,100 l
Nominal retention time	2.19 d
Inlet/outlet	Top surface

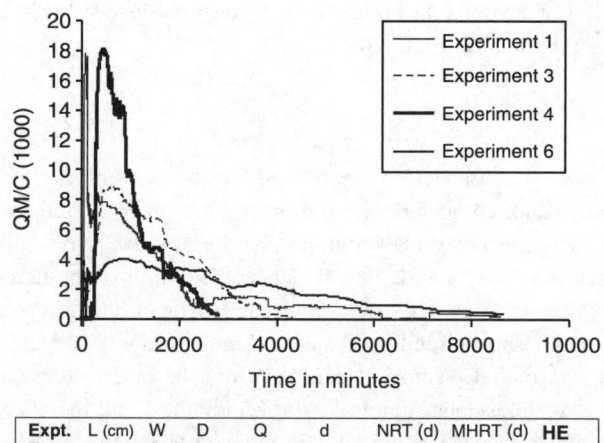

Expt. Trace No.	L (cm)	W (cm)	D (cm)	Q ml/s	d	NRT (d)	MHRT (d)	HE (%)
1	660	86	37.1	12.00	1.8	2.03	1.16	**57.1**
3	660	86	36.8	11.83	0.3	2.03	1.02	**50.2**
4	746	86	37.0	12.54	0.48	2.04	1.40	**68.6**
6	746	86	38.3	12.04	0.36	2.09	1.56	**74.6**

Figure 3 Comparison of tracer age distribution in the physical model open and baffled pond layouts: 1) open centrally aligned in/out, 3) open diagonally opposite in/out, 4) baffle placed near inlet, and 6) baffle near outlet

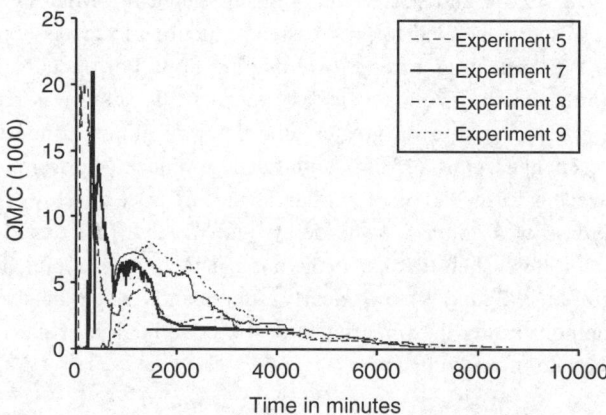

Expt.T Trace No.	L (cm)	W (cm)	D (cm)	Q ml/s	d	NRT (d)	MHRT (d)	HE (%)
5	832	86	37.	13.23	3.4	1.84	1.24	**67.4**
7	1880	30	34.1	11.41	0.78	1.96	1.36	**69.4**
8	1930	30	34.0	12.19	0.21	1.83	1.64	**89.6**
9	1970	30	32.3	11.02	0.18	1.93	1.78	**92.2**

Figure 4 Comparison of the age distribution in the physical model layouts: 5) double baffle, and three channels with 7) 50 cm, 8) 25 cm and 9) 5 cm gaps in the bends at the end of each channel

An application of HYDRO-3D was developed jointly by CEHE (University of Surrey) in collaboration with Mott MacDonald specifically for WSPs (Guganesharajah, 2001). HYDRO-3D is a CFD model, which has been successfully applied to water bodies including WSPs. The model can simulate impacts of wind on hydraulic conditions, and temperature on water quality parameters. HYDRO-3D was used to simulate the physical model 1 shown in Figure 1 as a baseline for comparison with various subsequent interventions. Simulations included the production of hydrodynamic vector maps at 4 depths and the curves of tracer concentration against time. However, due to space limitations, only the simulation of layout 8 in Figure 1 (three channel lagoon with 25 cm gaps) with opposing wind, is presented in Figure 6.

Results

Physical model without wind

Baffled model pond. The impact of one baffle placed 1 m from the inlet-edge is shown in trace **4** (Figure 3) and, compared with the open pond (traces **1** and **3**), is characterised by a substantial increased delay (414 minutes) before any dye leaves the model reactor. The MHRT is increased to 1.4 days and hence the HE is also increased, to 68.6%. However, the first dye fraction leaves the model in less than one third of NRT indicating that there is still major short-circuiting. Trace **5** (Figure 4) shows the effect of the second baffle in the physical model reactor, placed 1 m from the outlet. Surprisingly the double baffle reduced the dye exit delay time to about 85 minutes, and there was no increase in hydraulic efficiency; in fact it was marginally reduced to 67.4%, probably due to higher flow (13.23 ml/s). Trace **6** (Figure 3) shows the effect of the baffle placed 1 m from the outlet. This produced a MHRT of 1.56 days, and hence an HE of 74.6%, the best of the three, although all three baffle configurations are superior to the open pond model.

Three-channel model pond. The traces (**7–9**) shown in Figure 4 for the three channels constructed in the physical model, showed progressive and significant improvement for the two end of channel gap sizes, 50 cm, 25 cm and 5 cm. The hydraulic efficiencies were respectively, 69.4%, 89.6% and 92.2%; this reflects increasing MHRTs of 1.36 d, 1.64 d and 1.78 d, and also corresponded to progressive reductions in dispersion numbers, from 0.78, to 0.21, to 0.18, reflecting a trend towards plug flow. For trace **7** the first dye peak exited the pond after 4.30 h (258 min), and as expected this delay was much greater than in layout 1. Trace **8** shows an even greater time delay of almost 9 hours, demonstrating that the reduced channel gaps (25 cm) significantly reduce short-circuiting. The age-distribution showed in trace **9** is very similar to that of trace **8**. However, in trace **9,** a further large increase was noticed in the delay, and the dye first exited at 12.25 hours. Overall the curve is less skewed than in layout **8** and closer to plug flow. The last 2 channel configurations (**8** and **9**) are clearly significantly superior to all three baffle configurations, demonstrating the importance of the combination of narrow channels and the gap size at the end of each channel.

Effects of wind action on models and full-scale ponds

Wind effects on open pond configuration. It is noteworthy that the HE of all configurations of the physical model (in the absence of wind) were much higher than those of the prototype full-scale open pond and three-channel Lidsey ponds under natural conditions. However, all the physical model tests discussed thus far were conducted in still air, it is therefore essential to assess the impact of a similar, opposing wind condition as indicated in layout 2. The results from the full scale open pond and layout 2 (both with wind opposite to the inlet) are presented in Figure 5. They demonstrate similar

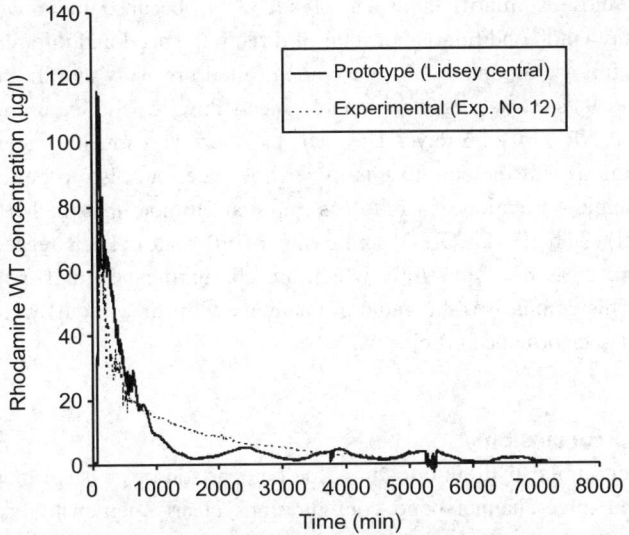

Figure 5 Agreement between full scale prototype open pond and experimental physical model with opposing wind

curves and similar MHRTs of 18.4 h (full scale) and 21 h (physical model), and similarly low HEs of 35% (at full scale) and 40% (physical model), reflecting high short-circuiting with the peak tracer concentrations leaving the ponds in <1.5 h (<90 min). Similar reductions in hydraulic efficiency were demonstrated using HYDRO-3D simulations.

Wind effects on three-channel configuration. In Figure 6 the tracer curves from 1) the Lidsey full-scale south three-channel configuration, 2) the physical model (layout 8), and 3) a HYDRO-3D simulation are compared. All three traces present a similar type of skew distribution and overall curve shape, however the similarity between 2) the physical model (layout 8) and 3) the HYDRO-3D simulation is the most impressive. The main

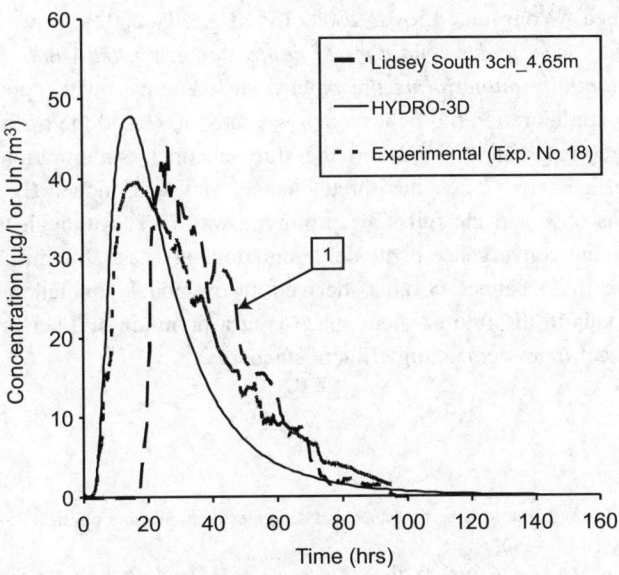

Figure 6 Agreement between 3 channel methods with opposing wind

reason for the close similarity between 2) and 3) is because the flow rate (9 l/s) is constant, and the wind condition is constant in direction, speed and duration. By contrast, the flow in the full-scale pond, could not be controlled precisely and the flow rate varied between 4.5 and 9 l/s, so taking a reasonable mean flow for the experimental period as 8 l/s, this gave a NRT of 2.36 days. The fact that there is a much greater delay time at full scale is primarily attributable to this lower flow rate, but also to periods of lower or no wind, or changing direction of wind. It is important to note that the MHRTs were 33 h (physical model), 34 h (HYDRO-3D) and 39.35 h (full scale). Their respective hydraulic efficiencies were 83%, 68% and 70%, which are all significantly higher than open pond configuration. This emphasises the value of channels with (or without) wind conditions in raising hydraulic performance efficiency.

Discussion and conclusions

In still air, under similar hydraulic loading, the comparison of physical model open pond, with baffled and three-channel pond configurations clearly demonstrated that the open pond configuration produced the lowest hydraulic efficiency (HE 50–57%), baffles produced a significant increase to 67–74.6%, whilst the narrow three-channel configuration produced the highest efficiency (69–92.2%). The best configuration (HE 92.2%) was with the smallest (5 cm) gaps between each of the three channels. Delay times, for the first detectable dye to leave the three-channel model, ranged from a minimum of 4.3 hours to a maximum of 12.25 hours, which represented a large improvement in hydraulic efficiency within the physical model. It is fair to recommend channels for use in WSPs in order to increase hydraulic performance.

There are, however, a number of factors, in addition to pond configuration, which influence hydraulic performance and can influence advective (short-circuiting plume) flow paths in full-scale ponds under natural conditions: these include wind stress, temperature, viscous effects, boundary shear, inlet discharge, and during this study it was observed that *wind aggravates short-circuiting and reduces HE*. In numerical terms it was demonstrated that wind could reduce MHRT by more than 25%. This adds significantly to the growing body of evidence that wind is damaging rather than enhancing pond performance (Wong and Lloyd, 2004, Lloyd *et al.*, 2002). However, it was also demonstrated that, *even under conditions of natural or generated wind, the channel configuration substantially outperforms the open pond*. Whereas, with opposing wind, the HE of the open configuration physical model was only 40%, and the open full-scale pond was 35%, the corresponding winds on the three-channel configuration still produced much higher efficiencies. Thus the three channel physical model HE was 83%, the HYDRO-3D was 68% and the full-scale prototype was 70%. Although further increases in the precision and convergence of model simulations are desirable, in practical terms it is already clear that channel designs, derived from model simulations, are likely to replace open ponds in the future where space is at a premium and performance must be maximised to meet more demanding effluent standards.

References

Agunwamba, J.C. (1992). Field pond performance and design evaluation using physical models. *Water Research*, **26**(10), 1403–1407.

Aldana, G.J., Bracho, N.R. and Esteves, J. (1999). Hydraulic parameters analysis in facultative ponds. *Revista Tecnica de Ingenieria Universidad del Zulia*, **22**(2), Agosto 106–117.

Aldana, G.J. (2004). Hydraulic behaviour and performance improvement of waste stabilisation ponds (WSPs) using a computational fluid dynamic (cfd) and a physical model. PhD thesis, University of Surrey, UK, 167–260.

Camp, T. (1946). Sedimentation and the design of settling tanks. *ASCE*, **111**, 895–958.

Finney, B.A. and Middlebrooks, E.J. (1980). Facultative waste stabilisation ponds design. *Journal of the Water Pollution Control Federation*, **5**(1), 134–147.

Gloyna, E.F. (1971). *Waste stabilisation ponds*. Monographic series No 60. WHO, Geneva Switzerland.

Guganesharajah, R.K. (2001). Numerical aspects of computational hydraulic and water quality models for rivers, estuaries, reservoirs and aquifers with particular reference to waste stabilisation ponds. PhD thesis University of Surrey, UK, 90–120.

Lloyd, B.J., Vorkas, C.A. and Guganesharajah, R.H. (2002). Reducing hydraulic short-circuiting in maturation ponds to maximise pathogen removal using channels and wind breaks. *Wat. Sci. Tech.*, **48**(2), 153–162.

Marais, G.R. (1974). Faecal bacterial kinetics in stabilisation ponds. *Journal of the Environmental Engineering Division*. February **EE1**, 119–139.

Middlebrooks, E.J. (1987). Design equations for BOD removal in facultative ponds. *Wat. Sci. Tech.*, **19**(12), 187–193.

Persson, J. (2000). The hydraulic performance of ponds of various layouts. *Urban Water*, **2**, 243–250.

Polprasert, C.H. and Bhattarai, K.K. (1985). Dispersion model for stabilisation ponds. *Journal Env. Eng.*, **111**(1), 45–59.

Water Pollution Control Federation (1990). Manual of practice, natural system, wetlands chapter, February 1990, *MOP FD-16 WPCF, 270–275*, USA.

Wong, S. and Lloyd, B.J (2004). An experimental investigation of the impact of wind shielding on hydraulic retention times in waste stabilisation ponds. Paper OW19 presented at the IWA *6th Specialist Conference on WSPs*. Avignon, France, Sept 2004, pp 167–175.

Variations in BOD, algal biomass and organic matter biodegradation constants in a wind-mixed tropical facultative waste stabilization pond

C.G.R. Meneses*, L.B. Saraiva*, H.N. de S. Melo*, J.L.S. de Melo* and H.W. Pearson**

*Department of Chemical Engineering, Federal University of Rio Grande do Norte, Natal, Brazil
(E-mail: carlagracy@eq.ufrn.br)

**Department of Civil Engineering, LARHISA, Federal University of Rio Grande do Norte, Natal, Brazil
(E-mail: howard_william@uol.com.br)

Abstract This study considered the impact of wind mixing on the efficiency of BOD removal and the first order biodegradation constant for organic matter in a primary facultative pond. Wind speeds of 1–4 m/s blowing from the effluent end of the pond towards the influent created surface-water flows of up to 0.94 m/s as determined by orange and coconut drogues moving in the opposite direction to the bulk hydraulic flow of 0.217 m/s. This was sufficient to cause mixing of the water column resulting in loss of stratification in terms of chlorophyll a, temperature and dissolved oxygen. BOD and chlorophyll a concentrations were spatially and temporally homogeneous throughout this large pond. BOD removal efficiency was only 50.3% as opposed to a projected value of 79% despite an acceptable surface organic loading of 350 kgBOD5/ha/d and an actual k value for BOD removal using influent sewage samples of $0.29\,d^{-1}$ close to the projected value of $0.30\,d^{-1}$. It would seem that wind mixing reduced pond efficiency by destroying stratification and thus reducing the microbial activity necessary to consume organic material. Mixing also increased the mean chlorophyll a concentration compared to stratified facultative ponds receiving similar loads and non-motile algae dominated the water column.

Keywords Algae; BOD removal; facultative pond; wind mixing

Introduction

Algae generate the bulk of the molecular oxygen required for aerobic bacterial degradation of organic material in waste stabilization ponds and are fundamental to the overall treatment processes therein. The total algal biomass concentration and species diversity depends on a range of factors including water temperature, solar radiation, nutrients, ammonia and sulphide toxicity and predation. However it is likely that under most conditions light and possibly ammonia toxicity are the controlling factors in ponds.

The high algal concentrations in ponds are likely to influence, the overall degradation constant (k) for organic material removal since they represent a significant proportion of the overall organic matter concentration. Bearing this in mind Weatherell et al. (2003), amplifying on the work of Polprasert and his fellow workers, developed an empirical regression equation expressing algal concentration as a function of temperature, average daily light intensity, the hydraulic retention time and BOD concentration.

When determining the kinetic constant k, in waste stabilization ponds it is necessary to take into consideration, the temperature, flow, pond geometry (depth, length and width), the direction and the speed of the wind and the hydrodynamic state. In ponds the hydrodynamic state can be one of either complete mixing, plug-flow or dispersed flow. Methods used to determine the hydrodynamic state in ponds include the use of chemical tracers and drogues. Frederick and Lloyd (1996) used oranges placed in the pond inlet to study dispersion at the surface.

The present study seeks to quantify the spatial and temporal variations in algal biomass concentration and correlate these with BOD concentrations and the biodegradation constants determining the rate of organic matter removal in a large tropical primary facultative pond operating under windy conditions.

The pond system

The pond system under study comprised a facultative pond 5.5 ha in area with a depth of 2.0 m followed by 2 maturation ponds in series each of 2.8 ha and 1.5 m deep. The ponds treat domestic sewage from the tourist resort of Ponta Negra in the coastal city of Natal, Rio Grande do Norte, Northeast Brazil. Pond dimensions and orientations are presented in Table 1 and Figure 1. The mean water temperature was 27 °C and the incident mean annual solar radiation at the surface was 5900 W/h/m^2.

During the period of this study the average influent flow was 5003 m^3/d (57.9 L/s) with a mean influent BOD_5 of 385.7 mg/L after de-gritting and screening. The surface organic loading on the primary facultative pond was 350 kg BOD_5/ha/d. The hydraulic retention time (HRT) of the facultative pond was 22 days, calculated on the basis of volume and influent flow (as opposed to the design value of 12.3 days) and 8 days in each of the 2 maturation ponds (as opposed to the design value of 5 days in each of 3 maturation ponds).

Methods
Sampling and analyses

Eighteen sampling stations distributed across the three sampling zones (A_1, A_2 and A_3) in the primary facultative pond were positioned by GPS (GPS II *Plus* – Garmin) and marked by polystyrene buoys secured by concrete bases to the pond floor thus forming a comprehensive sampling grid (see Figure 1). Water column samples were collected bi-weekly at each sampling station (during 07.00 h to 11.00 h) between September 2003 and March 2004 using column samplers (Pearson *et al.*, 1987). The water column samples were analysed for BOD_{20} using the OXITOP manometric method, COD using the reflux/colorimetric method, chlorophyll *a*, and phaeophytin were measured spectrophotometrically after 90% acetone extraction. All the methods were according to *Standard Methods* (APHA, 1998).

In situ depth profile measurements of the water column were also made for chlorophyll *a* to determine the degree of algal stratification and for temperature and dissolved oxygen to determine physico-chemical stratification. Light penetration was measured by secchi disc. In the absence of sufficient tracer dyes for such a large pond complex at the time of the study, simple drogue studies were performed using oranges (Lloyd *et al.*, 2003) and green coconuts to measure the impact of wind on surface hydraulic flow in the primary facultative pond.

Table 1 Dimensions of the Ponta Negra Waste Stabilization Pond System, Natal, Rio Grande do Norte, NE, Brazil (L = length, B = breadth, D = depth)

Pond	Dimensions (m)					Area (m^2)	Volume (m^3)
	L_1	L_2	B_1	B_2	D		
Fac	438.5	450.6	58.4	168.1	2.0	55,174.2	110,348.4
Mat 1	138.0	144.4	171.1	212.0	1.5	28,037.8	42,056.7
Mat 2	116.0	120.5	210.0	246.6	1.5	28,599.5	42,899.2

Source: CAERN (Companhia de Aguas e Esgotos do Rio Grande do Norte)

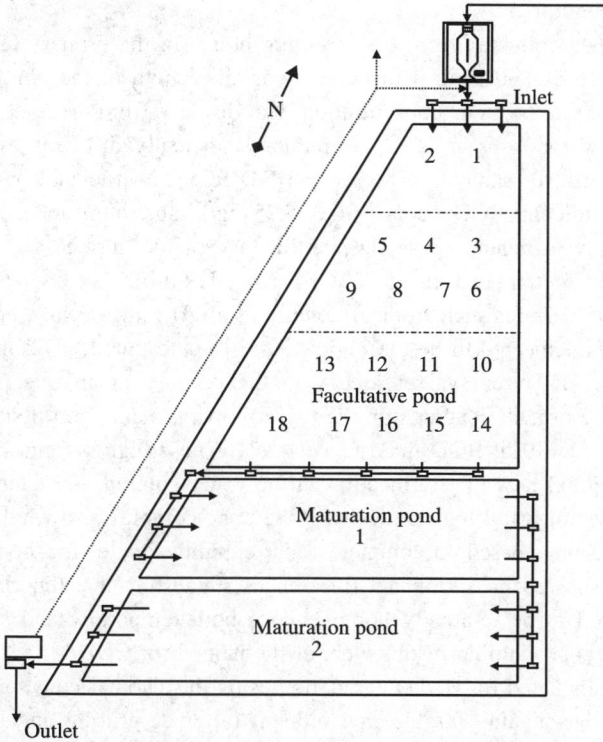

Figure 1 The layout of the Ponta Negra waste stabilisation ponds showing the sampling points in the trapezoidal primary facultative pond. Source: CAERN (Companhia de Águas e Esgotos do Rio Grande do Norte)

Calculation of *k* values for BOD degradation

The biodegradation constant (k) for BOD removal was calculated using the Thomas equation (see von Sperling, 1983) in the following rearranged form:

$$\left(\frac{t}{y}\right)^{1/3} = (2.3\,kL)^{-1/3} + \frac{k^{2/3}}{3.4\,L^{1/3}} t \tag{1}$$

where:
  t = time for biodegradation of organic material in days
  y = BOD produced in time t in mg/L
  L = the final BOD in mg/L
  k = the biodegradation constant for organic material

In certain cases the k value for BOD removal was also calculated using the equation of Mendonça (1990) to compare the data collected in this study with the theoretical k value used in calculations for the original Ponta Negra project:

$$k_1 = 0.796 \times (t_d)^{-0.355} \times (1.085)^{T-26} \tag{2}$$

where:
  k_1 = the constant for BOD removal
  t_d = the hydraulic retention time of the pond, in this case 22 days
  T = the temperature of the pond liquid, in this case 27 °C

Results and discussion

When the BOD data collected from the sampling points in the primary facultative pond (see Figure 1) were plotted against distance along the length of the pond there was no statistical difference in the BOD concentration with distance from the inlet (Figure 2).

Also the values for *in-pond* BOD_5 were not statistically different from the values recorded for the effluent samples. The mean BOD removal efficiency was only 50.3% given the mean influent BOD_5 conc. of 385.75 mg/L and a mean effluent value of 194 mg/L. This low efficiency was despite the acceptable surface BOD_5 loading of 350 kg.ha/d (mean air temperature of 27 °C) and an HRT of 22 days (see Mara *et al.*, 1992; Yanez, 1993). Under such tropical conditions BOD removal in a primary facultative pond would be expected to be the region 70–80% and thus BOD removal was suboptimal. This low BOD removal efficiency in the primary facultative pond adversely affected the surface organic loadings on the two following maturation ponds in the series which were 195 and 189 kg BOD_5/ha/d respectively. These high organic loadings on the maturation ponds (and lack of stratification in the water column, see later) contribute to the low FC removal of just 3 log removals for the pond complex as a whole.

Chlorophyll *a* values based on complete water column samples for the three sampling regions (A_1–A_3) showed no significant differences suggesting that the chlorophyll concentration appeared to be relatively homogeneous both temporally and spatially across the surface of the pond. Similar results were also obtained for COD.

The k values calculated for BOD degradation using the Thomas equation are presented in Table 2. The mean value for the raw influent (after de-gritting and screening) was $0.27\,d^{-1}$. The *k* values in the three sampling regions of the facultative pond were not significantly different from one another or from the effluent value but were less than half the influent value.

When the *k* values for BOD were plotted against chlorophyll *a* concentration obtained for the same water samples there was no significant relationship. This contrasts with the findings of El Hamouri *et al.* (2003) for high rate ponds and facultative ponds in Morocco but this might be explained by the homogeneity of the pond contents in this study.

Using the calculated theoretical retention time for the primary facultative pond of 22 days and a pond liquid temperature of 27 °C, a value of $0.29\,d^{-1}$ was obtained using the Mendonça equation which approximated very closely to the value of $0.27\,d^{-1}$ obtained

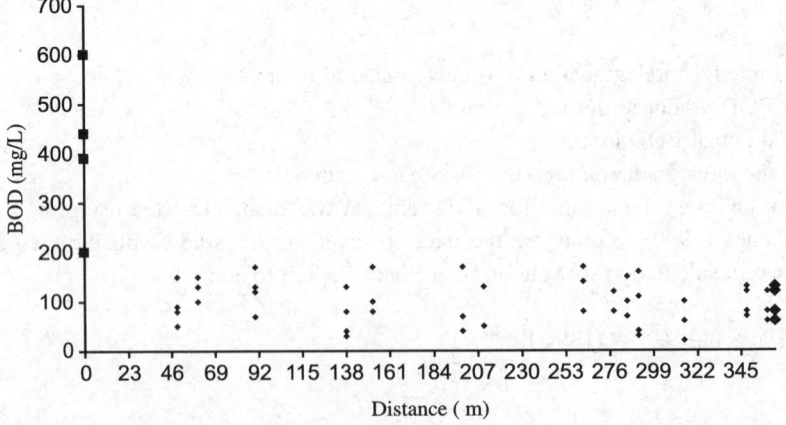

Figure 2 Variations in BOD_5 concentration with distance along the facultative pond from inlet to outlet. The solid squares on the left of the figure are influent values and the solid diamonds to the right are effluent values

Table 2 Variations in k values (d^{-1}) for BOD removal in the different zones along the facultative pond

	*Mean	Minimum	Maximum	Std. Dev.
k influent	0.27	0.21	0.35	0.06
$k\ A_1$	0.10	0.04	0.15	0.05
$k\ A_2$	0.13	0.08	0.21	0.05
$k\ A_3$	0.11	0.07	0.18	0.04
k effluent	0.12	0.10	0.14	0.02

*Based on 5 sets of samplings

using the Thomas equation. The theoretical project value for k BOD calculated using the Rolim equation was $0.30\,d^{-1}$. Thus although the actual k values calculated here closely agreed with the projected value, actual BOD removal was much lower at 50% compared to the predicted value of 79%. This calls into question the use of k values to predict pond efficiency and effluent BOD concentrations.

Wind data obtained from the meteorological station less than 1 km from the pond complex are presented in Figure 3 and show that the prevailing wind directions are from the SE quadrant (actually predominantly SSE) with velocities of between 1 and 4 m/s (i.e. $3.6-14.4\,km\,h^{-1}$) but with values reaching 10 m/s ($36\,km\,h^{-1}$) on occasions. Thus the prevailing wind blows along a diagonal from the outlet end of the pond towards the inlets, a situation favored by physical design manuals.

Oranges and coconuts were used simultaneously as drogues to calculate the counter flow of the surface water layer in the facultative pond. Ten oranges and 10 coconuts were released at the pond surface at the outlet end at 08.00 h and their movement towards the inlet end of the pond monitored. At the time of this experiment the mean wind speed

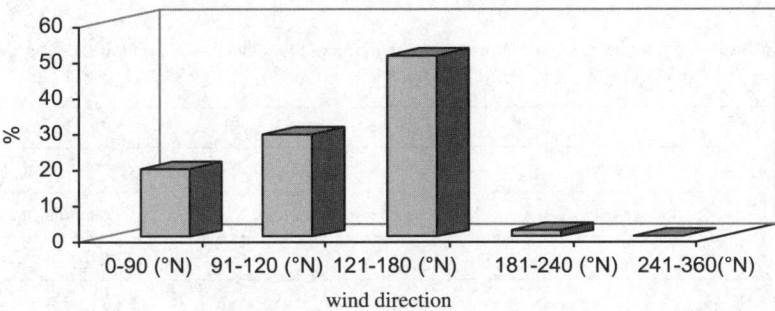

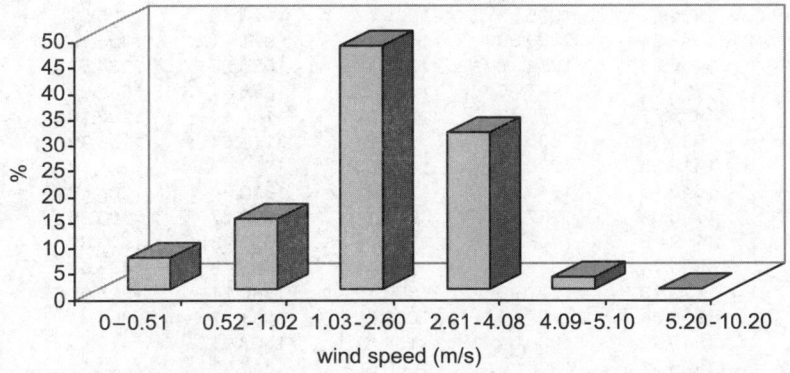

Figure 3 Prevailing wind directions and velocities at the Ponta Negra Waste Stabilization Pond Complex. Data based on the local meteorological station approximately 1 km from the site

blowing towards the inlet end of the pond was 8 m/s. The counter-flow of the oranges was calculated to be 0.068 m/s and the velocity of the coconuts even faster at 0.094 m/s. The experiment with oranges was repeated at 16.00 h on the same day when the wind speed had dropped to 1 m/s. In this case the oranges took over 14 h to reach the inlets corresponding to a counter-flow velocity of only 0.009 m/s. Both the oranges and the coconuts were approximately 98% submerged and thus based on the mean diameter of the oranges they represented a water surface layer of 4 cm depth whereas the coconuts corresponded to a water surface layer of 35 cm.

The impact of these prevailing wind velocities during the day was to mix the pond water column as is shown by the examples of typical depth profiles for chlorophyll a concentration at 2 sampling points P_5 and P_{18} representative of the zones A_2 and A_3 respectively (Table 3) and for temperature and dissolved oxygen (Table 4). The mean wind speed was 2 m/s on the day these data were collected.

Usually, facultative ponds show algal banding with the algae occupying the photic zone i.e. approximately the top 40 cm of a facultative pond water column. However the data here show no such surface stratification but high algal concentrations extending down through virtually the entire depth of the pond water column even though secchi disc values varied from just 10 to 20 cm on the sampling days. ANOVA analysis showed no significant difference (at $p = 0.05$) in the chlorophyll a concentration with depth through the water column irrespective of time or sampling location (although there was a significant difference between the mean column chlorophyll a values at certain different times). Similarly there was no significant difference (at $p = 0.05$) in temperature with depth in the water column (Table 4) and oxygen concentrations were also extremely low and showed little difference with depth.

Table 3 Chlorophyll a concentration at different depths in the primary facultative pond at different periods of the day

Depth (m)	Time					
	09.00		13.00		16.00	
	Chlorophyll a μg/L		Chlorophyll a μg/L		Chlorophyll a μg/L	
	P_5	P_{18}	P_5	P_{18}	P_5	P_{18}
0.1	1674.00	1380.74	–	1515.15	1678.07	1488.75
0.2	1612.91	1637.34	–	1472.39	1531.45	1527.37
0.3	1800.26	1710.66	1509.04	1574.29	1350.20	1893.94
0.4	1362.42	1203.57	1009.97	1020.28	1295.21	1930.59
0.5	1743.24	2309.39	1405.18	971.41	1435.73	2028.35
0.6	1056.94	1234.12	1014.18	1075.27	2264.58	2333.82
0.7	1551.81	1197.46	1044.72	1356.31	1692.33	2028.35
0.8	1655.67	1588.47	1185.24	1140.44	1906.16	1722.87
0.9	1930.60	1460.17	775.91	2211.64	1156.73	1887.83
1.0	1319.65	1399.07	1564.03	1344.09	1197.46	1270.77
1.1	1643.45	1240.23	268.82	1313.54	1264.66	2022.24
1.2	1728.99	1637.34	910.31	1270.77	2223.85	1751.38
1.3	1197.46	1405.8	1374.64	1337.98	1523.30	1678.07
1.4	989.74	1686.22	916.42	1533.48	1425.55	1906.16
1.5	1179.13	1221.90	1362.42	1380.74	1295.21	1796.19
1.6	1313.54	1704.55	1032.50	3781.77	1356.31	2382.70
1.7	1840.99	1930.60	1154.69	1507.01	1417.40	1289.10
1.8	1295.21	1325.76	1063.05	1447.95	1490.72	1796.19
1.9	103.86	1344.09	1368.53	–	1533.48	1252.44
2.0	568.18	65.17	912.35	391.01	643.53	1971.32

Table 4 Typical dissolved oxygen (DO mg/L) and temperature depth profiles in different zones of the primary facultative pond at 13.00 h

Depth (m)	Zone A_1		Zone A_2		Zone A_3	
	DO	Temp. °C	DO	Temp. °C	DO	Temp. °C
0.30	0.35	28.6	0.54	28.6	0.40	28.3
0.60	0.19	28.6	0.26	28.5	0.40	28.3
0.90	0.36	28.7	0.25	28.5	0.12	28.2
1.20	0.19	28.6	0.14	28.4	0.06	28.0
1.50	0.02	28.4	0.07	28.2	0.02	27.6
1.80	0.00	28.3	0.03	28.1	0.00	27.7

During daylight hours facultative ponds usually show thermal and oxygen stratification with super saturated oxygen concentrations (>20 mg/L) in the surface waters corresponding to the actively photosynthesizing surface algal band and with anoxic conditions prevailing below approximately 50 cm depth. The data for the Ponta Negra facultative pond would suggest that the pond water column is well-mixed during the day as a result of the windy conditions.

The mixing of the pond water column probably explains why a *Chlorella* species and a pennate diatom (non-motile species) were the dominant algal types and why there was a complete absence of flagellate genera (i.e. motile algae) such as *Euglena*, *Chlamydomonas* and *Pyrobotrys* which would be expected to dominate the algal population of a facultative pond at a surface organic loading of 350 kgBOD/ha/d. The mixing of the pond water column may also be responsible for the relatively high mean chlorophyll *a* concentrations measured in this primary facultative pond and thus for the high suspended solids in the effluent and the poor solids removal efficiency of 37.5% (see Saraiva *et al.*, 2005).

Pearson *et al.* (1988) made a distinction between gentle surface-mixing of the water column which optimizes the light contact for the algal population and evenly distributes oxygen and nutrients throughout the aerobic zone helping to optimize conditions for aerobic degradation and strong mixing, which destroys microbial and physico-chemical stratification deeper in the pond where anoxic, anaerobic and electronegative conditions are fundamental to microbes involved in anoxic and anaerobic processes important to the overall treatment process.

Conclusions

The data presented here suggest that the Ponta Negra facultative pond was functioning as a well mixed reactor as a result of the windy conditions. It would seem that de-stratification of the pond water column is reducing the efficiency of biodegradation (BOD removal) despite the correct organic surface loadings, and retention times and correct orientation and pond geometry. It would seem that highly mixed facultative ponds as a result of strong wind action during daylight hours (most ponds turnover at night), maybe less efficient than stratified ones. This calls into question the use of large ponds in windy regions in favour of smaller ponds where wind action is reduced permitting biological stratification and thus increased biological activity. It would be interesting to see if introducing stratified conditions by using wind breaks and surface baffles improves the efficiency of this complex as was the case in Mexico (Lloyd *et al.*, 2003). Studies on the impact of mixing in conventional ponds are called for.

References

APHA (1998). *Standard Methods for the Examination of Water and Wastewater*, 20th edition, American Public Health Association, Washington, DC.

El Hamouri, B., Rami, A. and Vasel, J. (2003). The reasons behind the performance superiority of a high rate algal pond over three facultative ponds in series. *Wat. Sci. Tech.*, **48**(2), 269–276.

Frederick, G. and Lloyd, B. (1996). An evaluation of retention time and short-circuiting in waste stabilization ponds using *Serratia marcescens* bacteriophage as a tracer. *Water Science and Technology*, **33**(7), 49–56.

Lloyd, B.J., Vorkas, C.A. and Guganesharajah, R.K. (2003). Reducing hydraulic short-circuiting in maturation ponds to maximize pathogen removal using channels and wind breaks. *Wat. Sci. Tech.*, **48**(2), 153–162.

Mara, D.D., Alabaster, G.P., Pearson, H.W. and Mills, S.W. (1992). *Waste stabilization ponds: A design manual for Eastern Africa*, Lagoon Technology International, Leeds, England.

Mendonça. S.R. (1990). *Lagoas de Estabilização e Aeradas Mecanicamente: Novos conceitos.*

Pearson, H.W., Mara, D.D. and Mills, S.W. (1988). Rationalizing waste stabilization pond design: The biological factor. In: *Water Pollution in Asia.* (Adv. Wat. Pollut. Control no. 6), Panswad, T., Polprasert, C. and Yamamotot, K. (eds), Pergamonn Press, Oxford, UK.

Pearson, H.W., Mara, D.D. and Bartone, C.R. (1987). Guidelines for the minimum evaluation of the performance of full-scale waste stabilization ponds. *Wat. Res.*, **21**(9), 1067–1075.

Saraiva, L.B., Meneses, C.G.R., Melo, H.N.S., Araújo, A.L.C. and Pearson, H.W. (2005). Determination of the sedimentation constants for total suspended solids and the algal component in a full-scale primary facultative pond operating at high wind velocities under tropical conditions. *Wat. Sci. Tech.* **51**(12), 213–216.

von Sperling, M. (1983). *Autodepuração dos Cursos d'água.* Dissertação (mestrado em Engenharia Sanitária), Departamento de Engenharia Sanitária, Universidade Federal de Minas Gerais, Belo Horizonte, Brazil.

Weatherell, C.A., Elliott, D.J., Fallowfield, H.J. and Curtis, T.P. (2003). Variable photosynthetic characteristics in waste stabilization ponds. *Wat. Sci. Tech*, **48**(2), 219–226.

Yanez, F. (1993). *Lagunas de estabilizacion. Teoria, diseno y mantenimiento. ETAPA*, Cuenca, Equador.

Control of chironomid midge larvae in wastewater stabilisation ponds: comparison of five compounds

R. Craggs*, L. Golding*, S. Clearwater*, L. Susarla** and W. Donovan***

*National Institute of Water and Atmospheric Research, P.O. Box 11-115, Hamilton, New Zealand
**URS, P.O. Box 821, Auckland, New Zealand
***Bioresearches, P.O. Box 2828 Auckland, New Zealand (E-mail: r.craggs@niwa.co.nz)

Abstract Chironomid midge larvae are a valuable component of wastewater stabilisation pond (WSP) ecology. However, in high numbers, adult midge swarms can be a nuisance to near-by urban areas. Improving WSP treatment by incorporating aerobic or maturation ponds or by the addition of pre-treatment to reduce organic loading also increases the availability of aerobic sediment (midge larva habitat) in the pond system and the potential for midge nuisance problems. The efficacy of Maldison, an organophosphate traditionally used to control midge larvae in New Zealand WSPs, was compared to *Bacillus thuringiensis* var. *israelensis* (Bti), Methoprene, Pyriproxyfen and Diflubenzuron which are all more specific to insects and have fewer adverse environmental effects. Initial laboratory trials established the concentration of each compound required to achieve 95% control of the midge population. During 21-day small-scale trials within the WSP, Bti, Diflubenzuron and Maldison reduced live larvae numbers substantially (80–89%) compared to controls and adult midge emergence was markedly reduced by all compounds (72–96%). Large-scale trials with Bti (Vectobac® WG) powder (1000 μg/L) only caused a slight reduction in midge larvae numbers compared to controls and had little effect on adult emergence, however, Methoprene (Prolink XRG granules) (50 μgAI/L) reduced midge adult emergence by ~80% over 25 days and has been used successfully to control several midge nuisance outbreaks.
Keywords Chironomid; control; larvae; midge; methoprene; nuisance; wastewater stabilisation ponds

Inroduction

Chironomid midge larvae are commonly found in aerobic bottom sludges of wastewater stabilisation ponds (WSPs) (Cranston, 1995) and are recognised as a valuable component of WSP ecology (Pinder, 1995; Tokeshi, 1995), contributing to treatment by consuming algae, bacteria and detritus (reducing pond TSS and nutrient levels), which are then completely removed from the system when the adult midge emerges from the pond. Midge adults swarm around pond margins at dawn and dusk, where they mate, and females lay eggs back into the pond (Tokeshi, 1995). When present in high numbers, and blown by the wind or attracted by light to near-by urban areas, swarms of adult midges can cause considerable nuisance (Ali, 1996). Common nuisance problems include invasion of buildings, disruption of outdoor activities and accumulation beneath exterior lighting.

Nuisance problems typically occur with newly commissioned WSP systems where aerobic pond sediments provide ideal midge larvae habitat. The nuisance is usually only short-term as the midge larvae population declines when anaerobic sludge accumulates on the pond bottom; restricting the available midge larvae habitat to the pond edges. However, pond sediments of established WSPs can temporarily become aerobic (due to favourable environmental conditions or reduced organic loading) and cause temporal midge nuisance problems (Spiller, 1964). Thus, a potential downside of improving the design of WSPs to provide higher levels of treatment (e.g. by incorporation of aerobic or maturation ponds) is the increase of available aerobic sediment (midge larvae habitat) in the pond system and the higher susceptibility of near-by urban areas to midge nuisance problems. Likewise, installation of a mechanical treatment plant ahead of a WSP system and the consequent improvement in pond water quality can result in a similar increase in midge larvae habitat and nuisance from adult swarms (e.g., Broza *et al.* 2000).

This paper compares the efficacy of Maldison (a compound that has traditionally been used to control midge larvae in New Zealand (NZ) WSPs) to four alternative compounds (*Bacillus thuringiensis* var. *israelensis* (Bti), Methoprene, Pyriproxyfen and Diflubenzuron) that are more specific to insects and have fewer adverse environmental effects. When comparing the efficacy of several types of control agent it is important to understand their different modes of action and the relevant '*endpoint*' for control.

Maldison is a fast acting, broad-spectrum organophosphate insecticide that is effective upon external contact or after ingestion. Maldison causes disruption of the central nervous system, but in insects it becomes a much stronger inhibitor when it is metabolised to malaoxon (Extoxnet, 2000). *Endpoints:* larvae mortality, reduced pupation and reduced adult emergence. *Bacillus thuringiensis var. israelensis* (Bti) is a naturally occurring soil bacterium that contains a proteinaceous toxin which is activated by enzymes in the intestine of the larvae. The protoxin disintegrates the intestine causing death within a few hours up to a few weeks depending on the dose ingested and chironomid species (Liber *et al.* 1998). *Endpoints:* larvae mortality, reduced pupation and reduced adult emergence.

Methoprene, Pyriproxyfen and Diflubenzuron are insect growth regulators (IGRs) that are specific to insects and tend to sorb to organic matter thereby targeting benthic organisms (including chironomids). An advantage of IGRs is that the treated larvae are not killed immediately and remain in the aquatic food chain (Ali *et al.*, 1993). Methoprene and Pyriproxyfen are both juvenile hormone analogues (JHAs). Methoprene interferes with the insect's life cycle preventing metamorphosis of larvae to adults but is not toxic to pupae. Therefore adults will still emerge from pupae present at the time of application, but larvae will pupate then fail to emerge (Extoxnet, 2000). *Endpoint:* reduced adult emergence. Pyriproxyfen interferes with larval development of pupae by disrupting the sequence of gene activation. Therefore adults will emerge from pupae present at the time of application but larvae will fail to pupate. *Endpoints:* reduced pupation and reduced adult emergence. Diflubenzuron is a chitin synthesis inhibitor (CSI) that prevents exoskeleton hardening after moulting and is effective through contact or ingestion (Extoxnet, 2000). Therefore midge larvae are most susceptible prior to moulting. *Endpoints:* larvae mortality, reduced pupation and reduced adult emergence.

Experiments were conducted at the North Shore WWTP, Auckland, NZ, where there was concern that an upgrade of the mechanical treatment plant upstream of the two-pond WSP system and consequent reduction in the organic load would increase the available habitat (aerobic sediment) for chironomid midge larvae in the ponds and the potential for midge nuisance problems.

Methods
Small-scale field trials
The results of initial laboratory trials (96-hour acute static tests for the fast acting toxins Maldison and Bti, and 10-day chronic static tests for the IGRs Methoprene, Diflubenzuron, and Pyriproxyfen using midge larvae (*Chironomus zealandicus*) cultured from egg masses collected from the North Shore WSPs) were analysed and compared with literature values to derive the following concentrations of each compound for use in field trials: 30 µg active ingredient (AI)/L Maldison (Malathion 50EC); 1000 µg/L Bti (Vectobac®WG, 3000 international toxic units (ITU)/mg); 50 µgAI/L Methoprene (Prolink); 10 µgAI/L Pyriproxyfen (Admiral 10EC); and 50 µgAI/L Diflubenzuron (Dimilin

25W™). Unfortunately, Pyriproxyfen which has been found to provide better control of midges than both Methoprene and Diflubenzuron (Ali et al., 1999) could not be tested further as it was not currently approved for water-related applications in NZ.

Small-scale trials were conducted in enclosures at a site in Pond 1, where high numbers of midge larvae had routinely been recorded in the pond sediment. The enclosures (0.785 m^2, 471 L) were made from 1.6 m sections of 1 m diameter fibreglass pipe that were driven into the pond substrate. Triplicate enclosures for each treatment and 6 control enclosures were randomly positioned at 0.6 m water depth along a line parallel to the pond edge. The trials of each compound were conducted at two concentrations: (1) the concentration recommended from the laboratory trials, and (2) 10 × that concentration, to account for factors such as insecticide degradation and sediment adsorption which may decrease toxicity in the pond environment. Maldison (a liquid emulsion) was directly applied and mixed into the enclosure water. Bti and Diflubenzuron (both powders) were applied by premixing in water taken from the enclosure then mixing the suspension into the enclosure water. Methoprene pellets (42.5 gAI/kg) were evenly scattered over the water surface and sank to the bottom of the enclosure.

The enclosure water was aerated with bubble diffusers to ensure aerobic conditions (DO >3 g m^{-3}). Numbers of live larvae and pupae and emerging adults were monitored in the enclosures before and over 21 days after application of the control compounds. Larvae and pupae numbers were counted from a sieved (500 μm mesh) sediment sample taken from each enclosure using an Eckman grab sampler (15 cm × 15 cm). Adult emergence numbers over a 24 h period were monitored using an emergence trap in each enclosure that caught the adults on adhesive paper. Statistically significant differences between treatments were detected using a one-way ANOVA ($p < 0.05$) with post hoc analysis of least significant difference using DataDesk® software.

Large-scale field trials

Methoprene and Bti were tested individually and in combination at two areas in Pond 1 (that typically had high numbers of midge larvae) to determine how factors such as dilution and dispersion in the pond water or adsorption to the pond sediment influenced performance, and to test the ease of applying each compound. Area SH on the north-eastern side of Pond 1 stretched 160 m along the shore (at a depth of 0.5 m) and extended 30 m into the pond to a 2 m depth. Area SI on the southern side of Pond 1 stretched 80 m along the shore (at a depth of 0.25 m) and extended 20 m into the pond to a 2 m depth. Both trial areas were divided into four sites (Control, Methoprene, Bti/Methoprene and Bti) each 40 m × 30 m in Area 1 and 20 m × 20 m in Area 2, and separated to reduce water movement using polythene curtains extending to just below the water surface, that were weighted on the pond bottom and held in place with stakes.

Methoprene granules (15 gAI/kg) were used to achieve a more even distribution than the larger, higher strength pellets (42.5 gAI/kg) used in the small-scale trial. The granules were applied using a rotary dispenser (10 m swath) fitted to the front of a flat-bottomed boat at a rate of 32 kg total ingredient (TI)/ha to give a concentration of 50 μgAI/L in the water column (assuming a 1 m water depth). Bti (wettable powder) was mixed with 200 L of pond water in a plastic drum and the continuously mixed suspension was applied to the trial sites by pumping through a boom manifold (with 16 hoses that reached down to the pond sediment surface) that was fitted to the front of the flat-bottomed boat. An application rate of 10 kg TI/ha was used to give a concentration of 1000 μg/L in the water column (assuming a 1 m water depth). A hand-held boom manifold (2 m swath with 4 hoses), was used for inshore areas. Both compounds were re-applied on day 20.

Midge larvae were enumerated from two sieved (500 μm mesh) sediment samples that were collected using a 0.023 m² Eckman grab sampler from four monitoring stations within each trial site. Following the first application, sediment samples were randomly collected at pond depths between 0.25 m and 1 m, however, following the second application, all sediment samples were collected at a uniform pond depth of 0.75 m to control for a possible depth effect. Pupae were not enumerated because previously their numbers were highly variable. Sediment samples were collected on days 3, 5, 10, 13 and 17 after the first application of the compounds and on days 24, 27, 31 and 34 following the second application. Numbers of adults emerging from the trial sites were monitored using emergence traps placed within each monitoring station (water depths: 0.25–1 m after the first application; 0.75 m (above larvae sampling sites) following the second application). Emergence traps were deployed for 24 h periods, once prior to the first application of the compounds (to assess the initial populations); on days 3, 5, 10, 13 and 17 after the first application; and on days 24, 25, 26, 27, 31, 32 and 33 following the second application.

Results and discussion
Small-scale field trials

Small-scale field trials were conducted in enclosures within the WSP to determine the efficacy of each compound to control a sample of the midge population at 1 × and 10 × the concentration recommended by the laboratory tests. The average percentage change in live larvae and pupae (per m²) and adult emergence (per emergence trap) for each of the compounds compared to controls are given with 95% confidence limits in Table 1. Despite a gradual increase in the midge population within the control enclosures, all four compounds (Maldison, Bti, Methoprene and Diflubenzuron) were effective at reducing the midge population by the end of the 21-day experimental period.

Compared to controls, treatment with Maldison reduced live larvae numbers by 80% and 92% by day 11 at the 30 μgAI/L and 300 μgAI/L concentrations, respectively, and pupae by 100%. By day 14 numbers of emerging adults were significantly reduced by 88% and 83%, respectively. The 30–300 μgAI/L concentration used in this experiment is much lower than the 22,000 μgAI/L that was routinely used at the Mangere WSP, NZ to achieve 100% larvae mortality (Spiller, 1964), but falls within the ranges suggested by Ali (1981b) for effective control (4–56 μgAI/L) and for control of a Maldison-sensitive species in a 1 m deep pond (25–50 μgAI/L).

Bti significantly reduced live larvae numbers by 93% by day 4 at the 10,000 μg/L concentration compared to controls and by day 11, larvae numbers were reduced by 61% and 89% and pupae numbers by 100% and 100% at the 1000 μg/L and 10,000 μg/L concentrations respectively. By day 14, numbers of emerging adults were significantly reduced by 75% at the 10,000 μg/L concentration compared to controls and by day 21, adult emergence was reduced by 72% and 96% at the 1000 μg/L and 10,000 μg/L concentrations respectively. Previous field studies have shown that Bti is effective for midge larvae control e.g. at 1600 μg/L (1200 ITU) (Charbonneau et al. 1994), 88% larvae reduction at 2500 μg/L (1000 ITU) (Ali, 1981a) and 71% larvae reduction at 9000 μg/L (200 ITU) (Liber et al., 1998). Typical reapplication rates are 2–4 weeks (Ali, 1981a; Liber et al., 1998). Therefore, Bti would appear to be an effective replacement for Maldison.

As expected, Methoprene did not significantly reduce larvae numbers compared to controls, but by day 11, pupae numbers were significantly reduced by 100% at both concentrations, and by day 14 adult emergence was significantly reduced by 93% and 98% at the 50 μgAI/L and 500 μgAI/L concentrations, respectively. These results are in agreement with Ali (1991) who found that 22 μgAI/L (for a 1 m depth lake; 0.22 kgAI/ha) reduced adult emergence by 64–98% for 7 weeks.

Table 1 Results of small-scale field trials. Mean % change relative to controls in live midge larvae, pupae and emerging adults in enclosures treated with Maldison, Bti, Methoprene, and Diflubenzuron relative to the control (no treatment) over 21 days

	Post-treatment mean % change in												
	Live larvae				Live pupae				Adult emergence				
Treatment (AI) \ Day	4	11	21		4	11	21		2	4	7	11	14	21
Maldison 30 µg/L	−54	−80*	−33		−100	−100*	−100*		−52	−18	22	−89	−88*	−88*
Maldison 300 µg/L	−70	−92*	−96*		−100	−100*	−100*		−44	−29	11	−84	−83*	−97*
Bti 1000 µg/L	−41	−61*	−83*		767*	−100*	−91		188	994*	444*	189	63	−72*
Bti 10 000 µg/L	−93*	−89*	−100*		−100	−100*	−100*		4	912*	189*	−79	−75*	−96*
Methoprene 50 µg/L	13	−17	−33		−100	−100*	−93*		116	500*	67	−89	−93*	−95*
Methoprene 500 µg/L	42	−23	−31		33	−100*	−100*		364	112	−11	−84	−98*	−96*
Diflubenzuron 50 µg/L	116*	−7	−89*		−100	−100*	−64*		52	−88	−11	−95*	−80*	−96*
Diflubenzuron 500 µg/L	15	−25	−85*		−100	−100*	−96*		−68	−41	22	−89	−73	−95*

(negative % implies a decrease, positive % implies an increase, * statistically significant, $P < 0.05$)

Diflubenzuron performed more quickly and was slightly more effective than Methoprene; significantly reducing live larvae numbers by 89% and 85% compared to controls by day 21 at the 50 µgAI/L and 500 µgAI/L concentrations, respectively. By day 11, numbers of pupae were reduced by 100% at both concentrations and adult emergence was reduced by 95% at the 50 µgAI/L concentration. By day 21 numbers of emerging adults were significantly reduced by 96% and 95% at the 50 µgAI/L and 500 µgAI/L concentrations, respectively. Previous studies have demonstrated similar control of adult emergence by Diflubenzuron at concentrations of: 0.028–0.056 kgAI/ha for 24 d (Ali and Lord, 1980) and 30 µgAI/L (Liber et al. 1998). Our results also corroborate studies that have found Diflubenzuron to be more effective at controlling midges than JHAs such as Methoprene (Ali and Lord, 1980). However, for exposure periods greater than 21 days, the slow release pellet and granule formulations of Methoprene could be more effective than Diflubenzuron powder, the only formulation presently available in NZ. Diflubenzuron was not tested further because of concern over possible effects on non-target organisms at the marine outfall.

Large-scale field trials

Based on the small-scale trial results, large-scale field trials were conducted at two sites (SH and SI) within Pond 1 to determine the efficacy of Bti and Methoprene alone, and in combination, to control the midge population. The midge larvae population was highly variable at the control sites during the experiment (2,600–22,000/m^2 at Area SH, and 3,800–11,400/m^2 at Area SI (Figure 1), reflecting the inherent variability in the WSP midge population as cohorts of larvae overlap one another.

We expected Bti to greatly influence larvae numbers, however, only a slight reduction was observed compared to controls. A detectable reduction in live midge larvae numbers was not recorded at Area SH until day 24 (4 days after the second application) with numbers reduced from 17,000/m^2 (day 19) to 3,300/m^2, while numbers at the control sites increased from 7,800/m^2 to 21,500/m^2 over the same period (Figure 1). Larvae numbers at Area SI were no different to controls throughout the trial period (Figure 1).

Bti only had a slight effect on adult emergence at both trial areas; with numbers at Area SH varying between 2–26/trap (control 5–36/trap) and those at Area SI varying between 2–15/trap (control 1–26/trap) (Figure 2). Field trials previously conducted at Mangere WSP, NZ using Bti (Vectobac 12AS) were also found to be ineffective (pers comm. Sam Tan, Mangere WWTP). Poor control by Bti in field trials has been attributed to dilution and dispersion of the Bti suspension in the water column (Ali, 1981a; Liber et al., 1998) and this probably occurred in the Mangere trial where the Bti was applied to the water surface. Charbonneau et al. (1994) found that larvae instar stage was important, with second and third instars more susceptible to Bti than the fourth-instar larvae, but this is an unlikely factor in the present study that was conducted over 35 days. Thus, dispersion of the Bti into the water column above the sediment or within the sediment itself is the most plausible explanation. We had hoped to use a concentrated Bti briquette in the large-scale trial, which could have alleviated this problem, but it was not approved for use in NZ at the time.

Compared to the control sites, Methoprene also had little control on live midge larvae numbers which varied between 3,400–10,400/m^2 at Area SH and 4,700–11,200/m^2 at Area SI (Figure 1). This result was expected as larvae exposed to Methoprene do not die immediately. The effect of Methoprene on adult emergence was clearly evident at both trial areas, with the numbers decreasing after both applications. On average, Methoprene reduced adult emergence by 80% compared to controls at both trial areas (Figure 2). Methoprene appeared to have greatest effect on adult emergence 8–10 days following

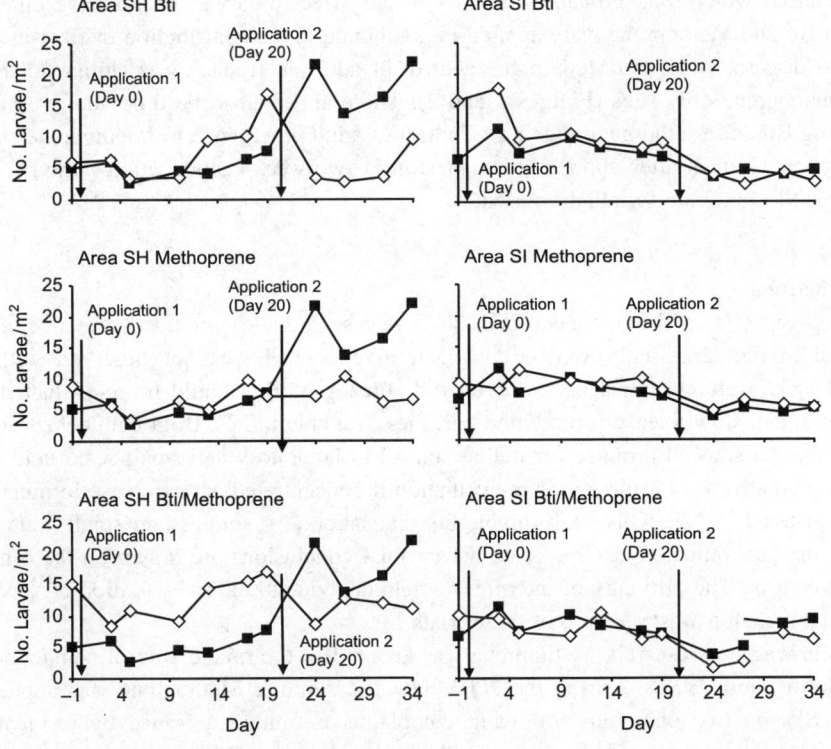

Figure 1 Live larvae numbers at sites in large-scale field trial areas (SH and SI) treated with Bti, Methoprene, or Bti/Methoprene combined (all white symbols) compared to control sites (black symbols) over 35 days

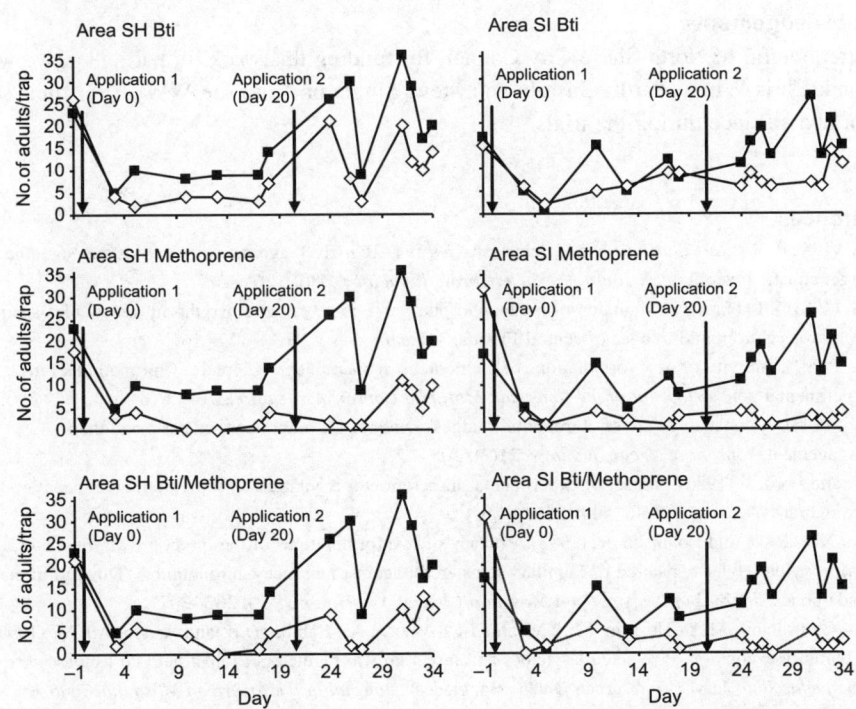

Figure 2 Adult emergence at large-scale field trial areas (SH and SI) treated with Bti, Methoprene, and Bti/Methoprene combined (all white symbols) compared to control sites (black symbols) over 34 days

application, which was probably the time when affected larvae would have emerged. When Bti and Methoprene were applied in combination, the contribution of Bti was difficult to discern, while Bti/Methoprene control of adult emergence was similar to that at the Methoprene only sites (Figures 1 and 2). There appeared to be little additive benefit of using Bti and Methoprene together. Control of adult emergence by Methoprene started to decline 17 days after application suggesting a two-week reapplication period for the Prolink XRG granule formulation used.

Conclusions

Methoprene (50 µgAI/L, concentrated, slow release granule) provided effective midge control in the large-scale WSP trials. Definitive control was not observed with Bti (1000 µg/L, wettable powder); however, the efficacy of Bti should be re-evaluated if a concentrated, slow-release formulation becomes available in NZ. Both Diflubenzuron and Pyriproxyfen showed promise for midge control in laboratory and small-scale field trials and should also be considered for re-evaluation if concentrated, slow-release formulations are registered in NZ. This study highlights that laboratory studies and small-scale field trials must be validated at large-scale before final conclusions are drawn on the efficacy of the control. The difficulty of interpreting field data due to natural variation of the WSP midge population was a feature of these trials.

Following this research, Methoprene was adopted as the midge control compound for the North Shore WSPs. During the 2003/2004 NZ summer Methoprene was applied to the WSPs on two occasions following complaints of midge nuisance by residents of urban areas close to the pond system. On both occasions the WSP midge population was successfully controlled, and no further nuisance complaints were received.

Acknowledgements

We are grateful to North Shore City Council for funding this work. In particular, we wish to thank Chris Wium, Partha Susarla and Steve Singleton from the WWTP for their support and assistance during the trials.

References

Ali, A. (1981a). *Bacillus thuringiensis Serovar. israelensis* (ABG-6108) against chironomids and some non target aquatic invertebrates. *Journal of Invertebrate Pathology*, **38**(2), 264–272.

Ali, A. (1981b). Laboratory evaluation of organophosphate and new synthetic pyrethroid insecticides against pestiferous chironomid midges of central Florida. *Mosquito News*, **41**(1), 157–161.

Ali, A. (1991). Activity of new formulations of methoprene against midges (Diptera: Chironomidae) in experimental ponds. *Journal of the American Mosquito Control Association*, **7**(4), 616–620.

Ali, A. (1996). A concise review of chironomid midges (Diptera: Chironomidae) as pests and their management. *Journal of Vector Ecology*, **21**(2), 105–121.

Ali, A. and Lord, J. (1980). Impact of experimental insect growth regulators on some non target aquatic invertebrates. *Mosquito News*, **40**(4), 564–571.

Ali, A., Xue, R.-D. and Lobinske, R. (1993). Efficacy of two formulations of the insect growth regulator, pyriproxyfen (Nylar registered or Sumilarv registered), against nuisance Chironomidae (Diptera) in manmade ponds. *Journal of the American Mosquito Control Association*, **9**(3), 302–307.

Ali, A., Chowdhury, M.A., Hossain, M.I., Mahmud-Ul-Ameen, A., Habiba, D.B. and Aslam, A.F.M. (1999). Laboratory evaluation of selected larvicides and insect growth regulators against field-collected *Culex quinquefasciatus* larvae from urban Dhaka, Bangladesh. *Journal of the American Mosquito Control Association*, **15**(1), 43–47.

Broza, M., Halpern, M. and Inbar, M. (2000). Non-biting midges (Diptera; Chironomidae) in waste stabilization ponds: an intensifying nuisance in Israel. *Water Science and Technology*, **42**(1-2), 71–74.

Charbonneau, C.S., Drobney, R.D. and Rabeni, C.F. (1994). Effects of *Bacillus thuringiensis var. israelensis* on nontarget benthic organisms in a lentic habitat and factors affecting the efficacy of the larvicide. *Environmental Toxicology and Chemistry*, **13**(2), 267–279.

Cranston, P. (1995). Introduction. In *The Chironomidae: the Biology and Ecology of Non-Biting Midges*, Armitage, P., Cranston, P.S. and Pinder, L.C.V. (eds), Chapman and Hall, London, pp. 1–11.

Extoxnet (2000). http://ace.orst.edu/info/extoxnet. Extension Toxicology Network.

Liber, K., Schmude, K.L. and Rau, D.M. (1998). Toxicity of *Bacillus thuringiensis var. israelensis* to chironomids in pond mesocosms. *Ecotoxicology*, **7**(1), 343–354.

Pinder, L.C.V. (1995). Biology of the eggs and first-instar larvae. In *The Chironomidae: the Biology and Ecology of Non-Biting Midges*, Armitage, P.D. *et al.* (eds), Chapman and Hall, London, pp. 87–106.

Spiller, D. (1964). *A* submission on the chironomid midge problems at Manukau Purification Plant, Mangere, DSIR, Auckland.

Tokeshi, M. (1995). Life cycle and population dynamics. In *The Chironomidae: the Biology and Ecology of Non-Biting Midges*, Armitage, P., Cranston, P.S. and Pinder, L.C.V. (eds), Chapman and Hall, London, pp. 225–268.

Mosquito development and biological control in a macrophyte-based wastewater treatment plant

I.M. Kengne Noumsi*, A. Akoa*, R. Atangana Eteme*, J. Nya*, P. Ngniado*, T. Fonkou** and F. Brissaud***

*Wastewater Research Unit, Faculty of Science, University of Yaounde I, P.O. Box: 8404 Yaounde, Cameroon (E-mail: ives_kengne@yahoo.fr)
**Faculty of Science, University of Dschang, P.O. Box: 67 Dschang, Cameroon (E-mail: tfonkou@yahoo.fr)
***Hydrosciences, University of Montpellier II, 34095 Montpellier Cedex 05, France
(E-mail: brissaud@msem.univ-montp2.fr)

Abstract A one-year study of the proliferation of mosquito in a *Pistia stratiotes*-based waste stabilization ponds in Cameroon revealed that *Mansonia* and *Culex* were the main breeding genera with about 55% and 42% of the total imagoes respectively. Though the ponds represent a favorable breeding ground for mosquitoes, only 0.02% of captured imagoes was *Anopheles gambiae*, suggesting that this wastewater treatment plant does not significantly contribute to the development of the malaria vector in the area. *Gambusia* sp. introduced to control mosquito population in the ponds acclimatized relatively well in most of the ponds (B3–B7) and their feeding rate without any diet ranged from 15.0 to 50.2 larvae/day for a single fish.
Keywords Biological control; *Gambusia* sp.; macrophyte lagoons; mosquito; wastewater treatment

Introduction

Many advantages plead in favor of waste stabilization ponds (WSP) technology in developing countries which lack capital and qualified manpower to run sophisticated systems: i) its simplicity of construction and operation, ii) low operation and maintenance costs and iii) capability to withstand excess organic and hydraulic loads. Furthermore, most developing countries have a warm tropical and subtropical climate that allows high biological activity year-round, which means high efficiency of the system (Denny, 1997; Kivaisi, 2001). Unfortunately, large water bodies covered by macrophytes may represent a favorable breeding ground for mosquitoes, thus providing nuisance pests and or disease vectors to the nearby communities (Dill, 1989; Russell, 1999). Methods to control mosquito breeding at acceptable level include periodic spraying of insecticides (expensive and sometimes ecologically dangerous), periodic macrophyte harvesting and/or the use of mosquito predators such as larvivorous fishes (Charudattan, 1987; Eldridge and Martin, 1987).

This work was undertaken in order to seek how to minimize the proliferation of such disease vectors or nuisance pests in a *Pistia stratiotes*-based treatment plant devoted to treat approximately 45 m^3/day of domestic sewage from the Biyem-Assi (Yaounde) residential quarter. The first step of the investigation aimed at assessing the biodiversity of mosquitoes breeding in this plant. The second step focused on the biological control of mosquitoes with *Gambusia* sp.

Methods

This work was carried out in a 0.1 ha WSP system comprising a decantation–digestion pond (B0) followed by a series of 7 rectangular ponds vegetated mainly with the floating macrophyte *Pistia stratiotes*. Details of its construction and operation are given in Kengne Noumsi (2000).

Biodiversity of the culicids was assessed five times a month from November 1997 to October 1998 taking into account immature and adult mosquito stages breeding in all the macrophyte ponds (B1–B7). Concomitantly, the following physicochemical characteristics of water were measured at the outlet of each pond: pH, temperature, conductivity, dissolved oxygen, BOD_5, SS and turbidity.

Immature free living mosquitoes were sampled by randomly dipping at 5 stations in each lagoon with a 250 mL dipper while, for attached ones, 5 macrophyte roots were washed in a white basin by gently swirling them in water to release the mosquito larva and nymphs. Adult mosquitoes were trapped using a $0.25\,m^2$ trap derived from the model proposed by Aubin et al. (1973). In each pond, 3 traps were randomly located and inspected after 24 hours. Identification was done according to the immature mosquito determination keys of the Ethiopian region (Hopkins, 1952) and the adult mosquito key of the Ethiopian region (Edwards, 1941).

Prior to the assessment of their feeding rate, 50 *Gambusia* sp. fishes were allowed for a week to acclimatize to the rough conditions of the different macrophyte ponds. In each pond (except B1 which had an excess of sludge), 50 individuals ($25 \leq size \leq 45\,mm$) collected in the surrounding natural wetlands were introduced in a $0.4\,m \times 0.4\,m \times 0.4\,m$ wooden box wrapped in a net. After this period, 5 fishes without any diet and 300 mosquito larvae at 3 to 4th stage belonging both to the genus *Culex* and *Mansonia* (most frequent species in the plant) were introduced in aquariums containing 3.5 L of water and their number evaluated after 24 hours. Mosquito fish and water were from each pond where the fish had survived. The feeding capacity of *Gambusia* was evaluated using this formula: $Nle = Ni - (Nlp + Nn)$, with Nle the number of mosquito larvae eaten, Ni the number of mosquito larvae introduced (Ni = 300 larvae), Nlp the number of mosquito larvae present in the aquarium after 24 hours and Nn the number of nymphes.

Results and discussion

A total of 17,568 adult mosquitoes were captured during the investigation period among which 54.8 and 41.9% belonged respectively to the genus *Mansonia* and *Culex* (Table 1). Identification of species both at the immature and mature stages indicated that most of the *Culex* belonged to *C. quinquefasciatus*, *C. decens* and *C. tigripes,* while for *Mansonia* it was both *M. africana* and *M. uniformis*. About 3% of captured imagoes were *Coquillettidia* (especially *C. metallica*), *Ficalbia* and *Anopheles*. Though not found at adult stage, a few *Aedes* larvae were found in ponds B1 and B2 only during the month of December 1997. Taking into consideration the whole plant, approximately 43 mosquitoes emerged on an

Table 1 Biodiversity of adult mosquitoes in the plant

Immature stage characteristics	Genus	Number of imagoes captured during the whole period of study			% with respect to the total number of imagoes
		Male	Female	Total	
Living attached to *Pistia stratiotes* roots	Mansonia	3,969	5,654	9,623	54.78
	Coquillettidia	94	107	201	1.14
	Ficalbia	46	57	103	0.59
Free living	Culex	3,839	3,529	7,368	41.94
	Anopheles	0	3	3	0.02
	Aedes	0	0	0	0
	Unidentified *	–	–	270	1.53
	Total	7,948	9,350	17,568	100

*Parts essential for identification were lost during collection.

average basis per m²/day. Luckily, bearing in mind the endemic malaria situation of the region (Manga et al., 1992), this plant can be considered as not favoring the development of the malaria vector, since only 0.02% of the total capture was *Anopheles gambiae*.

The larvivorous fishes acclimatized relatively well in the latest ponds (B5–B7) since no death was encountered during the one week period of study (Table 2). On the contrary, a stabilization of the death rate was observed the 3rd and 4th day in ponds B3 and B4 (respectively 35 and 46 fishes alive). In pond B2, no fish survived after the 2nd day.

The death rate of fishes could be due to the rough conditions of the milieu (Table 3). In pond B2 for example, the low level of dissolved oxygen (0.8 mg/L) and the high level of SS (149 mg/L) certainly hampered the survival of fishes. The increase of their acclimatization could be due to the progressive improvement of the water quality. In pond B5 where no death was encountered despite a still high NH_4 content, the level of D.O increased up to 4 mg/L, while SS dropped to 64 mg/L. These conditions, though remaining difficult, seemed to permit their survival.

The feeding rate of *Gambusia* ranged from 15.0 to 50.2 larvae/day for a single fish (Table 4). The highest average rate was recorded in aquarium receiving water from pond B6 (39.4 larvae/day) and the lowest in water from pond B4 (27.4 larvae/day). However, analysis of variance of Kruskal–Wallis revealed no significant difference between the ponds ($H = 5.71$, $df = 4$, $p = 0.223$). The statistical insignificance of the number of larvae consumed between the various pounds could be explained by the fact that once acclimatized, fishes were no longer influenced by the water quality. The average number of mosquito larvae consumed for all the ponds was estimated at approximately 31 larvae/fish/day. This could be attributed to the fact that no fish was submitted to a diet prior to the experiment.

Table 2 Number of fishes alive in the various ponds in function of time ($n = 5$)

	Number of fishes alive (mean ± SEM)					
	B2	B3	B4	B5	B6	B7
Day 1	9 ± 6	41 ± 3	47 ± 2	50 ± 0	50 ± 0	50 ± 0
Day 2	0 ± 0	36 ± 4	47 ± 2	50 ± 0	50 ± 0	50 ± 0
Day 3	0 ± 0	35 ± 4	47 ± 2	50 ± 0	50 ± 0	50 ± 0
Day 4	0 ± 0	35 ± 4	46 ± 2	50 ± 0	50 ± 0	50 ± 0
Day 5	0 ± 0	35 ± 4	46 ± 2	50 ± 0	50 ± 0	50 ± 0
Day 6	0 ± 0	35 ± 4	46 ± 2	50 ± 0	50 ± 0	50 ± 0
Day 7	0 ± 0	35 ± 4	46 ± 2	50 ± 0	50 ± 0	50 ± 0
% survival after one week	0	70	92	100	100	100

Table 3 Physicochemical characteristics of the wastewater in the ponds

Ponds	Physicochemical properties of wastewater (mean ± SEM)						
	pH	Temperature (°C)	Conductivity (μS/cm)	D.O (mg/L)	BOD_5 (mg/L)	SS (mg/l)	$N\text{-}NH_4^+$ (mg/L)
B1	7.07 ± 0.04	26.4 ± 0.2	1,292 ± 26	0.3 ± 0.0	308 ± 12	277 ± 13	89.2 ± 4.6
B2	7.12 ± 0.04	26.6 ± 0.2	1,258 ± 27	0.8 ± 0.1	212 ± 13	149 ± 9	85.6 ± 4.7
B3	7.18 ± 0.03	26.7 ± 0.2	1,215 ± 22	1.7 ± 0.2	173 ± 10	103 ± 7	79.1 ± 3.8
B4	7.22 ± 0.04	26.7 ± 0.2	1,147 ± 23	2.2 ± 0.2	173 ± 9	70 ± 13	71.4 ± 3.7
B5	7.26 ± 0.04	26.8 ± 0.2	1,094 ± 22	3.6 ± 0.3	135 ± 8	64 ± 4	65.4 ± 3.2
B6	7.30 ± 0.04	26.7 ± 0.2	1,045 ± 23	4.4 ± 0.2	101 ± 7	45 ± 3	60.2 ± 3.3
B7	7.63 ± 0.06	26.7 ± 0.3	903 ± 28	9.6 ± 2.0	83 ± 2	30.1 ± 2	52.3 ± 3.2

Table 4 Mean number of mosquito larvae eaten by a single fish in 24 hours ($n = 5$)

Origin of water	Number of larvae eaten/day	
	Mean	Minimum – Maximum
B3	30.3	18.2–50.2
B4	27.9	15.0–45.0
B5	29.4	15.4–39.6
B6	39.4	24.0–47.2
B7	25.9	16.0–35.4

Conclusion

As any other standing pool, this *Pistia*-based treatment plant represents a favorable breeding ground for mosquitoes. However, it can be considered as not favoring the development of the malaria vector, since *Anopheles gambiae* was found only accidentally and in small number. These results highlight the dilemma of sanitary engineers and local public health authorities that have to treat water at low-cost while reducing the nuisance pest.

Successful trials of the acclimatization of *Gambusia* in most of the treatment ponds and their feeding rate without any diet lower than 25 mosquito larvae/day are viewed as very promising. *Gambusia* are present in streams and marshes of Yaounde, and could be domesticated for biological control of mosquitoes in such treatment plants.

Acknowledgements

This work was supported by the International Foundation for Science, IFS (grant n° 1580/3F) and the French Ministry of Foreign Affairs (Programme CAMPUS, grant n° 94 016 400).

References

Aubin, A., Bourassa, J.P. and Pellisier, M. (1973). An effective emergence trap for the capture of mosquitoes. *Mosquito News*, **33**(2), 250–252.

Charudattan, R. (1987). Impact of pathogens on aquatic plants used in water treatment and resource recovery systems. In *Aquatic Plants for Wastewater Treatment and Resource Recovery*, Reddy, K.R. and Smith, W.H. (eds.), Magnolia Publishing Inc., Orlando, Florida, pp. 795–803.

Denny, P. (1997). Implementation of constructed wetlands in developing countries. *Wat. Sci. Tech.*, **35**(5), 27–34.

Dill, C.H. (1989). Wastewater wetland: User friendly mosquito habitats. In *Constructed Wetlands for Wastewater Treatment: Municipal, Industrial and agricultural*, Hammer, D.A. (ed.), Lewis Publishers, Chelsea, MI, pp. 665–668.

Edwards, F.W. (1941). *Clé des Culicinae adultes de la région éthiopienne*, ORSTOM, Paris.

Eldridge, B.F. and Martin C.V. (1987). Mosquito problems in sewage treatment plants using aquatic macrophytes in California. In *Proceed. & Papers of the 55th Ann. Conf. of the California Mosquito and Vector Control Association*, pp. 87–91.

Hopkins, G.H.E. (1952). *Mosquitoes of the Ethiopian Region. Part 1: Larval bionomics of mosquitoes and taxonomy of culicine larvae*, Brit. Mus. Nat. Hist. 2nd edn., London.

Kengne Noumsi I.M (2000). *Evaluation d'une station d'épuration des eaux usées domestiques par lagunage à macrophytes à Yaoundé: Performances épuratoires, développement et biocontrôle des Diptères Culicidae*. Thèse Doctorat 3e Cycle, Univ. Yaoundé I, Cameroon.

Kivaisi, A.K. (2001). The potential for constructed wetland for wastewater treatment and reuse in developing countries. A review. *Ecol. Eng.*, **16**, 545–560.

Manga, L., Robert, V., Messi, J., Desfontaine, M. and Carnavale, P. (1992). Le paludisme urbain à Yaoundé Cameroun: Etude entomologique de deux quartiers centraux. *Mem. Soc. R. Belge Ent.*, **35**, 155–162.

Russell, R.C. (1999). Constructed wetlands and mosquitoes: Health hazards and management options – An Australian perspective. *Ecol. Eng.*, **12**, 107–124.

Reduction of odors from a facultative pond using two different operating practices

A. Truppel*, J.L.M. Camargos*, R.H.R. da Costa*** and P. Belli Filho***

*Fundação Nacional de Saúde de Santa Catarina. FUNASA-SC Rua Marinheiro Max Schramm, 2179 - Estreito – Florianópolis - SC - CEP: 88095-001 – Brazil (E-mail: atruppel@terra.com.br)
***Department of Sanitary and Environmental Engineering, Federal University of Santa Catarina, Brazil
(E-mail: rejane@ens.ufsc.br; belli@ens.ufsc.br)

Abstract This paper presents the results of a proposed intervention to deal with the odor problems of a sewage treatment works (STW), which is located near a populated area. The STW consists of a facultative pond. Since this pond functions under close to anaerobic conditions, unpleasant odors are emitted. In this respect, two possible ways to deodorize the pond were evaluated. Firstly, the recirculation of effluent using 1/6 of the flow stream followed by aeration of the pond with a reduced power aerator. In order to study the efficiencies of the deodorization methodologies chemical analyses of the gases NH_3 and H_2S, olfactometric analyses and evaluation of the environmental perception of the population in relation to the odors originating from the STW, were carried out for each experimental situation. The results showed a significant reduction in odors when aeration with reduced power equipment was utilized in combination with recirculation of effluent in the pond. Reductions in emissions of H_2S from 0.1345 mg/m^3 to 0.0083 mg/m^3 and of NH_3 from 0.021 mg/m^3 to 0.0073 mg/m^3 were obtained. To analyze the behavior of the pond, its planktonic community was investigated, with a difference in species for the situations with and without odor being observed.
Keywords Facultative pond; odor; olfactometry; recirculation; surface aeration

Introduction

Odors originating from wastewater treatment systems present undesirable situations and conflict between sanitary institutions and communities. Among the potential sources of emission of these nuisances are sewage treatment works (STW). Among the various types of treatment processes, stabilization pond systems are commonly used in Brazil. The ponds may emit odorous gaseous compounds when anaerobic, or when they present some instability in their functioning, if they are facultative or maturation ponds. Most of the compounds responsible for the bad odors from this source, belong to the sulfur (hydrogen sulfide and mercaptanes) and nitrogen (ammonia and ammonium) families. Second in importance are some volatile organic compounds such as phenols, organic acids, esters, aldehydes and alcohols. Various methodologies may be adopted for the control or reduction of odors. Belli *et al.* (2001) and Stuetz and Frechen (2001) present several possibilities for intervention to reduce the odors emitted by a STW. The measures may be implemented to control and/or treat odors through corrective and/or preventative action at the source of the emissions.

The odors can be reduced by simple operating practices. Surface aeration has been shown to be efficient at laboratory-scale (Zhang *et al.*, 1997) and at large-scale (Schulz and Barnes, 1990) for anaerobic ponds treating agricultural wastewater. The recirculation of an effluent rich in dissolved oxygen and algae from a facultative or maturation pond located late in the pond series has previously been used to overcome odor problems associated with primary anaerobic ponds (Pescod, 1996). However, recirculation has been more frequently used to solve odor problems and increase efficiency in primary facultative ponds (Shelef and Kanarek, 1995).

This paper presents evaluations of alternatives to control the odors emitted from a sanitary sewage treatment system, consisting of a facultative pond. Two operating practices were tested: the recirculation of the pond effluent, and the recirculation of the combined effluent with the use of low density aeration at the pond surface.

Materials and methods

Characterization of the study area

This study was carried out at the sanitary sewage treatment station of the São Ludgero municipality, Santa Catarina, in the south of Brazil, which serves a population of 3,200 people. The station comprises pre-treatment followed by a facultative pond, with baffles, with an area of 7,000 m^2 and a depth of 1.20 m. Figure 1 shows a treatment system plan with the position of the aerators and an indication of the recirculation of effluent.

Methodologies of deodorization

For the deodorization two alternatives were utilized: (a) recirculation of the facultative pond effluent to the entrance of the pond itself was introduced, using 1/6 of the treatment flow stream; and (b) recirculation of effluent was then combined with the installation of two mechanical aerators, placed at the points with the greatest odor problems and the formation of dead zones, determined *in loco*. In order to recirculate the effluent to the pond entrance a pump was used, with a pump flow of 1/6 of the total station flow stream and with a power of 3CV. The equipment was installed above the pond exit flow meter and the effluent was recirculated to the pond entrance. In the second stage, introducing aeration, two surface aerators, with paddles, with an aeration rate of 16 kg air/h and power of 2 HP were installed.

Monitoring of the system operation

The system was monitored for a period of eight months, when evaluations of the liquid phase of the pond influent and effluent, odorous gas evaluation and application of questionnaires to identify the perception of the community who live around the STW in relation to the nuisances were carried out.

Evaluation of the pond liquid phase. In this study only the results obtained for the behavior of the BOD$_5$ in the pond influent and effluent and the planktonic community are presented. The BOD$_5$ analyses were carried out by the manometric method, using Hach

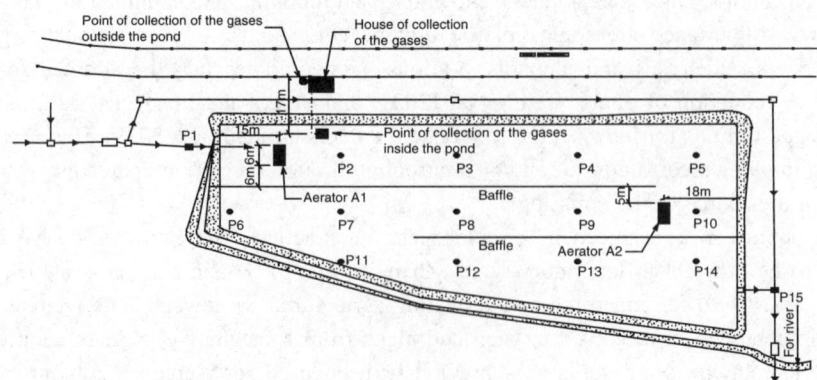

Figure 1 Treatment system plan

equipment, and following the *Standard Methods* (1998) recommendation. The analyses of the planktonic community were carried out utilizing a Olympus BX-50 microscope.

Evaluation of the deodorization methods. In order to evaluate the efficiency of the deodorization for each operating practice, techniques for the quantification of the chemical constituents (H_2S and NH_3) and evaluation of odors were employed. The gases H_2S and NH_3 were sampled at specific points in the STW, through absorption in $HgCl_2$ and HCl solutions, for the respective analyses, following the recommendations of Belli (1995). The H_2S was determined through the quantification of precipitated mass with mercury, while the NH_3 was evaluated through the distillation and titration expressed in the form of NH_4^+.

For the evaluation of odors, an olfactometric study was developed with the application of two analytical methodologies, following the recommendations of the standards AFNOR NF X 43 (1989) and VDI (1993). Samples were collected in 45-litre Tedlar bags, at a point above the pond surface and another at the extremity of the STW area. The samples were analyzed by an 'olfactometry jury', comprised of people who were trained to provide olfactory responses regarding each deodorization process, utilizing solutions of 1-butanol as a reference for different levels of odor. Another methodology for evaluating odors used a permanent olfactometric jury composed of people from the community neighboring the STW. Questionnaires with the aim of evaluating the hedonic characteristics of the population near the STW, totally approximately 30 residences, were applied every five days.

Results and discussion
BOD_5 behavior

The treatment with recirculation of pond effluent showed an average efficiency for the reduction of BOD_5 of 77% for $BOD_{5,total}$ and of 74% for $BOD_{5,filtered}$. In this experimental stage, the pond was operating with an average superficial organic load of 270 kg BOD_5/ha.day and a maximum of 440 kg BOD_5/ha.day. In the treatment with aeration combined with effluent recirculation the results obtained for the BOD_5 removal were 81% for $BOD_{5,total}$ and 77% for $BOD_{5,filtred}$. In this stage, the pond operated with an average superficial organic load of 336 kg BOD_5/ha.day and a maximum of 583 kg BOD_5/ha.day. These results indicate that the aeration aided the elimination of the carbonaceous fraction, along with contributing to the deodorization, even when the applied organic loads were at the maximum limit of the working conditions for the facultative pond not to enter into an anaerobic state (Mara and Pearson, 1998).

Treatment of odors

Hydrogen sulfide (H_2S). In the operational period with recirculation of effluent the emissions of hydrogen sulfide (H_2S) showed greater concentrations than those shown with the deodorization which employed aeration. Figure 2 shows this behavior. For the first experiment, average concentrations of H_2S of 0.054 mg/m^3 and 0.1345 mg/m^3 were obtained inside and outside the pond, respectively, while during the period with aeration the average emissions of H_2S were 0.0158 mg/m^3 and 0.0083 mg/m^3 inside and outside the pond, respectively. The values showed that the deodorization with a low rate of aeration was efficient in the reduction of H_2S. The concentrations of H_2S obtained only with the recirculation of effluent were above the perception limits, whereas those determined during the stage with aeration had values below or near the olfactory detection limits. According to Belli *et al.* (2001), in this situation an oxidation of H_2S to SO_4^{2-} occurs. The ambient operational temperature could be another factor which

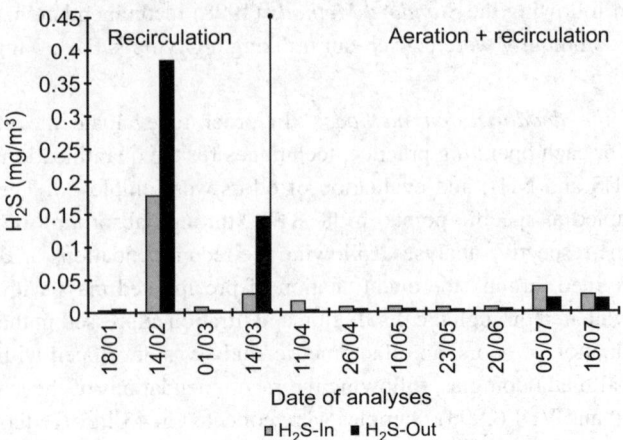

Figure 2 Results of H_2S

contributed to the better aeration efficiency, since it was applied at a time of the year when temperatures were not so high ($\sim 15\,°C$).

Ammonia (NH_3). The behavior for the ammonia (NH_3) emissions of the pond during the application of the two practices is shown in Figure 3. An evolution over a time similar to that for the H_2S is observed. The NH_3 concentrations emitted were higher during the application of recirculation of the pond effluent during the recirculation with aeration. In this period the average concentration inside the pond was $0.2903\,mg/m^3$, whereas outside was $0.0208\,mg/m^3$. During the aeration ammonia concentrations inside and outside the pond of $0.0957\,mg/m^3$ and $0.0073\,mg/m^3$, respectively, were obtained. The factors responsible for the reduction in NH_3 emissions during the aeration may be related to the reduction in pH (in relation to the experiment with only effluent recirculation), lower ambient temperatures and the maximization of NH_4^+ transformation into NO_3^- in relation to the NH_3 conversion. The efficiencies obtained in the H_2S and NH_3 reductions were 77% and 98%, respectively, outside the pond area (extremity of the STW area), at the point of interest for environmental legislation.

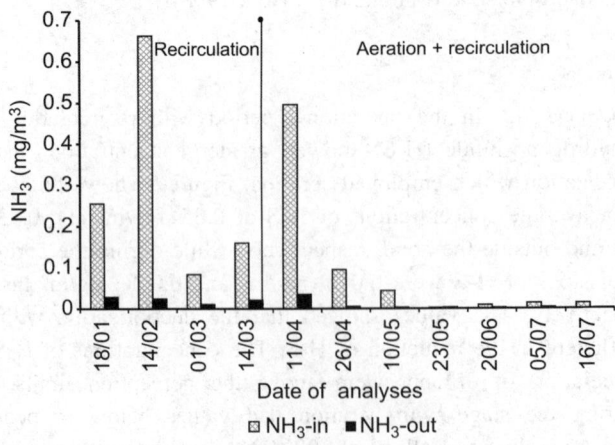

Figure 3 Results of NH_3

Olfactometric evaluation

The olfactometry confirmed the results obtained for H$_2$S and NH$_3$. Figures 4 and 5 show the responses for the odor intensities, obtained through an olfactometric jury trained in deodorization detection techniques. During the deodorization with effluent recirculation, the jury indicated the predominance of very strong to strong odor intensities inside the pond and strong intensity at the extremity of the STW area. For the application of recirculation and aeration odor intensities with a tendency towards strong to medium for the inside of the pond were obtained, and odor levels of medium to weak were obtained outside the pond.

Figure 6 shows the results of the community response to the odor intensity during the experimental period. For the experiment with effluent recirculation, 18% of the residents felt that the odor was extremely inconvenient and for 26% it was not inconvenient. During the experiment with aeration, only 3% of the residents responded that the odor

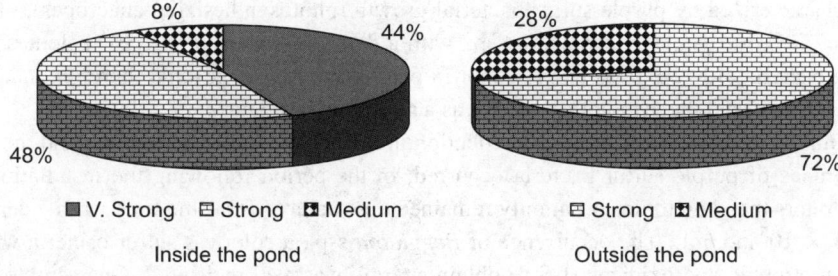

Figure 4 Odor intensity with effluent recirculation

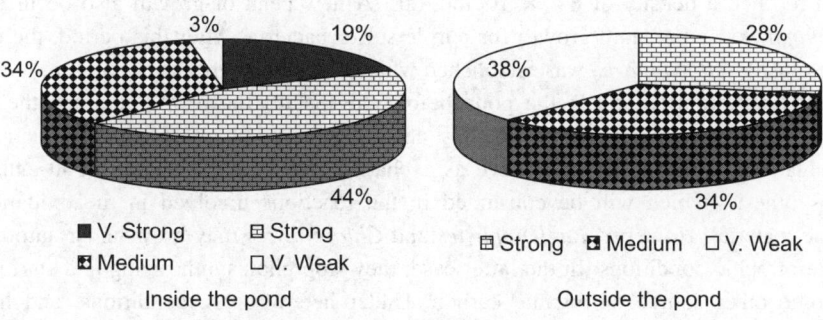

Figure 5 Odor intensity with effluent recirculation and pond aeration

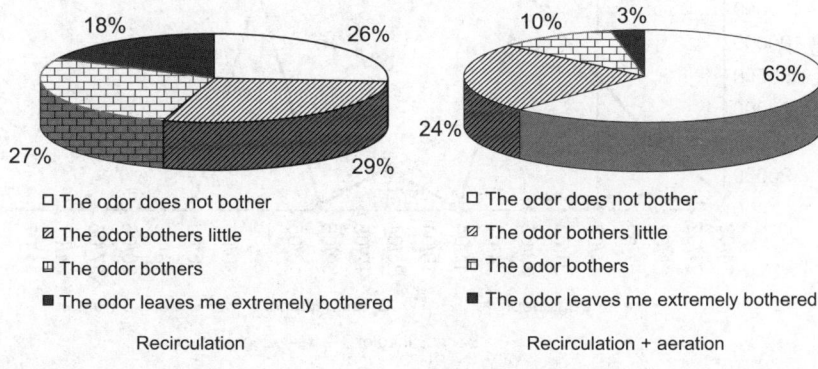

Figure 6 Responses of the community during the two experimental phases

was extremely inconvenient and 63% of them suggested that the odor didn't bother them at all. According to the perception of the community the application of aeration provided an improvement to their comfort. The results indicated the potential for applying aeration, with reduced power aerators, for the elimination of odors in situations such as those analyzed in this study. Paing *et al.* (2001) showed that the recirculation seem to be the more interesting solution to reduce odor emission in anaerobic ponds, with a good efficiency, low cost and simplicity of operation.

Dynamics of the planktonic community

The dynamics of the planktonic community of the stabilization pond studied were related to the instability of its functioning. The plankton was dominated by *Chlorella vulgaris* (Chlorophyte) with densities varying between 2.6×10^5 and 7.4×10^6 ind./mL. The period before the installation of the aerators was characterized by the occurrence of a great density of *Euglenophytes*, with values reaching 1.3×10^5 ind./mL. This period was also characterized by purple sulfur bacterial growth (photosynthesizing anaerobes), of the genus *Thiocapsa* ssp. and *Thiopedia* ssp., with a density of up to 9.0×10^3 colonies/mL. These bacteria utilize H_2S as an H_2 donor in photosynthesis suggesting, therefore, anaerobic conditions and the availability of H_2S as a substrate, Madigan *et al.* (1997).

With the installation of aerators a reduction in the density of *Euglenophytes* and the disappearance of purple sulfur bacteria occurred. In the period following the installation of the aerators the planktonic community remained destructured for some time, with a density of 3.1×10^5 ind./mL. The occurrence of *Beggiatoa* ssp., a colorless sulfur bacteria which is chemotrophic and oxidizes H_2S to obtain energy, was also registered. One sample was shown to be atypical, suggesting a sudden new change in the environment. In this period the plankton was dominated (Figure 7) by *Polytoma* sp, an achlorophylic chlorophyte which reached a density of 8.4×10^6 ind./mL. A new peak of growth also occurred for *Euglenophytes* and a small growth for purple sulfur bacteria. After this period, the dominance of *Chlorella vulgaris*, was established which reached its growth maximum.

In the conditions found in the pond before the installation of the aerators, the algae present were not able to supply the O_2 necessary to maintain a healthy functioning. The algae in stabilization ponds have as a characteristic the production of O_2, through photosynthesis, which will be consumed in the reactions involved in the oxidation of organic material. However, *Euglenophytes* and *Chlorophytes* may grow under autotrophic or heterotrophic conditions. In the latter case, they stop photosynthesizing and start utilizing some other source of organic carbon. Under heterotrophic conditions, and having

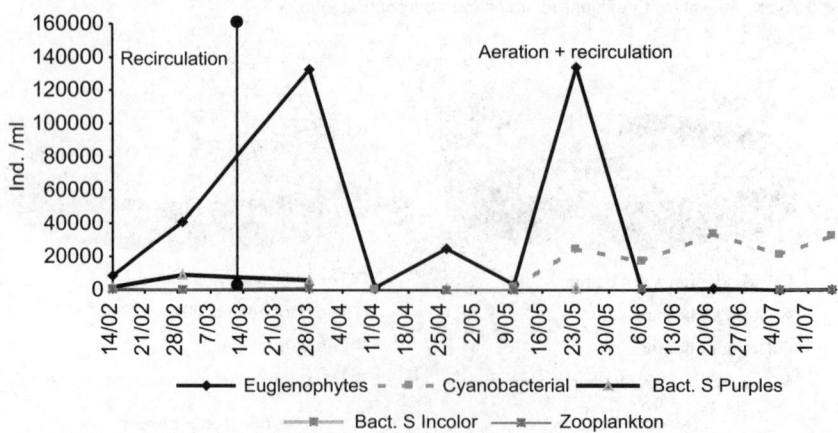

Figure 7 Results for the planktonic community

acetate as a substrate, Cohen and Post (1993) observed experimentally for *Chlorella vulgaris,* an increase in the respiratory activity and reduction in the photosynthetic potential, with an increase in the demand for oxygen. Some algae found in the environment studied did not contribute more to the O_2 production since they definitively lost the photosynthetic capacity, such as *Polytoma* sp and *Hyaloraphidum contortum* (chlorophytes).

Conclusions

Based on the analysis and discussion of the results the following can be concluded.

The pond effluent recirculation alone, with a flow stream of 1/6, was not sufficient for its deodorization.

The aeration integrated with effluent recirculation provided efficient reduction of H_2S and NH_3 emitted from the pond, in the order of 94% and 65%, respectively.

The perceived evaluation of the community in relation to each technology tested was fundamental to defining the effluent aeration/recirculation with the best deodorization option.

The planktonic community showed different behaviors in the periods before and after the installation and functioning of the aerators. The presence of purple sulfur bacteria and the high density of *Euglenophytes* were related to the period of inefficient functioning of the pond (occurrence of odors). On the other hand, high densities of *Chlorella vulgaris* and a reduction in other groups were related to the efficient functioning of the pond (absence of odors).

References

AFNOR (1989) *NF X 43-101: Qualité de l'air. Mesure de l'odeur d'un effluent gazeux. Détermination du facteur de dilution au seuil de perception.*

Belli, F.P. (1995). *Stockage et odeurs des déjections animales – cas du lisier de porc.* PhD thesis, Université de Rennes, France.

Belli, F.P., Costa, R.H.R., Gonçalves, R.F., Couracci, F. and Lisboa, H.M. (2001). Tratamento de Odores em Sistemas Sanitários. In *Pós-Tratamento de Efluentes de Reatores Anaeróbios*, PROSAB 2, FINEP, Brazil, pp. 455–488 .

Cohen, I., and Post, A.F. (1993). The heterotrophic connection in a photoautropic *Chlorella vulgaris* dominant in wastewater oxidation ponds. *Wat. Sci. Tech.*, **27**(7–8), 151, 155.

Madigan, N.M.T., Martinko, J.M. and Parker, J. (1997). *Brock Biology of Microorganisms*, 8th edn., Prentice Hall, Upper Saddle River, NJ, USA.

Mara, D.D. and Pearson, H.W. (1998). *Design Manual for Waste Stabilization Ponds in Mediterranean Countries*, European Investment Bank, Mediterranean Environmental Technical Assistance Programme. Lagoons Technology International, Ltd., Leeds, UK.

Paing, J., Sambuco, J.P., Costa, R.H.R. and Picot, B. (2001). Reduction of odor from anaerobic pond with different operating practices. *Preprints of 1st IWA International Conference on Odour and VOCs: Measurement, Regulation and Control Techniques.* Sydney, Australia, March 25–28, pp. 579–587.

Pescod, M.B. (1996). The role and limitations of anaerobic pond systems. *Wat. Sci. Tech.*, **33**(7), 11–22.

Schulz, T.J. and Barnes, D. (1990). The stratified facultative lagoon for the treatment and storage of high strenght agricultural wastewaters. *Wat. Sci. Tech.*, **22**(9), 43–50.

Shelef, G. and Kanarek, A. (1995). Stabilisation ponds with recirculation. *Wat. Sci. Tech.*, **31**(12), 389–397.

Standard Methods for the Examination of Water and Wastewater (1998). 20th. Edn, American Public Health Association/American Water Works Association/Water Environment Federation, Washington, DC, USA.

Stuet, R. and Frechen, F.B. (2001). *Odour in Wastewater Treatment, Measurement, Modelling and Control*, IWA Publishing, London, UK.

VDI 3883 part 2 (1993) *Effects and Assessment of Odours- Determination of Annoyance Parameters by Questioning- Repeated Brief Questioning of Neighbour Panellists.*

Zangh, R.H., Dugba, P.N. and Bundy, D.S. (1997). Laboratory study of surface aeration of anaerobic lagoons for odor control of swine manure. *Am. Soc. Agr. Eng.*, **40**(1), 185–190.

Determination of the sedimentation constants for total suspended solids and the algal component in a full-scale primary facultative pond operating at high wind velocities under tropical conditions

L.B. Saraiva*, C.G. Ribeiro Meneses*, H.N. de Souza Melo*, A.L. Calado Araújo** and H. Pearson**

*Department of Chemical Engineering, University of Rio Grande do Norte, Brasil
(E-mail: *libertalamar@eq.ufrn.br; carlagracy@eq.ufrn.br; henio@eq.ufrn.br*)
**LARHISA, Civil Engineering, Federal University of Rio Grande do Norte, Brasil
(E-mail: *howard@ct.ufrn.br; acalado@cefet.rn.br*)

Abstract This study evaluated the amount, distribution and sedimentation constant of solids in a full-scale primary facultative pond operating mostly under high wind conditions and the contribution made by the algal biomass. Solids deposition rates were measured using sedimentation traps placed in the inlet and outlet zones of the pond. Most sludge accumulation occurred, not surprisingly, in the inlet zone A_1 with a sludge volume of 9072 m^3 accumulating over an operating time of approximately 3 years. However, sludge deposition within this zone was uneven and affected by wind action. Mean proportionality constant (K) values for solids sedimentation were 3.02 and 5.70 for depths of 50 cm and 100 cm respectively for A_1. In contrast in zone A_3, (the outlet zone), reduced K values of 1.38 and 3.22 were obtained for depths of 50 cm and 100 cm respectively. The algal sedimentation constant varied from 0.8 d^{-1} in zone A_1 to 0.02 d^{-1} in A_3. These data suggest that in this large facultative pond the wind, blowing predominantly from the direction of the outlets towards the pond inlets, had a greater influence on solids deposition than the bulk hydraulic flow and also kept the pond completely mixed for most of the time.
Keywords Algal biomass; facultative ponds; sedimentation constants; suspended solids

Introduction

The accumulation of organic solids in the bottom sediments of waste stabilization ponds represents a reservoir of organic substances that can either be anaerobically digested away with the release of methane from the pond surface and or solubilized and recycled into the water column to act as substrates and nutrients for further microbial activity.

According to Nelson (2002), to determine the amount of sediments, the solids in the influent wastewater, the mass of bacteria and the maximum algae mass per unit area occurring in the surface layers of the pond must all be considered.

The methods currently used to determine the amount and rate of sedimentation of suspended solids in ponds are based on the concept of sedimentation in an ideal tank or by using computational models of flow dynamics. However these do not properly evaluate the intervening factors that affect sedimentation at full-scale (Krishnappan and Marsalek, 2002).

The present study seeks to determine the quantity, distribution and rate of accumulation of solids and the sedimentation constants for both total suspended solids and that of the contributing algal biomass by using sedimentation traps in a full-scale, tropical primary facultative pond operating most of the time under high wind conditions.

Methods

The pond system (total area 11 ha) comprises a facultative pond (5.5 ha, depth 2.0 m) and 2 maturation ponds (2.8 ha each, with depths of 1.5 m) in series. It treats domestic sewage from the tourist resort of Ponta Negra, in the coastal city of Natal, Rio Grande Do Norte, Northeast Brazil. The climatic conditions are tropical, with incident mean annual solar radiation at the surface of 5900 W/h/m^2, a mean water temperature of 27 °C but also with wind speeds reaching 3.3 m/s during daylight hours but greatly reducing at night, blowing in the direction 130° north i.e. in the case of the facultative pond (but not the maturation ponds), from the direction of the outlets towards the pond inlets.

For the purpose of this study the trapezoidal facultative lagoon was divided into three sections or zones along its length starting from the inlet end namely A_1, A_2, and A_3 with areas of 14175, 17168 and 25831 m^2 respectively. The rate of deposition of particulate matter in the pond was determined every 15 days in the A_1 and A_3 zones (i.e. in the inlet and outlet zones) using sedimentation traps suspended vertically in the water column at depths of 50 cm and 100 cm at 12 m intervals across the complete width of the pond in the two zones. The depth of the sludge layer was determined using a portable echo sounder and by the empirical "White towel test" method (Mara and Pearson, 1998) at 30 points distributed across all three sections of the facultative pond (see Meneses et al., 2005), located using a Global Positioning System (GPS). Sludge core samples were also collected and subsequently sectioned in the laboratory for analysis for total, fixed and volatile solids (APHA, 1998).

Sedimentation rate was calculated from the suspended solids concentration in the traps in relation to trap area. As pond solids sedimentation is proportional to the inlet solids load, the proportionality constant called the sedimentation constant was calculated from the equations:

$$Sa = K.Q.C_{SSTi} \qquad (1)$$

Sa is the solids sedimentation rate (kg.d^{-1}); Q is the inflow (m^3.d^{-1}); C_{SSTi} is the total suspended solid concentration that enters the lagoon and K is the constant of proportionality.

Chlorophyll a and phaeophytin concentrations were determined for water and sludge column samples taken from six locations equidistant along the length of the pond (APHA, 1998). The algae sedimentation rate was calculated according to Nelson (2002):

$$Sa = \chi A K \qquad (2)$$

χ is the maximum concentration of chlorophyll a in the active band in mg/L, in this case in the water column of the lagoon assuming complete mixing; A is the area of influence in m^2; K is the algal sedimentation constant in d^{-1}.

Results and discussion

Sludge distribution in the pond showed that sludge accumulation was, not surprisingly, greatest near the inlet region i.e. in the first 100 m of the pond (zone A_1) (Figure 1), with the thickness of the sludge layer varying from 0.01 to 1.10 m and with a mean sludge concentration of 94.63 g/L in zone A_1. However sludge accumulation was uneven with most accumulation occurring in the bottom right-hand quadrant of Figure 1, i.e. corresponding to the area of pond zone A_1 receiving the prevailing wind. Surface water movement (>30 cm in depth) was estimated to occur at a velocity of 182 m/h from the outlets towards the inlets, that is in a counter flow direction compared to the main body of the water as determined in simple drogue experiments with oranges and coconuts which

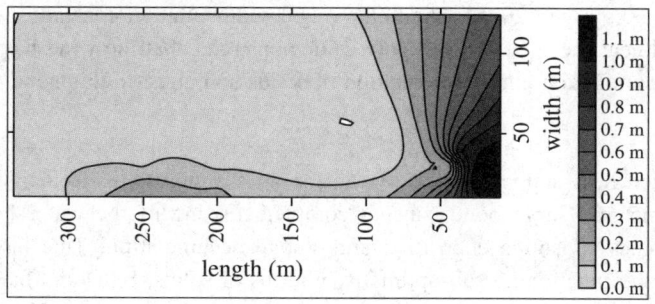

Figure 1 Sludge layer distribution along the primary facultative pond. The inlets are marked on the right-hand side of the figure

accumulated in this region (see Meneses et al., 2005). This suggests that sludge accumulation was more affected by wind speed (1–4 m/s) and direction than by bulk water flow entering the pond through the equal and efficient splitting of the influent flow between the inlets discharging 50 cm above the pond base and extending some 20 m into the pond. The calculated sludge volume for zone A_1 was of 9072 m^3 which had accumulated over 36 months of operation.

The proportionality constant (K) for solids sedimentation was determined using the mean flow of 5000 m^3/d with an influent mean total suspended solids concentration of 435 mg/L of which 362 mg/L were volatile suspended solids.

Using equation 1 the mean K values for zone A_1 i.e. in the first third of the pond were 3.02 for a depth of 50 cm and 5.70 for a depth of 1.0 m. Sedimentation measurements in zone A_3 near the outlet zone gave lower mean values of K of 1.38 and 3.22 for depths of 50 cm and 100 cm respectively. In the outlet section (A_3) the value for total suspended solids in the water column was 272 mg.L^{-1} and for volatile suspended solids 242 mg.L^{-1}.

The algal sedimentation constant was determined from the chlorophyll a concentration using equation 2, and was 0.80 d^{-1} in area A_1 reducing to 0.03 d^{-1} for A_3. Figure 2 shows the distribution of the algal K values along the pond.

The values for the proportionality constant (K) for the sedimentation of solids were similar at depths of 0.5 and 1.0 m since the values approximately doubled for a doubling of the depth. This tends to strengthen the theory that the hydraulic regime in the pond is one of complete mixing and suggests that in this large pond the speed and direction of the wind were important factors, possibly more so than geometry and inlet positions, in influencing the hydraulic regime. The lack of stratification was confirmed by the homogeneous nature of the data for chlorophyll a concentration and oxygen with depth in the

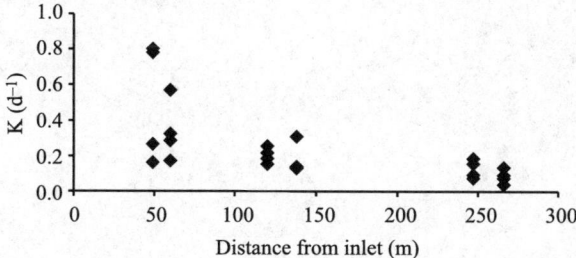

Figure 2 Variation in the algal sedimentation constant with distance from the inlet, based on chlorophyll a values for algal biomass

water column (Meneses *et al.*, 2005). The higher algal sedimentation constant in the first section of the pond near the inlet zone (Figure 2) demonstrates that the wind has a direct influence on the accumulation and sedimentation of solids and algae in this pond.

Conclusions

The results presented here support the notion that relatively high wind velocities affect the hydraulic regime in a large pond causing complete mixing of the water body and affecting the efficiency of solids deposition and sludge accumulation. This large pond system is known to be functioning sub-optimally in terms of solids, BOD and faecal coliform removal despite a correct surface organic loading of approximately 350 kg BOD/-ha/d on the primary facultative pond and a retention time of 22 days. In fact total suspended solids removal was only 37.5%. It would seem that there is a case for using smaller ponds in parallel rather than large single facultative ponds to reduce the adverse impact of the wind on treatment efficiency in such situations.

References

APHA (1998). *Standard Methods for the Examination of Water and Wastewater*, 20th edition, American Public Health Association, Washington, DC.

Krishnappan, B.G. and Marsalek, J. (2002). Modeling of flocculation and transport of cohesive sediment from an on-stream stormwater detention pond. *Wat. Res.*, **36**(15), 3849–3859.

Mara, D.D. and Pearson, H.W. (1998). *Design Manual for Waste Stabilization Ponds in Mediterranean Countries*, Lagoon Technology, Leeds, UK.

Meneses, C.G.R., Saraiva, L.B., Melo, H.N.S., Melo, J.L.S. and Pearson, H.W. (2005). Spatial and temporal variations in BOD and algal concentration and total organic matter biodegradation constants in a facultative tropical waste stabilization pond system mixed by wind action. *Wat. Sci. Tech.*, **51**(12), 183–190.

Nelson, K.L (2002). Development of a mechanistic model of sludge accumulation in primary wastewater stabilization ponds. *5th International IWA Specialist Group Conference on Waste Stabilization Ponds. Auckland, NZ.* Conference Papers. v 2, pp 551–560.

Some observations on the effects of accumulated benthic sludge on the behaviour of waste stabilisation ponds

C.J. Banks*, S. Heaven* and E.A. Zotova**

*School of Civil Engineering and the Environment, University of Southampton, Southampton SO17 1BJ, UK
(E-mail: cjb@soton.ac.uk, sh7@soton.ac.uk)
**BG Chair of Environmental Technology, AIPET, 126 Baytursynov Street, Almaty 480013, Kazakhstan
(E-mail: bgchair@aipet.kz)

Abstract The effect of accumulated bottom sludge on water column characteristics was studied in two pilot-scale ponds. Parameters measured were ammonia, nitrate, phosphate, COD, suspended solids, dissolved oxygen (DO), temperature and light intensity. The de-sludged pond showed a stronger correlation between DO, light intensity, nutrients and suspended solids with the controlling factor being availability of nitrogen. This was less apparent in the pond with sludge where nutrient levels were higher and more complex mechanisms controlled biomass concentration. Water column characteristics in the two ponds converged rapidly in 7–10 weeks, however, due to accumulation of fresh sludge.
Keywords Benthic feedback; sludge; waste stabilisation ponds

Introduction

One of the prime functions of a facultative waste stabilisation pond (WSP) is to act as a reservoir for the slow accumulation of sediments resulting from death and deposition of the pond flora and fauna and the influent suspended solids. In an ideal design the rate of sediment accumulation would be balanced by that of anaerobic decomposition in the sludge layer, with both soluble organic and inorganic nutrients returning to the water column. This return imposes an additional biochemical oxygen demand on the pond, and affects its performance in terms of nutrient, suspended solids and dissolved oxygen concentrations. Some of the products of anaerobic stabilisation, such as methane and sulphides, may also affect the microbiology of the water column.

A great deal of fundamental research has been carried out to elucidate the mechanisms and quantify the parameters associated with benthic feedback. Much of this has been applied to modelling natural water bodies such as lakes and estuaries. In their review of modelling techniques, Reckhow and Chapra (1999) point out the inappropriateness of simple zero and first-order approaches for systems with a high nutrient loading. Research specifically on WSP systems has been more limited. Bryant and Bauer (1987) reviewed factors affecting feedback of phosphorus and nitrogen, investigated the behaviour of sludge from an aerated stabilisation basin, and produced a model simulating feedback to the water column. Lumbers and Andoh (1987) modelled data from a pond system in the USA and concluded that in the hottest months of the year the benthic feedback was equal in magnitude to the incoming load. Giraldo and Garzon (2002) developed a compartmental model to predict soluble BOD concentrations in facultative ponds. The model was calibrated with data from a pilot-scale system in Colombia and results suggested that solubilisation and return of organic matter from pond sediments to the aerobic layer has a significant influence on effluent BOD. Di Toro *et al.* (1990) carried out extensive work on modelling sediment oxygen demand due to fluxes of methane, nitrogen and ammonia: it is estimated that loss of carbon as methane can account for as much as 30% of influent biochemical oxygen demand.

An experiment was set up to compare water column characteristics from two pilot-scale ponds, one of which had been de-sludged. The work aimed to gather data on the scale of the effect, and its significance in practical terms for WSP operation.

Materials and methods
Small-scale ponds construction and environmental conditions

The experiment was carried out using two ponds each with a volume of 550 litres, a water column depth of 0.6 m and a surface area of 0.9 m^2. These consisted of pre-fabricated semi-translucent polypropylene tanks that were externally insulated with 50 mm of polystyrene foam, preventing any light entering through the tank walls. To protect them from rain and provide a more stable temperature profile they were housed in a greenhouse with a southerly aspect to maximise the incident sunlight in Southampton, UK where the ponds were located. Because of the low light intensities and short daylight hours during certain months each pond also received supplemental illumination from an array of halogen floodlights providing a source wattage rating to each pond of 1600 W. These gave a surface illumination of the ponds equivalent to 300 W m^{-2}. To prevent excessive localised surface heating of air above the ponds they were ventilated continuously using a 375 mm blade diameter oscillating fan. During the brightest days of summer the combined artificial and natural irradiance could reach 1000 W m^{-2}.

Continuous measurement of light, dissolved oxygen and temperature. Light intensity and temperature were recorded at the surface of each of the ponds and light intensity, temperature and dissolved oxygen concentration were measured at 160, 345 and 545 mm below the surface in each of the ponds. Light intensity was measured using photodiodes (Siemens, type BPW 21) calibrated against a standard photovoltaic cell with an output of 71.4 µV W^{-1} m^{-2}. The photodiode output was recorded in mA as this is more stable under temperature change.

Dissolved oxygen (DO) was measured using a galvanic cell type with a zinc anode, silver cathode and Teflon membrane (Dryden Aqua, UK). The response of the probes is around 6 mV mg^{-1} l^{-1} of DO. The accuracy is usually better than ± 0.2 mg/l and they are self-temperature compensating from 0 to 40 °C. The calibration of the DO probes was checked daily in air by removing the probes from the pond, washing them and suspending them in air for a period of approximately 10–15 minutes. Output was recorded in mV and converted to DO concentration by a single point calibration. Temperature was measured using a type K fine wire exposed junction thermocouple offering a fast response over the temperature range 0–100 °C. Sensor output was internally configured to a direct temperature reading by the data logging equipment software. For temperature, light, and DO readings the probes were continuously sampled using a Data Taker D500 data logger and expansion unit. Under normal operation readings were averaged over a 30 second period and then further averaged to give a stored value for each of the sensors every 10 minutes. During the daily calibration of the DO probes readings were averaged and recorded each minute.

Routine pond feeding and maintenance. The experimental ponds had been running for a period of two years and received a synthetic wastewater that contained (g l^{-1}) semi-skimmed milk, 1.44; freeze-dried blood, 0.057; sterilized bakers' yeast, 0.23; sugar, 0.115; K$_2$HPO$_4$, 0.0056. This produced a feed with a COD of approximately 380 mg l^{-1}, BOD 160 mg l^{-1} and suspended solids of 190 mg l^{-1}. Before feeding each day a volume of 25 litres of pond water was siphoned off, in a period of approximately 15 minutes, from a point near the centre of the water body. Each pond was then batch fed 25 litres of the synthetic wastewater over a feeding period of approximately 1 hour. To account for evaporation, the level of the pond was topped up with clean tap water. This method of feeding resulted in a 22-day hydraulic retention time and a surface loading of 45 kg BOD ha^{-1} day^{-1}, while the batch feeding guaranteed a minimum retention period of 23 hours.

Sampling and analysis. During the experimental period samples were taken on alternate days from both ponds. Suspended solids (SS) were measured by filtration through a pre-dried and weighed GFC filter (Whatman, UK). Ammonia, nitrate, phosphate, alkalinity were measured with a Bran & Luebbe Autoanalyser model 3; filtered Chemical Oxygen Demand (COD) was measured by the closed tube digestion method (*Standard Methods*, 1998). Chlorophyll was determined by filtering a sample through a GFC filter (Whatman, UK) previously dosed with 1 ml of a saturated solution of MgSO$_4$, followed by grinding and treatment with acetone. The resultant colour was measured at 664 and 665 nm using a Cecil Instruments spectrophotometer (3000 series). Absorbance was measured at 664 nm in a colorimeter (Camlab DREL/5). Samples were analysed in duplicate or triplicate. Floating sludge were estimated on a scale of 0–5 where 0 = not present, 1 = isolated small pieces, 2 = <500 ml, 3 = < 1.5 l, 4 = >1.5 l and 5 = surface covered. A similar index was used for scum.

Experimental procedure. At the start of the experiment the top water from each of the ponds was pumped out, using a peristaltic pump and taking care not to disturb any of the accumulated bottom sediment. In this way it was possible to remove and mix together the top waters from the two ponds into a temporary holding tank leaving only 45 litres (approximately) of sludge in each of the ponds. The sludge itself was dark green/black in colour with a gelatinous nodular texture and had accumulated to a depth of 50–75 mm. The sludge was removed from one pond (Pond 2) and left undisturbed in the other (Pond 1). The mixed top water was then returned in equal volumes to each pond, taking care not to disturb the remaining sludge layer in Pond 1. At the beginning of the experiment conditions in both ponds were therefore equal in all respects apart from the presence or absence of a sludge layer. Feeding, monitoring and sampling of the ponds continued as before.

Results and discussion

Apart from some short-term transitional differences, in the two years of operation before the reported experiment the water column characteristics of the ponds were similar. Table 1 gives comparative values for some key parameters in the experimental period, the same periods in year 1 and year 2, and for the whole time of operation excluding any periods in which the ponds were treated differently. Figure 1 shows phosphate concentrations in the 16 months preceding the experiment: while one pond sometimes leads, the two clearly respond in similar ways to changing conditions. It was therefore assumed that any differences observed from the start of the experiment onwards could be attributed to the presence or absence of an accumulated benthic sludge layer.

Immediately after mixing, the suspended solids concentration of the top water from the two ponds was 160 mg l^{-1}. In the eight days following this fell sharply in both ponds to around 110 mg l^{-1}. The reason for this is unclear, but it is possible that after mixing the algal species balance was out of equilibrium. During this period there was good correlation between values of SS, absorbance, phosphate, and maximum and average DO in the two ponds, as shown in Table 2. COD concentrations were similar throughout,

Table 1 Comparative performance of ponds

In mg l^{-1}	SS	COD	NO$_3$	NH$_4$	PO$_4$
Experimental period (July–Nov 2003)					
av P1	106	55	0.10	1.65	1.71
av P2	102	60	0.05	0.13	1.19
R^2	0.34	0.21	0.00	0.27	0.01
Year 1 (July–Nov 2001), before experiment					
av P1	46	74	0.12	0.35	0.34
av P2	50	55	0.11	0.35	0.32
R^2	0.74	0.61	0.64	0.45	–
Year 2 (July–Nov 2002), before experiment					
av P1	61	50	0.26	2.63	1.34
av P2	51	53	0.23	2.28	1.33
R^2	0.75	0.12	0.75	0.48	0.60
Whole period (2001–2003)					
av P1	67	61	0.19	1.63	1.34
av P2	71	55	0.17	1.61	1.25

although the correlation was low. Both ponds showed a strong correlation between average light and average DO concentrations ($R^2 = 0.95$ in Pond 1 and 0.8 in Pond 2). The dominant trend was the fall in SS, but there were slight differences between the ponds: for example SS and chlorophyll were related in Pond 2 ($R^2 = 0.77$) but not in Pond 1 ($R^2 = 0.06$). Small quantities of floating sludge appeared in Pond 1 on six of the eight days (Figures 2 and 3).

From day 10 to day 35 suspended solids concentrations appeared relatively stable with an average of 117 mgl^{-1} SS in Pond 1 and 80 mgl^{-1} in Pond 2. During this period the soluble COD, measured on filtered samples taken from the ponds before the daily feed was added, showed low residual values of 65 and 60 mg l^{-1} in Ponds 1 and 2, respectively. While these values cannot be compared directly with those from the same period in previous years, due to changes in pond operating regime, the treatment efficiency is clearly similar to that of the overall period as shown in Table 1.

The major difference between the ponds during the first 35 days was in concentrations of nitrate, ammonia and phosphate which were significantly lower in Pond 2 than Pond 1 (see Table 2). It is clear that during this time the nutrient concentration in Pond 2 was very low, with available soluble nitrogen in the form of nitrate and ammonia at a level

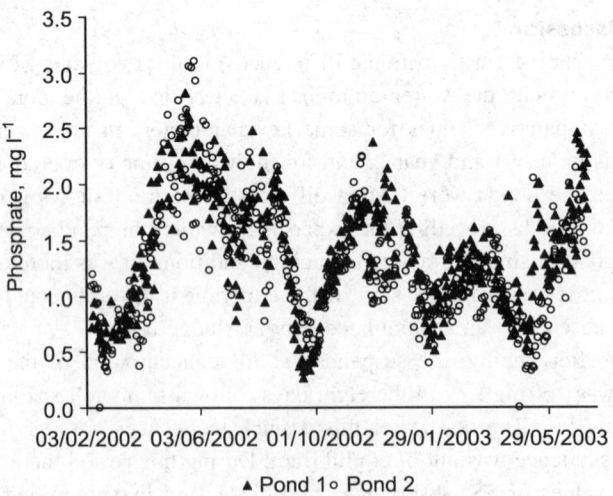

Figure 1 Phosphate concentrations in Ponds 1 and 2 for 16 months before the experiment

Table 2 Average parameters and correlation between ponds during phases of experimental period

mg l^{-1}	SS	Chloro	Abs	COD	NO$_3$	NH$_4$	PO$_4$	Alk	Scum[1]	Sludge[1]	DOmax	DOave
Day 1–9												
av P1	148	0.53	0.32	49	0.10	1.27	1.48	123	1.50	1.43	116%	43%
av P2	148	0.29	0.29	55	0.04	0.17	0.74	177	1.00	0.00	205%	97%
R^2	0.74	0.24	0.98	0.06	0.12	0.04	0.85	0.00	0.00	0.00	0.88	0.89
Day 10–35												
av P1	117	0.53	0.30	67	0.14	2.45	2.02	178	2.00	1.36	82%	28%
av P2	80	0.11	0.16	62	0.06	0.18	0.77	216	1.75	0.09	82%	27%
R^2	0.36	0.10	0.27	0.72	0.00	0.34	0.63	0.43	1.00	0.05	0.39	0.19
Day 36–59												
av P1	83	0.21	0.21	76	0.05	5.79	2.62	223	1.75	2.08	89%	31%
av P2	97	0.22	0.20	80	0.10	0.26	1.39	196	1.58	0.25	81%	26%
R^2	0.00	0.71	0.06	0.11	0.04	0.14	0.55	0.04	0.73	0.09	0.37	0.44
Day 60–143												
av P1	104	0.27	0.25	46	0.09	0.19	1.36	280	1.03	0.30	–	–
av P2	102	0.20	0.23	55	0.04	0.06	1.32	291	1.03	0.10	–	–
R^2	0.39	0.22	0.53	0.01	0.12	0.28	0.00	0.74	0.00	0.21	–	–
Whole experiment												
av P1	106	0.32	0.25	55	0.10	1.65	1.74	243	1.34	0.89	–	–
av P2	102	0.20	0.22	60	0.05	0.13	1.19	252	1.19	0.11	–	–
R^2	0.34	0.03	0.21	0.21	0.00	0.27	0.01	0.62	0.67	0.09	–	–

[1] Values for sludge and scum are based on index 0–5; DO is % saturation; all others mg l^{-1}

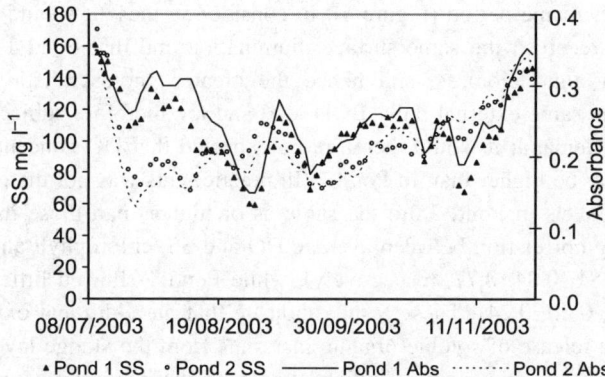

Figure 2 Ponds 1 and 2 suspended solids and absorbance during the whole experimental period

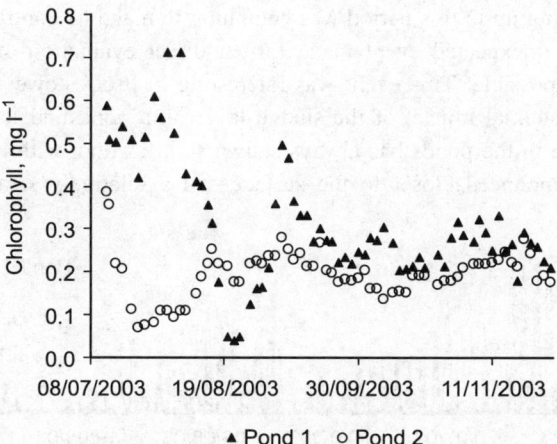

Figure 3 Ponds 1 and 2 chlorophyll during the whole experimental period

likely to be growth-limiting to planktonic algal species. This was not the case in Pond 1 where the average concentration of available soluble nitrogen from day 10–35 was 11 times greater than in Pond 2, and sufficient in itself to explain the difference in suspended solids. The chlorophyll concentration in Pond 1 during this time ranged from 0.4–0.7 mg l^{-1} with an average value of 0.5 mg l^{-1}, whilst in Pond 2 it remained between 0.1–0.2 mg l^{-1} indicating that at least a proportion of the difference in suspended solids between the two ponds could be attributable to algae. Further evidence that ammonia may be the limiting nutrient in Pond 2 comes from the relatively low correlation between levels of ammonia and phosphate ($R^2 = 0.65$ and 0.35 for days 1–9 and 10–35, respectively); the corresponding values in Pond 1 are higher ($R^2 = 0.93$ and 0.55) and may indicate that neither nutrient has been reduced to a limiting value but remains proportional to the amount supplied. Ammonia and phosphate showed negative correlations with SS in Pond 1 in this period ($R^2 = 0.69$ and 0.62), but little relation in Pond 2 ($R^2 = 0.07$ and 0.19). In both ponds the concentration of phosphates was positively correlated to water temperature, especially temperature in the bottom layer ($R^2 = 0.51$ and 0.86 in Pond 1 and 2, respectively).

As the feed to each pond was identical during this period, and as the potential for nutrient uptake was greater in Pond 1 because of the higher biomass density, the difference in soluble nutrient level can only be explained by release from the bottom sediment in Pond 1. Further evidence that the benthic deposits are in a dynamic state of interaction with the water column can be seen by reference to the levels of dissolved oxygen measured over the same period (Figure 4). In considering these it should be remembered that both ponds received the same surface illumination and that Pond 1 has the greater concentration of algal biomass, and hence the greater photosynthetic capacity. Both ponds receive the same external daily BOD load and are therefore subjected to the same external oxygen demand: it would therefore be expected that DO concentrations in Pond 1 would generally be higher than in Pond 2. In practice this was not the case: throughout this period DO levels in Pond 2 are the same as or higher than those in Pond 1. There was also a strong correlation between average DO and SS, chlorophyll and absorbance in Pond 2 ($R^2 = 0.81$. 0.84, 0.77, respectively), while Pond 1 showed little or no relationship ($R^2 = 0.21$, 0.00, 0.24). These results indicate that an additional oxygen demand is being exerted by release of soluble organic materials from the sludge layer. The material is effectively degraded, however, as indicated by both the increased oxygen demand and the fact that the residual COD in the two ponds is almost equal. By day 35 values for most parameters in Pond 2 had risen to levels at or near those in Pond 1, suggesting that fresh sludge building up in this period was beginning to make its contribution.

On day 36 an unexpected event made further direct evaluation of the effect of the benthic sludge impossible. This event was interesting in itself, however, and is reported as it shows the potential impact of the sludge layer on a pond under certain conditions. Water temperature in the ponds has always shown stratification, with diurnal fluctuations that are more pronounced closer to the surface and a relatively steady temperature at

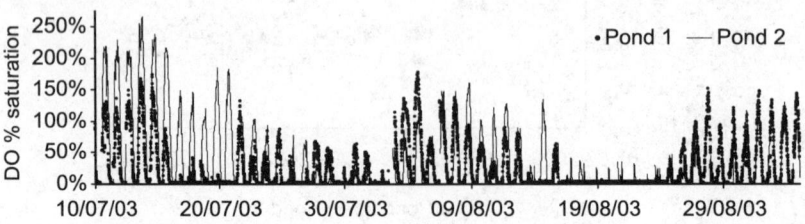

Figure 4 Dissolved oxygen concentrations in Ponds 1 and 2 for days 1–71

the bottom. Typically the surface water temperature rises in the day and then cools in the evening, when it may reach a point where the surface is slightly cooler than the layer below (see Figure 5). This results in a partial turnover, taking dissolved oxygen into the lower layers. During the experiment the weather was very hot, with daytime temperatures in the greenhouse reaching over 40 °C. In the early hours of day 36 the surface and middle layer of Pond 1 came to within 0.5 °C of the bottom temperature, resulting in a larger turnover that brought a significant amount of sludge to the surface (index 5). The conditions causing this were unusual for the ponds under study, despite their relatively shallow depth, and were a result of a prolonged series of hot days that raised the average water temperature. On day 38 the bottom temperature actually exceeded that in the upper layers (Figure 6), and a complete turnover occurred in which most of the sludge rose. The rising sludge in Pond 1 had a pronounced effect on a number of parameters:

- Soluble COD peaked at about $187 \, mg \, l^{-1}$, from previous values of about $70 \, mg \, l^{-1}$.
- Ammonia levels rose from $2-3 \, mg \, l^{-1}$ to $8-9 \, mg \, l^{-1}$ for 10 days before falling to $0.25 \, mg \, l^{-1}$.
- Suspended solids fell to around $60 \, mg \, l^{-1}$ from levels above $100 \, mg \, l^{-1}$.
- Chlorophyll levels averaging $0.5 \, mg \, l^{-1}$ declined very sharply to $0.1 \, mg \, l^{-1}$.

Similar complete turnovers of the water body also took place in Pond 2 but the effects were less marked as the quantity of sludge accumulated since de-sludging on day 1 was small. It was however detectable, with the concentration of ammonia in the water column briefly peaking at $0.86 \, mg \, l^{-1}$ from previous values ranging between $0.1-0.2 \, mg \, l^{-1}$. Soluble COD also rose briefly to $108 \, mg \, l^{-1}$ from $60-70 \, mg \, l^{-1}$. It should be noted that when there is an accumulated benthic sludge layer there is always the potential for rising sludge even when the conditions described above do not occur, although the event may be at a much reduced scale. Throughout the reported experimental period of 143 days there was some correlation between the index estimate for rising sludge in Pond 1, and temperature, especially temperature in the bottom ($R^2 = 0.45$). A very small quantity of floating sludge first appeared at the surface of Pond 2 on day 32 and after this there was a weak correlation between temperature and the rising sludge index, reflecting the consistency of the relationship but also the smaller sludge quantity in the benthic layer. Sludge that has risen tends to sink naturally during the daytime period and usually has no noticeable adverse effect on pond operation.

From day 36–59 the behaviour of the ponds reflected the different degrees of impact of pond turnover. As might be expected, while the COD and DO in Pond 2 were still largely determined by the presence of algae in the water column, Pond 1 was much more

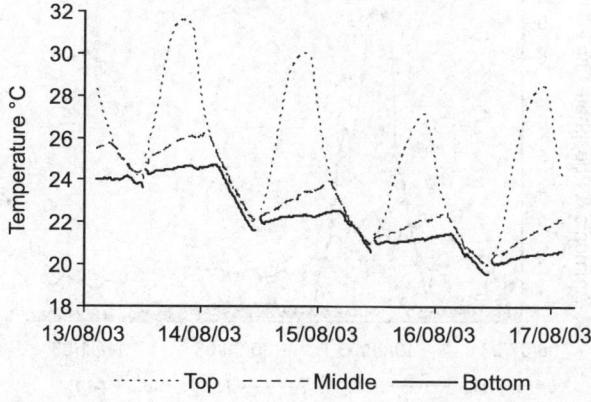

Figure 5 Pond 1 temperatures around turnover

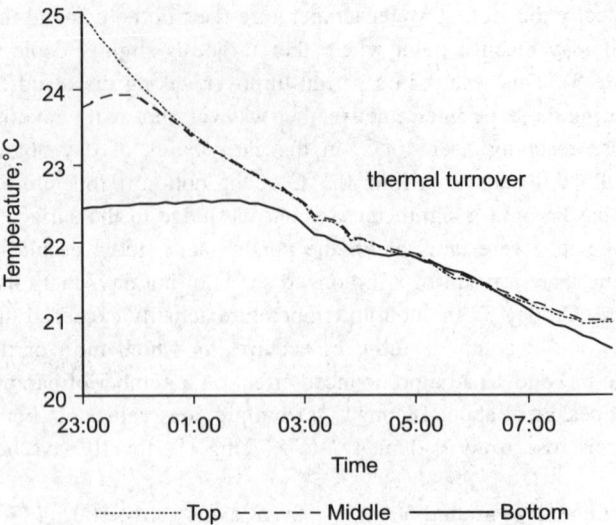

Figure 6 Pond 1 temperatures on turnover

affected by sludge. Pond 2 showed a reduced correlation between suspended solids and average DO ($R^2 = 0.62$), which disappeared in Pond 1 ($R^2 = 0.04$). COD in Pond 2 was negatively related to DO ($R^2 = 0.71$) and to SS ($R^2 = 0.81$), while Pond 1 showed no relationship ($R^2 < 0.02$). There was a weak negative correlation between phosphates and SS in Pond 2, ($R^2 = 0.54$), which may have been due to algal uptake; but no equivalent in P1. In general the correlations between related internal parameters in Pond 1 were weaker indicating that conditions in this period were more variable. The correlation between sludge and bottom temperature in Pond 1 was $R^2 = 0.72$, and there was zero correlation between SS in the two ponds in this period.

After falling from day 36–53 the concentration of SS in Pond 1 then began to rise, reaching 114 mg l^{-1} on day 61. Chlorophyll concentrations which fell even more rapidly also recovered to 0.5 mg l^{-1} on day 59. These results suggest that the rising sludge initially caused die-off or sedimentation of the algae through shading or toxicity, followed by recovery enhanced by the released nutrients. Chlorophyll concentrations in Pond 1 in this period were related to both phosphate ($R^2 = 0.81$) and ammonia ($R^2 = 0.52$). The relationship between chlorophyll and DO was weak but similar in both

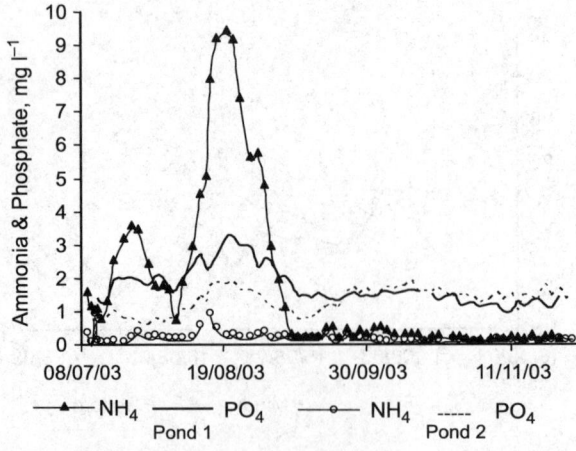

Figure 7 Ammonia and phosphate in Pond 1 and 2 around turnover

ponds ($R^2 = 0.27$ and 0.31). The average DO concentration in P1 was actually fractionally higher than in Pond 2 during this period (see Table 2); DO in Pond 1 fell almost to zero on pond turnover, but rose sharply as the algal population increased stimulated by the release of nutrients.

From day 60 onwards, the behaviour of the two ponds slowly equalised, with similar relationships between parameters in each pond, and individual parameter values moving closer, leading to increased correlation between the ponds (see Table 2). The remaining differences mainly concerned nutrient concentrations (Figure 7), with more nitrate and ammonia in Pond 1 than Pond 2 (Table 2). Bottom temperatures showed some correlation with nitrates in both ponds ($R^2 = 0.69$ in P1, 0.59 in P2), and with ammonia in Pond 1 only ($R^2 = 0.59$). Ammonia in Pond 1 is strongly linked with phosphate throughout, with a correlation coefficient $R2 = 0.75$ for the whole experimental period of 143 days.

An analysis was undertaken of the relationships between some key operational and discharge parameters (SS, DO, COD, nutrients) and driving factors such as light and temperature. Results for DO concentrations predicted by multiple regression from other parameters are shown in Table 3. In Pond 2 SS is an effective predictor of DO, and still more so when combined with light intensity. It is clear, however, that even before pond turnover Pond 1 is a more complex system influenced by many parameters: the most significant for DO concentration from Day 1–71 is ammonia followed by light, bottom temperature and phosphates with $R^2 = 0.34$, 0.26, 0.22 and 0.19, respectively.

The experimental period can therefore be split into four distinct phases: days 1–9 showed initial stabilisation following mixing and de-sludging of Pond 2; days 10–35 showed relatively stable operation and allowed direct comparison between the ponds to determine the influence of the sludge layer; days 36–59 showed the response of Pond 1 to a major rising sludge event; and days 60–143 showed conditions in the two ponds slowly equalising. The results show that the two ponds initially responded in a similar manner to ambient conditions, but the pond containing sludge had a significantly higher concentration of suspended solids and chlorophyll, indicating return of nutrients from the benthic sediments. During this time a fresh layer of sediment was building up in the previously de-sludged pond. The recovery of SS, chlorophyll and other parameters in the first 35 days after de-sludging indicates that the activity of the top layer of freshly deposited sludge contributes significantly to the effect on the water column. In general the results suggest that the presence of sludge does not have an inhibitory effect but in fact contributes to the growth of primary producers under normal conditions. Pond turnover had a dramatic short-term effect which gradually declined and the performance of the two ponds drew closer over a period of weeks. Effectively pond turnover fits within

Table 3 R^2 values for prediction of average DO based on other parameters

	Pond 1	Pond 2
Day 1–71		
Light only	0.26	0.29
SS only	0.02	0.73
Chlorophyll only	0.01	0.29
SS and light	0.25	0.78
Chlorophyll and light	0.25	0.43
Day 1–35		
Light only	0.64	0.46
SS only	0.21	0.81
Chlorophyll only	0.00	0.84
SS and light	0.72	0.91
Chlorophyll and light	0.61	0.87

the pond's self-adjusting system where the release of nutrients stimulates the growth of algae which then provide oxygen for breakdown of the associated COD.

Conclusions

Accumulated sludge contributes to the nutrient load in the water column in a way that significantly affects pond behaviour. In particular there is an increase in concentrations of SS and nutrients, which may be of importance if the pond is to be discharged to sensitive natural waters. This behaviour does not appear to be detrimental to overall performance in terms of COD removal or DO concentrations in the pond, however; and given that the nutrient contribution from freshly deposited sludge approached that from a mature sludge layer within 7–10 weeks it seems unlikely that changes in recommended design or de-sludging frequency can be used to regulate it in a practical manner.

Acknowledgements

Thanks are due to EU INCO-Copernicus Project CT98-0144 and the BG Foundation.

References

Bryant, C.W. and Bauer, E.C. (1987). A simulation of benthal stabilisation. *Wat. Sci. Tech.*, **19**(12), 161–167.

Di Toro, D.M., Paquin, P.R., Subburamu, K. and Gruber, D.A. (1990). Sediment oxygen demand model: methane and ammonia oxidation. *J. Environ. Eng. Div. ASCE*, **116**(5), 945–986.

Giraldo, E. and Garzon, A. (2002). Compartmental model for organic matter digestion in facultative ponds. *Wat. Sci. Tech.*, **45**(1), 25–32.

Lumbers, J.P. and Andoh, R.Y.G. (1987). The identification of benthic feedback in facultative ponds. *Wat. Sci. Tech.*, **19**(12), 177–182.

Reckhow, K.H. and Chapra, S.C. (1999). Modeling excessive nutrient loading in the environment. *Environ. Pollut.*, **100**(1–3), 197–207.

Standard Methods for the Examination of Water and Wastewater (1998). *American Public Health Association/American Water Works Association*, 20th edn, Water Environment Federation, Washington, DC, USA.

Wastewater stabilisation ponds: sludge accumulation, technical and financial study on desludging and sludge disposal case studies in France

B. Picot*, J.P. Sambuco**, J.L. Brouillet*** and Y. Riviere****

*Département Sciences de l'Environnement et Santé Publique, UMR 5569 Hydrosciences- UM1, Faculté de Pharmacie, BP 1149, 34093 Montpellier cedex 5, France (E-mail: *picot@univ-montp1.fr*)

**CEREMHER, Zone de Recherche du Lagunage, B.P. 118, 34140 Mèze, France
(E-mail: *jpsambuco@yahoo.fr*)

***Conseil Général de l'Hérault DARE-DEMA, rue d'Alco, 34087 Montpellier cedex France
(E-mail: *dema-gestionglobaleeau@cg34.fr*)

****SATESE Conseil Général de l'Hérault, (E-mail: *yriviere@cg34.fr*)

Abstract Waste stabilisation pond treatment was widely developed during the 1980s in France, where there are now over 3,000 plants. Desludging the ponds has now become essential. In 19 primary facultative ponds, in operation for 12–24 years, the net average sludge accumulation rate was 19 mm/yr. The average per capita accumulation rates ranged from 0.04–0.148 m^3/person.year (mean of 0.08 m^3/person.year). In primary facultative ponds the volume of sludge represented 15–39% of the total volume of the basin. A filling rate above 30% necessitates desludging. In France, a desludging interval of 15 years is recommended for primary facultative ponds. The cost evaluation of desludging and landspreading showed differences according to the desludging technique used. Desludging after emptying the water had an average cost of 38 €/m^3 of sludge with 10% dry solid (range from 20 to 83 €/m^3). Under-water desludging was 50% more expensive. Although desludging is carried out only after several years of operation, its cost must be allowed for in the annual operation and maintenance costs of the process. It can be estimated to be 3 €/person.year. Even with this additional cost, waste stabilisation pond treatment remains less expensive than other treatment processes.

Keywords Cost evaluation; desludging; landspreading; sludge accumulation; waste stabilisation pond

Introduction

Wastewater stabilisation ponds (WSP) provide a simple, low-cost, low-maintenance process for treating wastewater. However, a sludge layer forms in the bottom of the ponds, due to the sedimentation of influent suspended solids such as algae and bacteria which grow in the pond. Sludge accumulation is greatest in primary ponds and can affect performance by reducing the effective pond volume and shortening hydraulic residence time (Schneiter *et al.*, 1984). Therefore, periodic sludge removal is required and the long-term sustainability of WSP systems is dependent on the safe and effective management of their sludge. The accumulation rate of sludge must be estimated so that the frequency of sludge removal can be determined and integrated into the pond design, maintenance schedule, and budget (Nelson *et al.*, 2004). The sludge volume and characteristics change with time due to anaerobic degradation and compression. Little information is available on the per capita accumulation rate of sludge which depends on temperature, age and geometry of the pond etc. Thus more regional data are needed to determine the expected cost of sludge removal.

Wastewater stabilisation pond (WSP) treatment was widely developed during the 1980's in France, where it now accounts for over 3,000 plants. The design of the waste stabilisation pond system most often employed consists of three basins in series

(Racault, 1997). Currently, about 50% of the French WSPs are more than 15 years old and primary ponds now require desludging. The first objective of this paper is to determine the accumulation rates of sludge and the reductions of the initial pond volume in several primary ponds in order to anticipate periodic desludging. The second aim of this work is to evaluate the cost of desludging in order to allow for this additional cost in the annual operation and maintenance costs of WSP treatment.

Materials and methods

19 WSPs located in the south of France were selected for this research (Table 1). All the treatment plants were treating municipal wastewater. Ponds 1, 6, 7, 8 and 18 received winery wastewater during a part of the year. These primary facultative ponds had been in operation for 12 to 24 years and their surface areas varied from 1,100 to 40,000 m^2. The pre-treatment in all these plants was a bar screen and grit chamber; moreover ponds 1, 2 and 9 had a primary settling tank, ponds 4 and 16 had a trickling filter. If one considers the normal load allowance for a first basin to be, 80–100 kg BOD$_5$/ha.d., 14 ponds were overloaded.

The accumulation rates and distribution of sludge were determined by measuring the thickness of the sludge layer at 10–140 locations over each pond. Bathymetric surveys were carried out, using a grid according to the size of the basin: area < 0.5 ha, 5 m × 5 m; 0.5 ha < area < 1 ha, 5 m × 10 m; >1 ha, 10 m × 10 m. Sludge and water depths were measured using a Secchi disk and a sounding bar. Sludge cores were collected from a boat at 3 different locations in each pond: near the entrance, middle and exit. Sludge depth was measured in different years on ponds 3, 6, 10, 11, 12 and 18. Total solid (TS) and volatile solid (VS) measurements were performed on well homogenised samples gravimetrically according to *Standard Methods*. Apparent sludge accumulation rates (mm/yr) were calculated for each pond by dividing the total sludge volume by the area of the pond and the number of years of operation.

Two methods of desludging are available: desludging without emptying out the water called "under-water" and desludging after emptying. Two case studies will be discussed

Table 1 Characteristics of the 19 facultative ponds

Pond (N°)	Location	Population (p.e.)	Operation period (yr)	Load λ_s (kgBOD$_5$/ha.d)	Surface area (m^2)	Pond depth (m)
1	Gigean	4,200	10.5	240	14,000	1.3
2	Lattes	7,500	21.5	189	24,000	1.1
3	Poussan *	5,800–7,200	12.5–19.5	173	18,200–29,000	1.45
4	Viols	500	15	142	1,300	0.9
5	Mireval	2,400	13.5	133	10,000	1.3
6	Mèze *	10,000	8–14	130	40,000	1.6
7	Capestang	3,700	18	117	11,200	1.2
8	Montbazin	2,300	16	100	14,000	1.1
9	Pinet	2,800	24	120	12,500	1.0
10	Vic *	1,610	12.5–13.5	95	6,000	1.45
11	Vérargues *	650	13.5–17.5	75–90	4,100	0.9
12	Claret *	600–800	16.5–18.5	120–185	3,350	1.0
13	Beaulieu	1,200	13	150	3,500	1.3
14	Creissan	900	19	90	5,500	1.2
15	Vendémian	700	19	130	2,600	1.0
16	St Paul	550	22	110	2,100	1.0
17	Ste Croix	300	18	140	1,700	1.3
18	Valfaunes *	370–500	13.5–21	90–125	2,400	1.1
19	St Jean	120	24	80	1,100	1.0

* Sludge in these ponds was measured after two different operation periods.

here: the first case is the Meze WSP under-water desludging operation, landspreading and sanitary impacts on the receiving environment (Crabos *et al.*, 1996). The second is a feasibility study on desludging and sludge disposal (Crabos *et al.*, 1998) for 9 primary facultative ponds in the Herault Department (ponds 1, 3, 5, 7, 10–12, 18 and 19) to characterise the technical and financial conditions for cleaning out the ponds. Desludging of several primary ponds has been carried out over the last few years, (ponds 6 and 9–19), so our projected technical and economical results will be compared to those actually obtained.

Results and discussion
Sludge distribution and rate of accumulation

The distribution of sludge was very uneven, the maximum sludge thickness occurring near the single pond inlet. For example in the primary pond of Mèze (Figure 1), at the entrance of the first pond there was a significant depth of sludge (more than 1 m), but 50 m from the input the thickness decreased rapidly and the remainder of the zone to be cleaned had depths of less than 0.6 m. Greater accumulation also occurred in the corners. The sludge becomes anaerobic and is buoyed up by the gaseous products of anaerobic decomposition; these floating masses are then blown into the corners by the wind. Again there was little difference in the sludge depth in the directions parallel or perpendicular to the prevailing winds (Middlebrooks *et al.*, 1982).

The sludge accumulation rates were determined both as the net average annual increase in sludge thickness and on a per capita basis as both are used in design (Table 2). It is better to give sludge accumulation rates in volume per person per year (or in dry weight per person per year) in order to compare several ponds which may have different area/volume ratios. The measured average annual increase in sludge thickness was 19 mm/yr (range 10–27 mm/yr). These rates are similar to values reported by Nelson *et al.* (2004), 19–21 mm/yr in 3 Mexican primary facultative ponds in operation for 6–12 years. Carré *et al.* (1990) reported higher accumulation rates in France (15–85 mm/yr, with a median value of 28 mm/yr) corresponding to 12 ponds in operation for 3–10 years. The accumulation rate is not constant and decreases with time due to anaerobic degradation and consolidation of sludge. Hammou *et al.* (1992) reported an accumulation rate of 43 mm/yr in the primary pond of Meze after 8 years of operation and we found 27 mm/yr after 14 years of operation in pond 6. Schetrite and Racault (1995) showed that, after two years of pond functioning, the deposited sludge became denser, the annual accumulation rate became linear and was less than that for the two proceeding years.

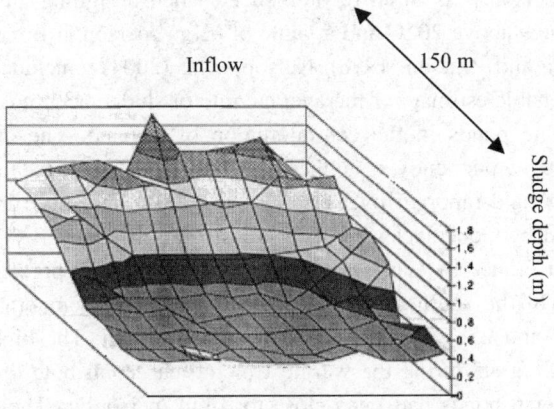

Figure 1 Sludge distribution in the primary facultative pond of Meze after being in operation for 14 years

Table 2 Mean concentration of solids in the sludge layer, sludge volume, filling rate and measured accumulation rates in primary ponds

Pond N°	Age Yr	Population p.e	Load kgBOD/ha.d	V_s m³	TS g/L	VS/TS %	Filling rate %	Accumulation rate mm/yr	m³ pers.yr	kgdw/pers.yr
1	10.5	4,200	190	2,950	220	47	18	20	0.067	14.9
2	21.5	7,500	189	10,300	63	53	39	20	0.064	4.0
3	12.5	5,800	173	3,921	119	49	17	17	0.054	6.4
3	19.5	7,200	149	6,958	–	–	17	12	0.050	–
4*	15.0	500	142	355	142	21	30	18	0.047	6.7
5	13.5	2,400	133	2,612	102	41	20	19	0.081	8.2
6	14.0	10,000	130	15,000	110	42	23	27	0.107	11.8
7	18.0	3700	117	5,000	60	59	37	25	0.075	4.4
8	16.0	2,300	100	4,650	–	–	30	21	0.126	–
9	24.0	2,800	120	4,268	89	34	34	14	0.064	5.7
10	13.5	1,610	95	1,730	80	55	29	21	0.080	6.4
11	13.5	650	75	723	122	39	25	17	0.082	10.1
11	17.5	650	87	1,444	87	63	38	20	0.127	10.5
12	18.5	800	185	1,200	80	37	36	19	0.081	6.5
13	13.0	1,200	150	1,115	59	46	25	25	0.086	5.1
14	19.0	900	90	1,323	160	–	20	13	0.077	12.4
15	19.0	700	130	750	92	38	29	15	0.056	5.2
16*	22.0	550	110	490	61	–	24	11	0.040	2.7
17	18.0	300	100	315	72	47	15	10	0.058	4.3
18	13.5	370	90	355	74	42	15	11	0.071	5.3
18	21.0	500	125	1,000	80	–	38	20	0.095	7.6
19	24.0	120	80	425	100	26	39	16	0.148	13.5
Mean								19	0.084	8

* Ponds 4 and 16 performed a tertiary treatment of trickling filter effluents; the corresponding data were not included in the calculation of mean values.
V_s volume of sludge; TS and VS: total and volatile solid concentration; dw: dry weight

Maybe this explains the low variability of accumulation rates observed in ponds aged between 10 and 24 years where sludge is well compressed and mineralised. For example in pond 11 the accumulation rate was the same after 13.5 and 17.5 years of operation. In this study, total dry solids (TS) concentration was found to be from 60 to 220 g/L; with a VS/TS of below 60%, sludge in all ponds was mineralised.

In contrast to the average annual increase in sludge thickness, the per capita rates of sludge accumulation varied widely from 0.04 to 0.148 m³/person.yr with a mean value of 0.084 m³/person.yr. Although published values determined from field data are few, a value of 0.04 m³/person.yr is often recommended when designing anaerobic ponds with average temperatures above 20 °C and a value of 0.1 m³/person.yr for winter temperatures below 10 °C (Mara and Pearson, 1998). Nelson et al. (2004) concluded that 0.04 m³/person.yr was a reasonable estimate of the average rate of sludge accumulation in both facultative and anaerobic ponds in the central region of Mexico. The average of 0.08 m³/person.yr observed in our study is higher than the rates measured in Mexico (0,021–0.036 m³/person.yr) and reported by Nelson et al. (2004) and those in Brazil (0.036 m³/person.yr) reported by Gonçalves (2000). It is less than the median value of 0.12 m³/person.yr observed in France by Carré et al. (1990) in 12 primary ponds under oceanic climatic conditions. In the south of France the average water temperature of the coldest month is 7–10 °C and 25–27 °C during the warmest month. The higher temperature in the Mediterranean region during the warmest month can contribute to better degradation of sludge; However 6 ponds had rates close to 0.1 m³/person.yr: The high rates in pond 11 and 19 were due to a growth of duckweed in spring and summer, the sedimentation of this biomass increased sludge. Sludge accumulation rates in dry weight per person per year ranged from 4 to 14.9 kg/person.yr with an average of 8 kg/person.yr.

In this study, 15–39% of the pond volumes were occupied by solids resulting in proportional decreases in the design hydraulic residence time. These high filling rates can affect pond performance. Namèche et al. (1997) also referred to the possibility of increasing sediment oxygen demand. We think that with a filling rate above 30% ponds require desludging. However, filling rate alone is not sufficient to decide on the requirements for desludging. It is also necessary to take into account the position of the deposits, the presence of sludge floating masses on the surface of the pond and unpleasant odours. The decision to clean out a pond was taken for varying thickness of sludge. In our case 9 ponds had a thickness of sludge corresponding to a volumetric loss in the pond of more than 30%, 6 of them have been desludged (ponds 4, 7, 9, 11, 12 and 19), one will be shortly (pond 18). On the other hand, 7 ponds (6, 10, and 13 to 17) have been desludged with filling rates 20–30%. The sludge was found to contain significant amounts of fertilising material (1.5–2.6% N, 1.9–6% P_2O_5) and levels of heavy metals and organic compounds which were below the limit set by the French authorities for agricultural reuse so could be used for landspreading.

Methods of desludging

Two methods of desludging are available: desludging without emptying out the water called "under-water" and desludging after emptying which first requires the by-passing and emptying of the pond that it is to be cleaned.

Desludging without emptying the pond. This method is chosen when the receiving water is sensitive to pollution and when it is not possible to bypass the pond. It was adopted for pond 6 and considered for ponds 1 and 3 due to the high sensitivity of the receiving water, the Thau lagoon, where oysters are cultivated. It was also chosen for ponds 15 and 17 due to the high sensitivity of underground water. We report here on the "under-water" desludging of the primary pond of the Meze WSP, carried out 14 years after its start-up. More detailed investigations on Meze WSP plant performances are given by Picot et al. (1992) and characteristics of sludge by Hammou et al. (1992). The sludge volume that should be extracted, 8,000 m^3 was determined after measuring sludge depth at 140 points. The zone to be cleaned extended over 1.5 ha, which represented 1/3 of the primary pond. The main physicochemical sludge characteristics were: 11% total dry solids (TS), 42% organic matter, 2.5% nitrogen and 1.9% phosphorus (P_2O_5), with a ratio C/N of 8.3. Heavy metal concentrations were lower than the those set by French standards for agricultural landspreading.

The administrative step required a spreading plan that included the following phases:
- *The choice of the land* which took into account the following criteria: pond proximity to reduce the costs of transport: far from dwellings in order to limit the unwanted effects of odours; far from the edge of the Thau lagoon because of sensitivity due to intensive shellfish breeding; sufficient flat land, far from rivers and with an unsaturated layer to avoid any transportation towards surface or underground waters.
- *The appropriateness of the land* for spreading was determined by analyses on soil samples and interpretation of pedological maps.
- *The calendar for spreading*: because of the climatic conditions this work should be carried out between August and October.
- *A temporary storage area* (1,200 m^3) was created to allow for the impossibility of land spreading on rainy days.

- *The sludge extraction* was carried out under water for 47 days using a raft-mounted sludge pump towed by a cable. This system had a pumping capacity of 300 m^3/h and the sludge concentrations were 40–60 g/L.
- *Dewatering the sludge*: this low concentration obliged us to dewater the sludge on-site using a centrifuge, with the capacity to treat 80 m^3/h. Since dewatering was not continuous the maximum production was 50 t of dewatered sludge per day. Thus the sludge volume extracted was reduced by centrifugation to 2,620 m^3. The dewatered sludge had an average of 27% dry solids. In total, 800 t of dry solids were therefore removed.
- *Transport* was carried out by conveyors to containers of 12 m^3. In total, the transport of dewatered sludge required 240 truck round trips. A sludge volume of 780 m^3 was stored, which represented 2/3 of the storage capacity.
- *Spreading* was carried out over three weeks by a dry sludge automotive spreader. The total surface area subjected to spreading was 90 ha. Sludge quantities taken to the land were between 24 and 31 m^3/ha. This amount represents the equivalent in available fertilising nutrients of 80–145 kgN/ha and 85–150 kgP$_2$O$_5$/ha.
- *A follow-up program* of WSP performance before, during and after desludging was established with the objective of monitoring the ponds and the sanitary impacts on the receiving water at the same time. WSP monitoring was carried out at both the entrance and the exit with a total of 6 sampling points; the physicochemical and bacteriological measurements of the WSP systems were carried out over 8 months (2 months before desludging, 2 months during the work and 4 months after desludging). The results showed that desludging did not affect the effluent in a significant way. The impact on the Thau lagoon was analysed by observation of mussels dispersed on 3 radials of 750 m starting from the point of the effluent discharge. These mussels were decontaminated before use by keeping them in bacteria-free water for 24 hours. The flesh of the mussels was analysed after exposure. No *salmonella* was detected in mussels in the receiving environment.

Desludging after emptying the pond. This method was considered for most of the ponds investigated. Desludging after emptying the pond involves passing the raw wastewater to another pond. The sludge of these ponds had heavy metal and organic compound concentrations (PCBs and PAHs) below the limits required by the French standards, therefore the appropriate solution was landspreading. When determining the area for landspreading, the soil characteristics, farming requirements and sludge quality must be taken into account (Table 3). Thus, integration of sludge into the fertilisation plan took into account the crops cultivated in this region (hard wheat).

If the dry solids content was around 6%, the sludge could be pumped directly into liquid manure hauling trucks and spread. At 6–10% TS it was extracted by a self-priming pump to a buffer pond and then conveyed to the spreading area by a liquid manure hauling truck. Spread sludge was dug into the ground each evening.

Financial evaluation of desludging and landspreading

Table 3 shows the comparative costs of the two desludging techniques. The cost has been reported as several ratios: €/person/year, €/m^3 of sludge (reported as 10% dry solids) and €/m^3 of invoiced drinking water. The cost varied according to land accessibility, area and availability of land in the neighbourhood of the ponds, the truck rotation time and the choice of desludging method.

The average cost for desludging after emptying the pond was 38 €/m^3 of sludge (reported as 10% dry solids), with a range of 20–83 €/m^3. The percentage for each step was 21% for preliminary work (spreading plan, bathymetric and installation work),

Table 3 Required area for landspreading and analyses of desludging costs

Ponds N°, towns	Land area ha	Costs		
		€/person.yr	€/m³ sludge (10% dry solid)	€/m³ drinking water
Desludging after emptying water				
1 Gigean **	135	4.2	29	0.077
3 Poussan **	100	1.8	28	0.04
5 Mireval **	55	2.1	25	0.039
7 Capestang **	35	0.9	27	0.017
9 Pinet *	65	1.1	20	0.02
10 Vic *	20	2.5 (1.7 **)°	39 (28 **)°	0.046 (0.03 **)
11 Verargues *	18	2.7 (2.8 **)	26 (28 **)	0.040 (0,05 **)
12 Claret *	–	3.2	50	0.059
13 Beaulieu *	22	2.8	55	0.042
14 Creissan *	26	2.7	22	0.050
16 St Paul *	15.4	1.9	83	0.034
18 Valfaunes *	6	4 (3.8 **)	52 (51 **)	0.069 (0.073 **)
19 St Jean *	6	4.2 (2.1 **)	31 (48 **)	0.077 (0.037 **)
Mean costs		2.6	38	0.05
Desludging without emptying the pond				
1 Gigean **	135	5.55	47	0.1
3 Poussan **	100	2.97	47	0.05
6 Mèze *	90	3.05	53	0.06
15 Vendemian *	14	2.86	60	0.052
17 St Croix *	9	4.2	108	0.077
Mean costs		3.7	62	0.068

* Real costs of desludging. ** Costs were estimated in the feasibility study.

30% for extraction of sludge, 16% for transport, 26% for spreading and 7% for the agronomic survey. The most expensive step was the extraction of the sludge. The real costs of desludging carried out over the last few years on 9 ponds (mean 42 €/m³ of sludge with 10% dry solids) confirm the results of our feasibility study, although. Ferry and Wiart (2000) reported a lower price of 25 €/m³ of sludge (with 10% dry solids) for one small facultative pond (500 inhabitants).

The average cost for 5 ponds with "under-water" desludging was 62 €/m³ with a range from 47–108 €/m³ The percentage for each step was 18% for preliminary work, 20% for extraction of sludge, 35% for dewatering, 11% for transport, 13% for spreading and 3% for the agronomic survey. The most expensive step in this technique was the dewatering of sludge. The results show that "under-water" desludging is about 50% more expensive than desludging after emptying the pond.

The cost of desludging and sludge disposal is high for small communities and should be allowed for in annual operation and maintenance costs. It can be estimated to be about 0.05–0.07 € for each m³ drinking water invoiced or 2.6–3.7 €/person.yr depending on desludging methods, after or without emptying water.

Conclusion

Sludge accumulation in wastewater ponds is uneven and depends on pond geometry. In 19 primary facultative ponds in the south of France the range in per capita sludge accumulation rates was between 0.04 and 0.148 m³/person.yr with a mean of 0.08 m³/person.yr. In a Mediterranean climate the net average annual increase in sludge thickness rate was 19 mm per year. This accumulated sludge can affect pond performance by reducing the basin volume and shortening the hydraulic residence time. A filling rate of over 30% necessitates pond desludging

The choice of desludging method with or without emptying water should take account of local constraints. Recovery of sludge after emptying out the water is the most used method in France and it is cheaper than the "under-water" procedure. The latter requires more operations such as sludge dewatering and temporary storage which increase the final cost, however, it must be chosen when the receiving body is a particularly sensitive to pollution. Although desludging and sludge disposal are carried out only after about 15 years of operation, the cost must be programmed in order to determine the annual operating costs of the process. This addition to the cost of the annual operation and maintenance of WSP treatment is 30%. In spite of this desludging cost, the waste stabilisation pond process is still competitive in comparison to conventional wastewater treatment.

Acknowledgements

The authors wish to acknowledge the support of the Conseil Général de l'Hérault. Our thanks to Jean Louis Crabos, Rejane Costa, Luis Philippi and Helen Burnett for their kind assistance and also Agrodeveloppement, Cabinet Merlin, Ecosite, Entech, SDEI, Terra-sol and the mayors of the towns concerned for data on desludging.

References

Carré, J., Laigre, M.P. and Legeas, M. (1990). Sludge removal from some wastewater stabilization ponds. *Wat. Sci. Tech.*, **22**(3–4), 247–252.

Crabos, J.L., Sambuco, J.P., Bondon, D., Brouillet, J.L., Lebec, C. and Bonal, O. (1996). *Lagunage de Mèze-Loupian. Curage, deshydratation et épandage des boues (Meze-Loupian lagoons. Clearing out, dewatering and spreading of sludges)*, Rapport CEREMHER, IFREMER, France.

Crabos, J.L., Sambuco, J.P. and Couleuvrat, C. (1998). *Lagunes de l'Hérault – Curage et valorisation des boues – Etude de faisabilité (Herault lagoons – clearing out and valorisation of sludges – feasibility study)*, Rapport CEREMHER Mèze, France.

Ferry, M. and Wiart, J. (2000). Les coûts de traitement et de recyclage agricole des boues d'épuration urbaines (Costs of treatment and agricultural recycling of the urban clarification sludge). *Tech. Sci. Mun.*, **95**(9), 117–135.

Hammou, N., Picot, B. and Bontoux, J. (1992). Sedimentary deposits in a natural microphyte lagoon variation of quantities, physical-chemical characteristics and heavy metal loads. *Environmental Technology*, **13**, 647–655.

Gonçalves, R.F. (2000). *Gerenciamento do lodo de lagoas de estabilização não mecanizadas*, PROSAB, ABES, Rio de Janeiro.

Mara, D. and Pearson, H. (1998). *Design Manual for Waste Stabilization Ponds in Mediterranean Countries*, Lagoon Technology International, Leeds, England.

Namèche, T., Chabir, D. and Vasel, J.L. (1997). Characterization of sediments in aerated lagoons and waste stabilization ponds. *Int. J. Environ. Anal. Chem.*, **68**(2), 257–279.

Nelson, K.L., Jiménez, B., Tchobanoglous, G. and Darby, J.L. (2004). Sludge accumulation, characteristics, and pathogen inactivation in four primary stabilization ponds in central Mexico. *Wat. Res.*, **38**, 111–127.

Middlebrooks, E.J., Middlebrooks, Ch.H., Reynolds, J.H., Watters, G.Z., Reed, S.C. and George, D.B. (1982). *Wastewater Stabilization Lagoon Design, Performance and Upgrading*, Macmillan Publishing, USA, pp. 356.

Picot, B., Bahlaoui, A., Moersisik, S., Baleux, B. and Bontoux, J. (1992). Comparison of the purifiying efficiency of high rate algal pond with stabilisation pond. *Wat. Sci. Tech.*, **25**(12), 197–206.

Racault, Y., (coord.) (1997). *Le lagunage naturel. Les leçons tirées de 15 ans de pratique en France (Natural lagoonage. Lessons drawn from 15 years of practice in France)*, Rapport SATESE-CEMAGREF, France, p. 60.

Schetrite, S. and Racault, Y. (1995). Purification by natural waste stabilization pond: influence of ageing on treatment quality and sediments thickness. *Wat. Sci. Tech.*, **31**(12), 191–200.

Schneiter, R.W., Middlebrooks, E.J. and Sleter, R.S. (1984). Wastewater lagoon sludge characteristics. *Wat. Res.*, **18**(7), 861–864.

Evaluation of sludge from pond system for treatment of piggery wastes

C.T. Zanotelli*, R.H.R. Costa** and C.C. Perdomo***

*Universidade da Região de Joinville, Campus Universitário, s/n., Bom Retiro, Joinville, SC, Brazil
(E-mail: czanotelli@univille.br)

**Universidade Federal de Santa Catarina, Campus Universitário, Trindade. 88010-970, Florianópolis, SC, Brazil (E-mail: rejane@ens.ufsc.br)

***Universidade do Contestado- Concórdia- SC. Rua Vitor Sopelsa, 3000, 89700-000, Concórdia, SC, Brazil

Abstract Stabilization ponds used for the treatment of piggery wastes accumulate sludge over time, which is commonly used in agriculture. The objective of this study was to evaluate the agronomic potential of this kind of sludge. The samplings were collected in two different phases. The first in two anaerobic ponds (AP1 and AP2) and in one facultative pond with 5 transverse baffles and, the second in the same facultative pond with aeration. The removed sludge of AP1 and AP2 was characterized as rich sludge in volatile solids and with low stabilization, there was a great accumulation of the total phosphorus in the sludge of AP2. The facultative pond presented greater retention of nutrients in the sludge in relation to the anaerobic ponds. The annual accumulation of sludge was 13.3 cm/year in the AP1 and 6.70 cm/year in the AP2, while in the pond with aeration this was on the average of 0.5 cm/year, due to the aeration regime. The sludge can be used as a fertilizer in agriculture, if the chemical characteristics of the soil are taken into account so as to avoid the accumulation of nutrients and damage to plants.

Keywords Piggery wastes; sludge; stabilization ponds; treatment system

Introduction

The use of stabilization ponds for the treatment of piggery wastes has spread globally, in less developed countries or in areas of tropical and subtropical climate, mainly for its ability for the reduction of organic matter and of pathogenic microorganisms. However it should be observed that sludges accumulate in these ponds over time and that when removed they should have an appropriate destination. The sludge removed from the ponds is commonly used in agriculture; this is a positive practice, because it allies low cost and environmental impact, however it should be accomplished within safe approaches (Andreoli *et al.*, 1999). It is necessary to evaluate the potential pollutants of these sludges before using them as fertilizers. Tsutiya (2001) points out that despite the high agronomic value of nitrogen, the same is used as the factor for determining the maximum amount of sewer sludge to be applied, because in very high levels, the nitrogen can leach in nitrate form and contaminate the groundwater. With regard to phosphorus the author observes that its concentration is smaller than that of nitrogen, although the plants need smaller amounts of phosphorus for their development than nitrogen. The purpose of this work was to evaluate the potential pollutants of sludge removed from stabilization ponds in a treatment system for piggery wastes.

Methods

The work was performed in a experimental treatment system comprising an equalizer, a horizontal flow decanter (PD), two anaerobic ponds (AP1 and AP2), a facultative pond (FP) with 5 transverse baffles and a maturation pond (MP) with water hyacinth, in series.

The sampling sludge deposited in the ponds was carried out on three occasions. The first, done in the facultative pond, in June 2001 (sludge accumulated for 14 months), when the parameters: chemical oxygen demand (COD), total and volatile solids (TS and VS), total Kjeldahl nitrogen (TKN) and total phosphorus (TP) were analyzed. The sampling points were located before each compartment, where the collecting of the liquid samples was made. The second sampling was in January 2002 (sludge accumulated for 3 years), done in the anaerobic ponds (AP1 and AP2), with the following parameters being analyzed: total, volatile solids and fixed solids (TS, VS, FS), for AP1; and total and volatile solids, chemical oxygen demand, total Kjeldahl nitrogen, total phosphorus and potassium (K) for AP2. The third sampling, in the facultative pond, without the baffles, submitted to a system of aeration during the night, took place in February 2002 (sludge accumulated for 12 months). Chemical oxygen demand, total and volatile solids, total Kjeldahl nitrogen and total phosphorus were analyzed. Analyses were done according to *Standard Methods for the Examination of Water and Wastewater* (1998). The sludge samples were collected from the bottom of the ponds using a PVC tube with an "open-closed" system at one of the extremities. For the anaerobic ponds, the samples were collected from six points distributed in the bottom of the ponds and mixed to obtain one sample from each pond.

Results and discussion

Results for the anaerobic ponds

Table 1 presents the results for the sludge accumulated during the three-year period (January 1999 to December 2001) to the anaerobic ponds (AP1 and AP2). These ponds operated with a mean applied load of $468\,gCOD/m^3$.day and $65\,gCOD/m^3$.day, respectively.

It can be observed in Table 1 that the removed sludge of AP1 presented a high concentration of total solids (24.6%) and volatile solids (56% TS). These last ones indicate low stabilization of the sludge, or "primary" sludge, characteristic of this type of pond (Metcalf and Eddy, 1991). In the AP2, the values of TS and VS were smaller than in the AP1, even so they were still high. These values can be compared with those obtained with sludge of secondary raw sludge and dehydrated sewers (VS = 33–53%TS) presented by Nacheva et al. (2002). In relation to nutrients, the values found for AP2 (Table 1) for NTK and K are comparable to those obtained in different stations of treatment of sewers, shown in Fernandes (1999), Nacheva et al. (2002) and Tsutiya (2001). However, for the total phosphorus a great accumulation exists in these sludge, being two to three times higher than the data reported by Tsutiya (2001).

There was a great amount of sludge deposited in the anaerobic ponds, and for AP1 the medium height of the layer of sludge was approximately 40 cm, while for AP2 about 20 cm were obtained. Considering the three-year period between the cleanings, those values result in an annual accumulation of sludge of an amount of: AP1 = 13.3 cm/year and AP2 = 6.70 cm/year. According to Gonçalves et al. (1999), in the anaerobic ponds the lineal rates of sludge accumulation exceeded a value of 4 cm/year. In agreement with Picot et al. (2001), the time of operation of the pond is important to mention, because the

Table 1 Results for AP1 and AP2

Sample point	COD (g/L)	TS (%)	VS (%TS)	FS (%TS)	TKN (g/kg)	TP (g/kg)	K (g/kg)	TKN (%TS)	TP (%TS)	K (%TS)
AP1	–	24.6	56	44	–	–	–	–	–	–
AP2	88.50	12.6	49	51	3.0	8.3	0.5	2.4	6.6	0.4

annual rate of accumulation of the sludge is smaller with the age of the pond, as exemplified by the anaerobic ponds of Mèze, located in the Department of Hérault, in France, whose rate of sludge accumulation was of 62 cm/year after 7 months of operation of ponds, and decreased to 12 cm/year after 18 months of operation.

Results for the facultative pond with baffles

Table 2 presents the results of the sludge accumulated in the compartments of the facultative pond with baffles, after 14 months functioning. This pond operated with a mean applied load of 470 kg COD/ha.day.

The sludge from compartments 1 and 2 presented about 4% of Total Solids (Table 2), which, according to Metcalf and Eddy (1991), corresponds to sludge of anaerobic ponds or percolating filters submitted to dehydration for automated way (TS = 5–15%). The compartment 1, presented VS = 34%TS checking their characteristics of "stabilized" sludge, while in the other compartments VS ≈ 60%, the sludge can be characterized as "primary" sludge. The retention of nutrients was larger in the facultative pond than in the anaerobic ponds AP1 and AP2, a fact that was also verified by Gonçalves et al. (1999) and also reported in Tsutiya (2001). In Table 2 it can be observed that there was larger retention of nutrients in the sludge of the intermediary compartments of the pond: 3, 4 and 5. The obtained values were high, mainly the total phosphorus, when compared to those usually found in sludge of anaerobic ponds or primary facultative ponds with domestic sewers, mentioned in Gonçalves et al. (2000): NTK = 2.0%TS and PT = 1.0%TS. Still in agreement with these authors, the medium values found in fertilizers for the agriculture are of the order of NTK = 5.0%TS and PT = 10%TS. It can be verified, therefore, that sludge produced in a facultative pond, with piggery wastes, has a high fertilizer potential in relation to those parameters, demanding special cares for its application according to the type of cultivation and the characteristics of soils.

The depth of the measured sludge layer was between 5 and 10 cm, being thicker in the first compartments, corresponding approximately to a sludge accumulation of 4.3–8.6 cm/year, those results are comparable to those obtained by Picot et al. (2001), for primary facultative ponds after 8 years of operation. Although, in the same way as in anaerobic ponds, the annual rate of accumulation of the sludge is smaller with the age of the pond.

Results for the facultative pond with aeration

Table 3 presents the results for the sludge in the facultative pond with aeration, after 12 months of operation. It can be observed that there was a larger concentration of COD and of the solids in the sampling points placed in the middle of the pond (3, 4, 5 and 6), soon after the aerator. The retention of nutrients was high (TKN = 7.9%TS; TP = 3.1%TS; K = 4.9%TS), mainly for the phosphorus and potassium, when it is compared with biosolids produced in several treatment systems for domestic sewers (TKN = 2.2 to 5.5%TS; TP = 1.0 to 3.7%TS; K = 0.01 to 0.4%TS), shown by Tsutiya (2001) and by

Table 2 Results for the sludge from each FP compartment

Compartment	COD(g/L)	TS(%)	VS(%TS)	TKN(g/kg)	TP(g/kg)	TKN(%TS)	TP(%TS)
1	171	4.2	34	0.84	12.20	2	29
2	428	4.3	56	3.13	18.70	7	43
3	92	1.0	60	1.11	6.10	10	57
4	210	1.7	68	1.71	12.40	10	71
5	213	2.0	68	1.99	12.70	10	63
6	230	3.4	48	2.17	16.60	6	48

Table 3 Results of sludge in facultative pond with aeration

Point	pH	COD (g/L)	TS (%)	VS (%TS)	TKN (g/kg)	TP (g/kg)	K (g/kg)	TKN (%TS)	TP (%TS)	K (%TS)
1	7.19	8.14	0.76	67	0.65	0.20	0.37	8.5	2.6	4.8
2	7.16	8.10	0.75	65	0.68	0.24	0.41	9.0	3.2	5.4
3	7.25	11.41	1.11	64	0.90	0.39	0.47	8.1	3.5	4.2
4	7.21	11.41	1.03	66	0.82	0.32	0.45	7.9	3.1	4.4
5	7.27	16.41	1.66	63	1.17	0.50	0.49	7.0	3.0	3.0
6	7.24	10.47	1.05	68	0.78	0.32	0.44	7.4	3.0	4.2
7	7.31	6.56	0.67	68	0.54	0.22	0.43	8.0	3.3	6.4
8	7.33	6.36	0.66	62	0.50	0.17	0.42	7.6	2.6	6.4

Nacheva *et al.* (2002). The annual accumulation of sludge was on the average of 0.5 cm/year. This low value is due to the fact that the aeration favors the solids suspension in the pond.

Conclusions

The removed sludge of AP1 and AP2 was characterized as rich in volatile solids and with low stabilization. In relation to the nutrients, for AP2, the values found for total nitrogen and potassium were comparable to those obtained in different wastewater treatment plants. Even though, for the total phosphorus there was a great accumulation in the sludge.

For the facultative pond with baffles, the sludge of compartment 1 presented characteristics of "stabilized" sludge, while in the other compartments those had characteristics of "primary" sludge. The retention of nutrients was higher than in the anaerobic ponds. For the facultative pond with aeration, the retention of nutrients was high, potassium was the principal nutrient accumulated in the sludge. The annual accumulation of sludge in the facultative pond among the baffles varied between 4.3 and 8.6 cm/year, while in the pond with aeration this was on the average of 0.5 cm/year, due to the aeration regime in the pond.

The sludge can be used as fertilizer in agriculture, respecting the chemical characteristics of the soil, in order to avoid accumulation of nutrients and damage to the plants.

References

Andreoli, C.V., Ferreira, A.C. and Jurgensen, D.(1999). *Uso e manejo do lodo de esgoto na agricultura*. Rio de Janeiro: ABES, PROSAB. pp. 26–28.
Fernandes, F. (Coord.) (1999) *Manual prático para compostagem de biossólidos*. Rio de Janeiro: ABES.
Gonçalves, R.F., Lima, M.R., Passamani, F.R.F.(1999). Características físico-químicas e microbiológicas do lodo de lagoas. In Gerenciamento Do Lodo De Lagoas De Estabilização Não Mecanizadas, Rio de Janeiro, RJ: ABES, pp. 25–37.
Goçalves, R.F., Silva, V.V. and Taveira, E.J.A. (2000). *Algae and nutrient removal in anaerobic-facultative pond system with a compact physical-chemical process*. In Conferencia Latino Americana En Lagunas De Estabilizacion Y Reuso, 1Santiago de Cali, Colombia: [s.n.], pp. 68–176.
Picot, B., Costa, R.H.R. and Philippi, L.S. (2001). The desludging of waste stabilisation ponds and sludge disposal in France. IWA Conf. on Sludge Management, Acapulco, México.
Metcalf & Edddy (1991). *Wastewater Engineering: Treatment, Disposal and Reuse*, 3 edn. McGraw-Hill, New York, USA.
Nacheva, P.M., Moeller, G., Camperos, E.R. and Vigueros, L.C. (2002). Characterization and dewaterability of raw and stabilized sludge using different treatment methods. *Wat. Sci. Tech.*, **46**(10), 123–130.
Standard Methods for the Examination of Water and Wastewater (1998). 20th edn, American Public Health Association/American Water Works Association/Water Environment Federation, Washington DC, USA.
Tsutiya, M.T. (2001). Características de biossólidos gerados em estações de tratamento de esgotos. In *Biossólidos na Agricultura*, Sao Paulo: Ed. SABESP pp. 89–125.

Investigating helminth eggs and *Salmonella sp.* in stabilization ponds treating septage

G.S. Sanguinetti*, C. Tortul*, M.C. García*, V. Ferrer*, A. Montangero** and M. Strauss**

*Centro de Ingeniería Sanitaria, Facultad de Ciencias Exactas, Ingeniería y Agrimensura, Universidad Nacional de Rosario, Riobamba 245 Bis, 2000 Rosario, Argentina (E-mail: cis@fceia.unr.edu.ar)
**Swiss Federal Institute for Environmental Science and Technology, P.O. Box 611, CH-8600 Duebendorf, Switzerland (E-mail: strauss@eawag.ch)

Abstract Sludge management arises as a relevant problem after being accumulated in primary ponds of septage treatment plants. One of the most attractive options for sludge disposal is its use in agriculture and then specific guidelines regarding hygienic quality must be fulfilled. This study aimed at evaluating the storage time needed to inactivate *Ascaris* eggs and *Salmonella* in sludge accumulated in a primary pond treating septage. Raw septage exhibited very low concentrations of viable *Ascaris* eggs, thus experiments with *Ascaris suum* eggs spiking were conducted. The concentration of *Ascaris* eggs in the solids accumulated at the bottom of the pond was 20 eggs/g of total solids (g TS) at the time of pond closure. Although it decreased, some eggs remained viable (0.59 mean viable eggs/g TS) up to 20 months of in-pond storage of the biosolids. *Salmonella* survival was studied after developing an analytical method that inhibited the native flora. Sludge was seeded with *Salmonella enteritidis*. An equation adequately describing *Salmonella* die-off in biosolids subjected to 115 days of in-pond storage/dewatering, was found to be represented by the regression: $y = \log \text{MPN Salmonella/g TS} = 6.67 \cdot t^{-0.086}$, with $t =$ storage time elapsed in days. The initial concentration was 7.0×10^6 MPN/g TS and the removal efficiency was 99%.
Keywords *Ascaris*; biosolids; helminth eggs inactivation; primary ponds; *Salmonella*; septage

Introduction

In Argentina, 89% of the 37,000,000 inhabitants are living in small and medium-size towns and in cities. 54% of the urban population are served by sewerage and the remaining 46% are using on-site sanitation systems, mostly septic tanks with soak pits. This paper reports investigations on helminth eggs and *Salmonella* inactivation in-situ and at ambient temperature in biosolids generated during pond treatment of septage (the pump-outs from septic tanks). The use of biosolids in agriculture appears as the most attractive solution as it contributes to recycling organic matter and nutrients. Hence, guidelines or standards on the hygienic quality of the solids must be satisfied. There exists, to date, insufficient published information about the survival of excreted pathogens in the biosolids derived from faecal sludge treatment. Argentina enacted guidelines for the use and disposal of biosolids by adopting US standards (Ministerio de Desarrollo Social y Medio Ambiente 2001; USEPA 1993). They stipulate <1 viable helminth egg/4 g TS and <3 MPN *Salmonella*/4 g TS. Xanthoulis and Strauss (1991) proposed a guideline value for biosolids (as produced in faecal sludge or in wastewater treatment schemes) of 3–8 viable nematode eggs/g TS. This recommendation is derived from the WHO guideline of ≤1 nematode egg/litre of treated wastewater used for vegetable irrigation (WHO, 1989), and based on an average manuring rate of 2–3 tons TS/ha year.

Ascaris lumbricoides eggs are particularly important as indicator of the hygienic quality of biosolids as Ascariasis is one of the most widespread excreta-related infections in low-income areas and as *A.* eggs are the most resistant among the gastro-intestinal pathogens. Their removal suggests that all other pathogens have also been inactivated (Feachem *et al.*, 1983).

Standardized methods for the isolation of *Salmonella* in excreta-derived sludges have not been available to date. Although a proven methodology does exist and is widely applied for blood, food, drinking water, and wastewater samples, it does not work likewise for isolation in biosolids. The reason lies in the co-existence of a native flora which masks *Salmonella* development, particularly if *S*. occur in low concentrations. This lead the authors in previous investigations to develop a methodology, which is suitable to detect *Salmonella* in sludges.

The aims of this study were: 1) To evaluate the required storage time for *Ascaris* eggs inactivation in accumulated sludge of a primary septage pond. 2) To evaluate the die-off of *Salmonella* in the same type of sludge.

The study comprised the following:

(i) Determining helminth eggs viability in biosolids stored and subjected to natural dewatering in-situ in a primary septage pond.
(ii) Determining the viability of *Ascaris suum* eggs spiked into septage pond biosolids which were subjected to natural dewatering/drying in a container and to further drying in an experimental drying bed.
(iii) Determining the *Salmonella* die-off in septage sludge.

The septage-cum-wastewater co-treatment scheme of the town of Alcorta (Province of Santa Fe, Argentina), which was used to conduct the in-situ experiments, is depicted in Figure 1. The town has a population of 5,000 inhabitants, of which 60% are served by septic tanks and 40% by sewerage. The system treats $25\,m^3$/day of septage and $200\,m^3$/day of wastewater (Ingallinella *et al.*, 2000; 2002).

Materials and methods

Method for detecting and enumerating helminth eggs and for determining eggs viability

The method used was the USEPA protocol (1992) as modified by Schwartzbrod (1998). The test uses zinc sulfate ($\rho = 1.3$) instead of magnesium sulfate ($\rho = 1.2$) for the flotation. Helminth eggs counts were made in a Sedgwick-Rafter counting chamber with an Olympus Bmax40 Microscope, at $100\times$ and $200\times$ magnification. The eggs were differentiated by genus. Egg viability was determined by incubating the processed samples at $20\,°C$ during 4 weeks in H_2SO_4 0.1 N. Viable eggs were those that had a larva inside after incubation. Total and volatile solids were additional parameters analyzed using *Standard Methods* (19th edition, 1995). Rainfall and air temperature were recorded.

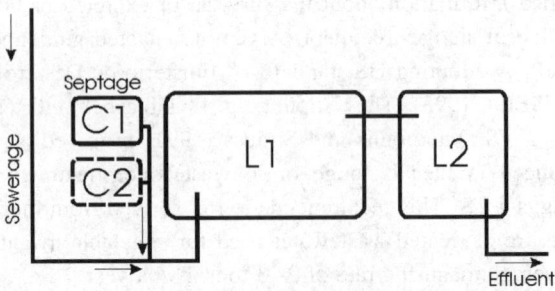

Figure 1 Alcorta (Argentina) pond scheme. C1 + C2 = alternatingly-operated septage sedimentation/digestion ponds; L1 + L2 = facultative and maturation ponds co-treating septage supernatant and raw wastewater

Determining helminth eggs viability in biosolids stored and subjected to natural dewatering in-situ in a primary septage pond (C1)

Sludge of the primary septage pond C1 was allowed to accumulate during the 12-month septage loading period. After this, the supernatant was pumped into the parallel pond C2, which was subsequently receiving the septage. The in-situ *Ascaris* eggs concentration at the beginning of the observation period amounted to 3 *Asc.* eggs/g TS. 42 composite samples of biosolids accumulated in pond C1 were taken over a period of 605 days. Each sample was composited from 6 sub-samples taken from different sites at a depth of 20 cm. The composite samples were collected in 6-L plastic recipients and stored at 4 °C. Sludge samples were taken every two weeks during 20 months.

Determining viability of *Ascaris suum* eggs spiked into septage pond biosolids which were subjected to natural dewatering/drying in a 70-L container (experiment IIA) and to further drying in an experimental drying bed (experiment IIB)

Container experiment (experiment IIA). The *Ascaris suum* eggs spiking experiment was conducted following the finding that helminth egg concentrations in the raw septage and in the accumulated biosolids were too low to conduct inactivation studies in a meaningful manner. *Ascaris suum* eggs were used because they are as resistant as those of *A. lumbricoides* (Carlander and Westrell, 1999), and as they have the advantage of being more easily obtained and less dangerous for humans.

Suspensions of *A. suum* eggs were prepared by removing the uterus of adult female worms, which were collected from the intestines of infected pigs in a slaughterhouse. The last centimetres of the uterus near the vulva containing the mature eggs were cut off. The uterus pieces were put in a test tube to which 15 ml of tap water were added. The uterus pieces were then squeezed with a glass stick to release the eggs. The suspension was then passed through a sieve of 100 μm into another test tube in order to get rid of large fragments. The suspension was concentrated through centrifugation for 10 minutes and the supernatant was withdrawn. In a drop of the pellet, a microscopic count (200 × magnification) of *Ascaris* eggs was made. Five drops of 0.05 ml were counted and the mean concentration of eggs was calculated. This result was extrapolated to the total volume of the pellet. Then the pellet was resuspended in a volume of physiological solution to have a final concentration of approx. 3×10^5 eggs/l.

A test tube with 10 ml *Ascaris* suspension was incubated at 20 °C in order to know whether *Ascaris* eggs in the suspension were potentially fertile. Microscopic observations were made twice a week during 4 weeks.

The container experiment. A 70-L plastic tank was set into the sludge accumulated in pond C1. A portion of sludge and a portion of *A. suum* egg suspension were added to the container simultaneously and homogenised with a stick. New portions of sludge and suspension were added and well mixed until the tank was full. The final concentration in the sludge was 20 *Ascaris* eggs/g TS. The sludge remained in the tank during 8 months, reaching a TS content of 55%. Samples of 400 g were collected at 20 cm depth every 15 days. They were refrigerated at 4 °C prior to analysis.

Viability of Ascaris suum eggs under drying bed conditions (experiment IIB). A plastic box of $40 \times 50 \times 20$ cm size with bottom drainage was used to simulate a drying bed. The sludge, which had previously been stored in an open container for 8 months, was subjected to further drying for 12 months. The sampling was similar to that for the container, yet the size of the composite sample was restricted to 100 g.

Die-off of Salmonella in septage sludge

A suitable methodology for *Salmonella* isolation was investigated in preceding experiments. The respective steps comprised: 1) Enrichment in Rappaport–Vassiliadis broth at 43 °C, 48 hours. 2) Isolation in XLD Agar at 40 °C, 24 hours. A 10-L container filled with solids accumulated in the septage pond was seeded with *Salmonella enteritidis* to study the die-off of *Salmonella*. The container was exposed to environmental conditions during 4 months.

Preparation of Salmonella suspension. Colonies from a pure culture of 24 hours were picked up and suspended in 1 litre of sterile peptoned water. The suspension was then incubated for 24 hours at 35 °C. After centrifugation at 3,500 rpm, the pellet was washed several times with sterile saline solution. After each washing, the suspension was centrifuged and the supernatant was discarded. The pellet was suspended in a sterile saline solution to yield 100 ml. The concentration of *Salmonella* in the suspension was determined by seeding 1 ml of several dilutions in nutrient agar plate by triplicate. The final concentration of the suspension was 7.0×10^9 CFU/ml. When preparing the experimental container experiment, portions of 1 litre of sludge were blended with 1 ml of the suspension of *Salmonella enteritidis* during 1 minute and then poured into the container until a volume of 10 litres was attained.

Sampling. Samples were collected every 3 days during the first 15 days and then once a week during 4 months. pH, total and volatile solids, and faecal coliforms were also analyzed.

Sample preparation for Salmonella detection. 50 g of wet sludge were blended with 500 ml of sterile saline solution + Tween 80 to a 0.1%v/v concentration, during 2 minutes (Method 1682, EPA-821-R-98-004 (USEPA, 1998)).

Isolation, identification and quantification of Salmonella. Samples were analyzed quantitatively (MPN/gTS) applying the method developed by the authors: 10 ml were inoculated in 5 tubes containing 10 ml of double concentration Rappaport–Vassiliadis (RV) broth, 1 ml in 5 tubes with 10 ml of simple broth, 0.1 ml in 5 tubes with 10 ml of simple broth and so on with series of higher dilutions. Control tests inoculating *Salmonella enteritidis* suspension in 5 tubes were done in parallel. The inoculated tubes were incubated at 43 °C for 48 hours. Positive tubes (turbidity) were streaked in XLD agar plates and they were incubated at 40 °C for 24 hours. Typical clear, pink-edged, black-centred *Salmonella* colonies were typified in TSI (alkaline/acid), LIA (positive) and Urease tests (negative). Then XLD plates with *Salmonella* colonies were correlated with the series of positives RV tubes in three significant dilutions (choosing the highest dilution that gives positive results in all 5 tubes), and according to 9221.IV. tables from Standard Methods, (19thedition) 1995, the most probable number were determined. Therefore, *Salmonella* MPN/gTS = MPN index/100 ml × 10 divided by the largest significant dilution volume × % TS (EPA/625/R-92/013 (USEPA, 1999)).

Results and discussion

Determining helminth eggs viability in biosolids stored and subjected to natural dewatering in-situ in a primary septage pond (C1)

The concentrations of total helminth eggs over the 22 months observation period varied considerably and ranged from 0.1 to 17 eggs/g TS. *Ascaris* eggs were prevalent and their concentrations varied between 0.1 and 16 eggs/g TS. *Trichuris* and *Hymenolepis* eggs were also detected. While the A. eggs concentration varied between 2.6 and 16 eggs/g TS during the first 330 days of in-pond storage, its concentration ranged from 0.1 to 1.4 eggs/g TS only between day 346 and 633, except for two samples with higher concentrations (Figure 2a).

The humidity remained constant at 80% during the first 535 days of storage, but gradually decreased to 53% during the subsequent 70 days. The biosolids were removed from the pond and arranged into open piles on day 636. There, the dehydration process continued. Pile samples collected from the piles on day 662 and day 697 exhibited humidities of 39% and 32%, respectively (Figure 2b). Viable helminth eggs were also detected.

Based on the results for the entire observation period of 697 days, there is no apparent direct relationship between sludge humidity and *Ascaris* eggs count. The decrease in eggs concentration from day 330 appears to be related more to storage duration rather than to decreasing humidity, since the humidity started to decrease around day 535 only (Figure 2). Volatile solids (VS) remained approximately constant at 38% of TS during the entire period of observation.

It must be noted that helminth eggs in the primary pond sludge became accumulated during 12 months prior to the beginning of this study. This may have contributed to the weakening of the external membranes of the eggs, which might had been degraded enzymatically by bacteria and fungi present in the sludge.

In the 45 samples analysed, the number of viable helminth eggs was lower than that stipulated by the Argentinian regulations (<1 viable egg/4 g TS). Hence, in this case, helminth eggs cannot serve as a reliable indicator to determine the storage period required to reach a satisfactory hygienic quality. This was the reason for continuing the investigations using artificial additions of *Ascaris* eggs. At the beginning of this experiment, 6 samples were extracted and analyzed in search of *Salmonella*, but all of them were negative.

Viability of *Ascaris suum* eggs spiked into septage sludge stored in a container

Sludge from primary pond spiked with *Ascaris suum* eggs was set into a plastic tank. The tank was located inside this pond and it remained here for almost 8 months. Then the sludge was transferred to a "drying bed" where it remained for the following 12 months.

The concentration of *Ascaris* eggs in the tank was 19 eggs/g TS at the beginning of the experiment. The *Ascaris* eggs concentration varied between 1.8 and 24 eggs/g TS during 213 days (7 months) of storage. This fluctuation might be attributable to a random distribution of the eggs in the sludge, which may have been caused by inadequate mixing during the filling and *A. suum* eggs admixing.

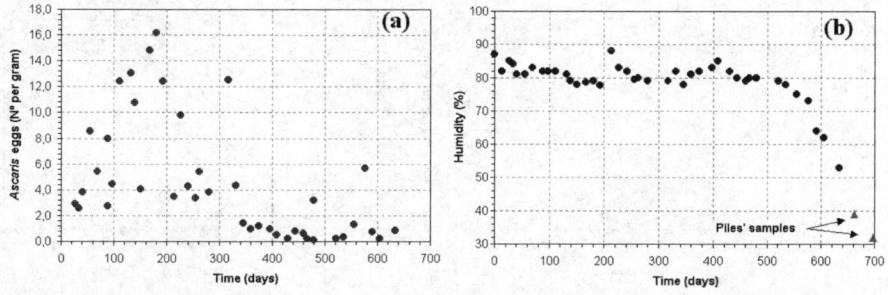

Figure 2 *Ascaris* egg concentration (a) and humidity (b) in pond C1 biosolids vs. storage time

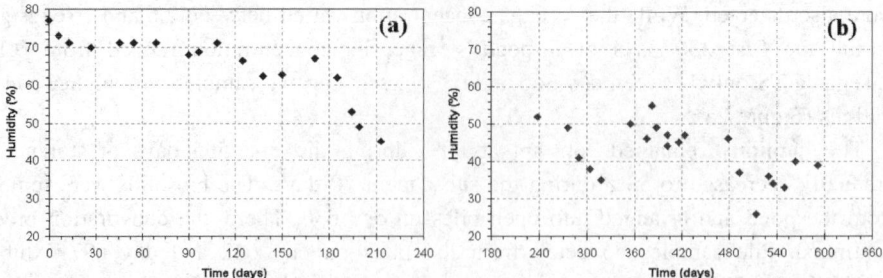

Figure 3 Humidity versus time in container experiment (a) and in drying bed (b)

Humidity ranged from 68% to 77% during the first 120 days (Figure 3a), while the *A. suum* concentration varied between 5.4 and 24 eggs/g TS (Figure 4a). No apparent relation between humidity and eggs concentration could be established during this period since the humidity remained constant. The eggs concentration decreased significantly from 15 to 1.8 *Ascaris* eggs/g TS max. and min., respectively between days 120 and 213. The humidity decreased from 67% to 45% simultaneously.

Viable eggs ranged from 0 to 5.4 viable eggs/g TS and the percentage of viability (ratio of viable vs. total eggs) was between 0% and 45%. However, higher percentages did not correspond to higher concentrations. Figure 4a shows that the total egg concentration varied considerably during the experimental period while the concentration of viable eggs remained almost constant at a relatively low level.

As in the in-situ experiment, the volatile solids (VS) content remained fairly constant at 29%. No correlation between VS and *A. suum* egg survival could be detected.

Viability of *Ascaris suum* under drying bed conditions

The sludges were moved to an experimental "drying bed" after 8 months of dewatering/drying in the experimental container (experiment IIA). The initial *Ascaris* egg concentration was of 13 eggs/g TS. The sludge remained on the drying bed during 12 months (Jan. – Dec. 2003). *Ascaris* egg concentrations ranged between 0.4 and 4.4 eggs/g TS during this period and were thus lower than in the preceding container experiment. The humidity varied between 55% and 26% during the 353 days of the drying phase (Figure 3b).

Figure 4b illustrates the gradual decrease of *Ascaris* eggs concentration. Observed egg viability fluctuated between 0.2 and 2.3 viable *Ascaris* eggs/g TS and the percentage of viable eggs ranged from 32% to 81%. There was no correlation between concentration and percentage, as low concentrations of viable eggs partly coincided with high percentages of viability. Even though the concentration of *Ascaris* eggs dropped in the drying bed relative to that obtained in the tank, the concentration of viable eggs did not decrease in the same proportion (Figs. 4a and b). This indicates that the viable eggs were more resistant than non-viable ones under the same adverse environmental conditions (Johnson

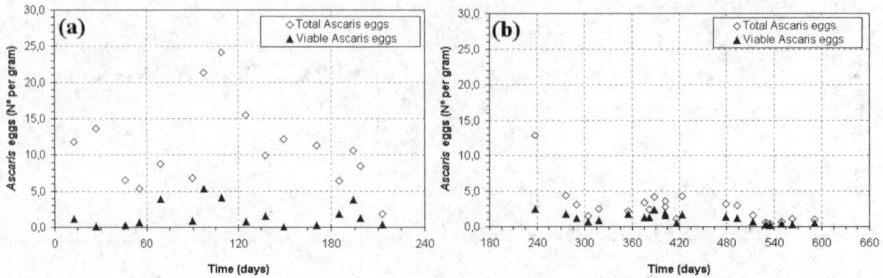

Figure 4 *Ascaris* concentrations vs. time: (a) container experiment (b) drying bed

et al., 1998). Hence, the *Ascaris* eggs that had disappeared could be those that did not have the capacity of developing into larvae because they were less evolved.

The humidity was amounted to 45% at the onset of the drying bed experiment and decreased to approx. 26% after 240 days. Figure 5 shows that the higher viability corresponded rather well with humidity values above 40%. At the end of the experiment, the concentration of viable eggs decreased to 0.2–0.4 viable *Ascaris* eggs/g TS while humidity decreased to below 40%.

A statistical study (ANOVA) was carried out in order to evaluate the relation of viability of eggs with humidity and time. The analysis shows that there was a statistically significant difference (at 95% confidence level) between the mean viable *Ascaris* eggs concentration (1.4 eggs/gTS) from one level of humidity range (40–55%) to another level (26–40%) with a mean viable eggs concentration of 0.59 eggs/g TS. The same statistical analysis showed that time is not an influential factor for viable *Ascaris* eggs counts in the same way as is humidity under these experimental conditions. Viable eggs concentrations, even though they exhibited highly variable values, decreased in parallel to a decrease of humidity to values in the range 26–40% along the whole drying period of 500 days (tank and drying bed).

Die-off of *Salmonella* in septage sludge

To minimize native flora development, the following method for *Salmonella* isolation technique was found most suitable after assaying different enrichment broths and isolation agars at different incubation times and temperatures: 1) Enrichment in Rappaport-Vassiliadis broth at 43 °C, 48 hours. 2) Isolation in XLD Agar at 40 °C, 24 hours. Identification of suspected colonies by biochemical tests: TSI, LIA, Urease and serological confirmation with Group O Antigen. To study the survival of *Salmonella* in the sludge, an experimental container with sludge seeded with *Salmonella enteritidis* was used. The proposed methodology mentioned above for *Salmonella* isolation was applied and the multiple test tubes method used for quantification. The results obtained during 4 months of experience are shown in Table 1.

The concentration of *Salmonella* was 7.0×10^6 MPN/g TS at the beginning of the experiment. The concentration decreased to 1.7×10^5 MPN/gTS within 6 days, remaining at a geometric mean of 1.4×10^5 MPN/gTS for the next 30 days. It can be remarked that the concentration increased in one log unit at approx. 50 days of storage possibly due to abundant rains during the previous days, which created an optimum humidity level for *S.* multiplication (Sidhu *et al.*, 2001). A similar behaviour was detected at 100 days of storage (Figure 6). The concentration of *Salmonella* decreased to a mean value of

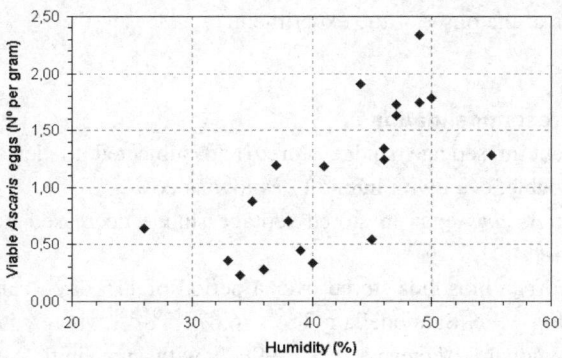

Figure 5 Ascaris viability versus humidity in drying bed

Table 1 Die-off of *Salmonella sp.* in sludge in experimental container

Days	pH	Hum %	TS %	FS %	VS %	F. Coliforms (MPN/gTS)	*Salmonella* (MPN/gTS)	Ambient Temp. (°C)	Sludge Temp. (°C)
1	7.3	71	29	78	22	3.1×10^4	7.0×10^6	15.0	14.5
3	6.9	71	29	75	25	1.7×10^4	2.1×10^6	20.0	20.0
6	7.3	70	30	75	25	1.1×10^4	1.7×10^5	13.0	15.0
10	7.5	68	32	75	25	2.1×10^4	3.4×10^5	14.0	14.5
16	7.5	63	37	78	22	1.2×10^4	1.8×10^5	20.0	19.0
23	7.5	63	37	78	22	3.5×10^3	1.3×10^5	15.0	17.0
30	7.5	52	48	77	23	7.9×10^2	2.8×10^4	20.0	18.0
37	7.2	39	61	77	23	1.0×10^3	2.0×10^5	28.0	26.0
51	6.5	38	62	78	22	1.2×10^4	1.9×10^6	28.0	29.5
65	6.5	27	73	79	21	1.6×10^3	5.6×10^4	27.0	26.5
79	6.5	20	80	79	21	1.8×10^3	4.2×10^4	26.0	26.5
100	7.0	47	53	80	20	7.5×10^2	3.2×10^5	23.0	25.0
115	7.5	44	56	82	18	3.8×10^2	3.0×10^4	27.0	29.0

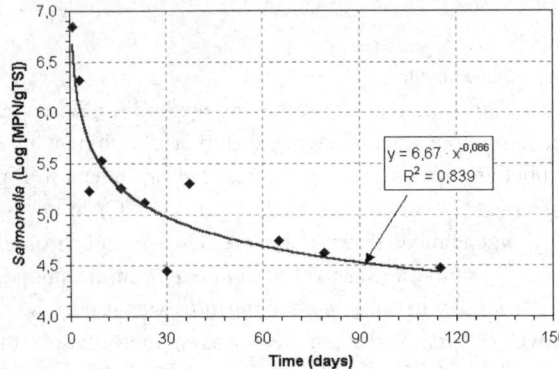

Figure 6 Salmonella die-off in biosolids

3.8×10^4 MPN/g TS after 65 days. In summary, the decay after 115 days of storage was 2 log units (from 10^6 to 10^4 MPN/g TS).

Humidity decreased from 71% to 20% during the first 79 days of the experiment while the *Salmonella* concentration decreased 2 log units during the same period. The concentration of faecal coliforms in the biosolids was 3.1×10^4 MPN/gTS initially. It decreased to 3.8×10^2 MPN/g TS by the end of the experiment. Note needs to be taken that faecal coliforms were already in the stored sludge while *Salmonella* were seeded at high concentrations at the onset of the experiment.

Conclusions and recommendations

Ascaris eggs present in septage sludge stored and subjected to dewatering during 20 months remained viable, even with humidity as low as 26%.

The viability of *Ascaris* eggs in stored septage sludge decreased when the humidity dropped to below 40%.

Salmonella die-off in biosolids stored over a period of 115 days, can be expressed by the equation: $y = \log \text{MPN Salmonella/g TS} = 6.67 \cdot t^{-0.086}$, where t is the elapsed time in days. The removal efficiency was 99%, with an initial concentration of 7.0×10^6 MPN/g TS.

A lower initial concentration of *Salmonella* ($1 \times 10^2 - 1 \times 10^3$ MPN/gTS) is recommended to be seeded in sludge to verify whether the survival curve matches with the one obtained at higher concentrations.

References

Carlander, A. and Westrell, T.A. (1999). *Microbiological and Sociological Evaluation of Urine – Deverting Double – Vault Latrines in Cam Duc, Vietnam*. Minor Field Studies No 91, Swedish University of Agricultural Sciences. International Office.

Feachem, R.G., Bradley, D.J., Garelick, H. and Mara, D.D. (1983). *Sanitation and Disease-Health Aspects of Excreta and Wastewater Management*, John Wiley & Sons, Chichester/NewYork.

Ingallinella, A.M., Sanguinetti, G., Fernández, R., Strauss, M. and Montangero, A. (2000). *Lagunas de Estabilización para Descarga de Líquidos de Camiones Atmosféricos. I Conferencia Latinoamericana sobre Lagunas de Estabilización y Reuso*, DINARA, IWA, Cali, Colombia, Noviembre 2000.

Ingallinella, A.M., Sanguinetti, G., Fernández, R., Strauss, M. and Montangero, A. (2002). Cotreatment of sewage and septage in waste stabilization ponds. *Wat. Sci. Tech.*, **45**(1), 9–15.

Johnson, P.W., Dixon, R. and Ross, A.D. (1998). An *in vitro* test for assessing the viability of *Ascaris suum* eggs exposed to various sewage treatment processes. *Int. J. of Parasitol.*, **28**, 627–633.

Ministerio de Desarrollo Social y Medio Ambiente (2001). *Reglamento para el Manejo Sustentable de Barros Generados en Plantas de Tratamiento de Efluentes Líquidos*. Resolución 97/2001. Argentina.

Schwartzbrod, J. (1998). *Method of Concentration of Helminth Eggs in Sludge, Report*, Laboratoire de Bactériologie-Parasitologie. Faculté de Pharmacie. Université Henri Poincaré, Nancy, France.

Sidhu, J., Gibbs, R.A., Ho, G.E. and Unkovich, I. (2001). The role of indigenous microorganisms in suppression of Salmonella regrowth in composted biosolids. *Wat. Res.*, **35**(4), 913–920.

USEPA (1993). United States Environmental Protection Agency. Part 503. *Standards for the Use or Disposal of Sewage Sludge*. 40 CFR Ch. I (7-1-93 Edition).

USEPA (1998). Environmental Regulations and Technology. Method 1682: *Salmonella spp. In Biosolids by Enrichment, Selection and Biochemical Characterization*. Office of Water, Washington, DC.

USEPA (1999). Environmental Regulations and Technology. *Control of Pathogens and Vector Attraction in Sewage Sludge*, Appendix F – EPA/625/R-92/013 - Office of Research and Development.

Xanthoulis and Strauss (1991). *Reuse of Wastewater in Agriculture at Ouarzazate, Morocco* (Project UNDP/FAO/WHO MOR 86/018). Unpublished mission/consultancy reports.

Isolation of *Salmonella* sp. in sludge from septage treatment plant

G.S. Sanguinetti*, V. Ferrer*, M.C. García*, C. Tortul*, A. Montangero**, D. Koné** and M. Strauss**

*Centro de Ingeniería Sanitaria, Facultad de Ciencias Exactas, Ingeniería y Agrimensura, Universidad Nacional de Rosario, Riobamba 245 Bis, 2000 Rosario, Argentina (E-mail: *cis@fceia.unr.edu.ar*)
**Swiss Federal Institute for Environmental Science and Technology P.O. Box 611, CH-8600 Duebendorf, Switzerland (E-mail: *strauss@eawag.ch*)

Abstract Waste stabilization ponds (WSP) are an often-used option to treat faecal sludges collected from on-site sanitation systems. Since agricultural use is one of the most attractive options for sludge disposal, specific guidelines on the hygienic sludge quality must be fulfilled, such as for viable helminth eggs and *Salmonella* sp. Although *Salmonella* isolation methods are well known for other types of samples, they are not suitable for faecal sludge. The reason can be attributed to the co-existence of a native bacterial sludge flora masking *Salmonella* development, especially if this bacteria is present at low concentrations. In order to select the best methodology for *Salmonella* recovery from septage sludge, different culture media were assayed at different incubation periods and temperatures. The proposed methodology for *Salmonella* recovery from sludge can be summarised as follows: (1) enrichment in Rappaport-Vassiliadis broth at 43°C, 48 hours, and (2) isolation in XLD agar at 40°C, 24 hours. Identification of suspected colonies by biochemical tests: TSI, LIA, urease and serological confirmation with Group O Antigen.
Keywords *Salmonella* isolation technique; septage; sludge

Introduction

Waste stabilization ponds (WSP) are an often-used option to treat faecal sludges collected from on-site sanitation systems. The Sanitary Engineering Center of the University of Rosario, Argentina, in collaboration with the Swiss Federal Institute for Environmental Science and Technology, Switzerland, has been investigating septage treatment in ponds since 1998 (Ingallinella *et al.*, 2002). Research is currently focusing on the inactivation of pathogens. Since agricultural use is one of the most attractive options for sludge disposal, specific guidelines on the hygienic sludge quality must be fulfilled, such as for viable helminth eggs and *Salmonella* sp. Although information on pathogen inactivation in fresh sludge is available, information on pathogen survival in sludges produced during septage treatment is scarce. Therefore, the effect of natural dewatering/drying under temperate climatic conditions on the hygienic quality of accumulated sludges was investigated.

What microbiological criteria are required for biosolids to be used in agriculture? Argentinian Regulations on the use and disposal of biosolids (Ministerio de Desarrollo Social y Medio Ambiente, 2001) stipulate < 3 MPN / 4 g Total Solid (TS) for *Salmonella* and <1 viable egg / 4 g TS for helminth eggs, thereby following USEPA standards (USEPA, 1993).

What are the difficulties encountered at the laboratory for Salmonella detection? Although *Salmonella* isolation methods are well known and widely applied in practice for blood, food, drinking water, and wastewater samples, they do not work likewise for faecal sludge. The reason can be attributed to the co-existence of a native bacterial sludge flora masking *Salmonella* development, especially if this bacteria is present at low concentrations.

The guidelines propose <3 MPN/4gTS. Does this value mean that Salmonella is actually lower than this concentration, or may it be present at higher concentrations and is unable to be detected due to the abundant competitive indigenous flora existing in faecal sludge? Consequently, a methodology must be developed which can be applied on sludge accumulated in WSP treating septage.

This study aimed at developing a suitable method to isolate and identify *Salmonella* sp. in faecal sludge containing a high concentration of competitive indigenous flora. The investigations were conducted with the sludge from a primary pond of the Alcorta treatment plant, Province of Santa Fe, Argentina (Ingallinella *et al.*, 2002).

Material and methods

Before investigating a suitable method for *Salmonella* isolation, sludge samples from a primary pond were analysed applying *Standard Methods* (20th edition, 1999) techniques. As *Salmonella* was not detected in any of the samples collected over a three-month period, other tests, such as seeding sludge with *Salmonella enteritidis,* were carried out to find an appropriate methodology.

Application of the Standard Methods technique

Seven sludge samples from a primary pond treating septage were collected during three months for *Salmonella* isolation. After appropriate homogenization of the sample, 30 g of sludge were blended with 270 ml of sterile physiological solution for one minute. The sample preparation steps recommended in EPA 625/R-92/013-Appendix F were used to determine *Salmonella* (USEPA, 1999). The processed samples were analyzed according to the *Standard Methods* (20th edition, 1999). The samples were seeded in the enrichment broth Selenite-Cystine (Merck) and positive cultures were then streaked for isolation in selective Xilose-Lysine-Desoxycholate agar (XLD, Merck). Suspected *Salmonella* colonies were picked to the following biochemical tests: Triple Sugar Iron (TSI), Lysine Iron Agar (LIA) and urease as well as serologically confirmed using the Antigen "O" Group. Additional parameters analysed comprised faecal coliforms; total (TS), fixed (FS), and volatile (VS) solids. The parameters were analyzed according to the *Standard Methods* (19th edition, 1995). Rainfall and ambient temperature were recorded during the experiment.

Application of the modified technique

Since the preliminary experiment (7 sludge samples) revealed a total absence of *Salmonella* in all samples, development of the influence of the indigenous bacteria masking *Salmonella* development could not be neglected. *Salmonella* is known to be strongly inhibited by native flora present in sludge (Sidhu *et al.*, 2001). It was therefore necessary to investigate another methodology which could better inhibit the competitive bacteria. Therefore, five different sludge samples were seeded with a suspension of *Salmonella enteritidis* (McFarland 0.5). A final concentration equivalent to MPN/g TS $1.5 \times 10^3 - 1.5 \times 10^5$ was obtained after blending. The seeded samples were processed according to EPA 625/R-92/013-Appendix F for qualitative analysis. In order to select the best methodology for *Salmonella* recovery, different culture media proposed by USEPA (1998), *Standard Methods* (20th edition, 1999) and other authors (Gibbs *et al.*, 1997) were assayed at different incubation periods and temperatures.

Enrichment broths:
(1) Selenite-Cystine broth (Merck) in 24, 48 and 72 hours at 41°C
(2) Rappaport-Vassiliadis (Merck) in 24, 48 and 72 hours at 43 °C

Isolation agars:
(1) XLD agar (Merck) in 24 hours at 35°C and 40°C
(2) SS (Salmonella-Shigella) agar (Merck) in 24 hours at 35°C and 40°C

Suspected colonies of *Salmonella* were typified in TSI, LIA and in a urease test. The positive colonies were confirmed by serological tests conducted by the Service of Enterobacteria of the Dr. Carlos Malbrán Institute (http://www.anlis.gov.ar/), Buenos Aires, Argentina.

Results and discussion
Application of the Standard Methods technique
Seven sludge samples from a primary pond treating septage were analysed. The results are given in Table 1.

Application of the modified technique
Based on the results obtained (absence) pertaining to the isolation of *Salmonella* in sludge and the significant development of indigenous flora in the cultures, it was necessary to develop a more accurate methodology to ensure *Salmonella* recovery. *Salmonella enteritidis* were seeded in five different sludge samples as previously described. The results obtained by triplicate tests are shown in Tables 2 and 3.

The abundant development of *Salmonella* at 43°C during 48 hours revealed that Rappaport-Vassiliadis was the best enrichment broth. When comparing the two isolation media, SS agar and XLD agar, better results in *Salmonella* recovery were attained with XLD at 40°C during 24 hours of incubation, as the sludge native flora was greatly inhibited.

The proposed methodology for *Salmonella* recovery from septage sludge can be summarised as follows: first, enrichment in Rappaport-Vassiliadis broth at 43°C, 48 hours;

Table 1 Physical and microbiological parameters of sludge accumulated in a primary pond treating septage

Sample	*Salmonella sp* (presence/absence)	Faecal coliforms (MPN/100 ml)	Faecal coliforms (MPN/g TS)	Humidity %	TS %	FS %	VS %
1	Absence	504	280	82	18	63	37
2	Absence	165	110	85	15	60	40
3	Absence	120	75	84	16	60	40
4	Absence	60	32	81	19	64	36
5	Absence	160	84	81	19	63	37
6	Absence	160	94	83	17	60	40
7	Absence	140	78	82	18	60	40

Table 2 Selection of enrichment broth and incubation time for *Salmonella* development

Sample	Selenite-Cistine broth (41°C)						Rappaport-Vassiliadis broth (43°C)					
	Salmonella			Indigenous flora			*Salmonella*			Indigenous flora		
	24 hs.	48 hs.	72 hs.	24 hs.	48 hs.	72 hs.	24 hs.	48 hs.	72 hs.	24 hs.	48 hs.	72 hs.
1	**	**	*	*	**	**	**	***	**	*	*	**
2	**	*	*	*	**	***	**	***	**	*	*	*
3	*	*	*	*	**	***	*	**	**	*	*	*
4	**	*	*	*	**	**	*	**	**	*	*	**
5	**	**	*	*	**	**	**	***	**	*	**	**

*Poor development ** Moderate development *** Abundant development

Table 3 Selection of isolation media and incubation temperature for *Salmonella* colonial growth

Sample	SS agar				XLD agar			
	Salmonella		Indigenous flora		*Salmonella*		Indigenous flora	
	35°C	40°C	35°C	40°C	35°C	40°C	35°C	40°C
1	*	**	**	*	**	***	**	*
2	**	**	**	*	**	***	*	*
3	*	**	**	**	*	**	*	*
4	**	**	**	*	*	**	**	*
5	*	**	**	**	**	***	**	*

*Poor development **Moderate development ***Abundant development

secondly, isolation in XLD agar at 40°C, 24 hours, and identification of suspected colonies by biochemical tests: TSI, LIA, urease and serological confirmation with Group O Antigen.

Conclusions and recommendations

The most appropriate methodology for isolating *Salmonella sp.* in sludges and inhibiting the development of indigenous flora is: enrichment in Rappaport-Vassiliadis broth, 48 hours at 43°C; then isolation in XLD agar, 24 hours at 40°C; and the use of identification tests: TSI, urease, LIA, serological tests.

Further studies are recommended to verify the effectiveness of the technique: first, using very low concentrations of *Salmonella* in sludge ($<10^2$ MPN/g TS); secondly, enlarging the number of samples to enhance statistical analysis; and finally, testing another inhibitory factor for indigenous flora.

References

Gibbs, R.A., Hu, C.J., Ho, G.E. and Unkovich, I. (1997). Regrowth of faecal coliforms and Salmonellae in stored biosolids and soil amended with biosolids. *Wat. Sci. Tech.*, **35**(11-12), 269–275.

Ingallinella, A.M., Sanguinetti, G., Fernández, R., Strauss, M. and Montangero, A. (2002). Cotreatment of sewage and septage in waste stabilization ponds. *Wat. Sci. Tech.*, **45**(1), 9–15.

Ministerio de Desarrollo Social y Medio Ambiente (2001). Reglamento para el Manejo Sustentable de Barros Generados en Plantas de Tratamiento de Efluentes Líquidos. Resolución 97/2001. Argentina.

Sidhu, J., Gibbs, R.A., Ho, G.E. and Unkovich, I. (2001). The role of indigenous microorganisms in suppression of Salmonella regrowth in composted biosolids. *Wat. Res.*, **35**(4), 913–920.

Standard Methods for the Examination of Water and Wastewater (1995, 1999). 19th and 20th editions. American Public Health Association/American Water Works Association/Water Environment Federation, Washington DC, USA.

USEPA (1993). United States Environmental Protection Agency. Part 503.*Standards for the Use or Disposal of Sewage Sludge*. 40 CFR Ch. I (7-1-93 Edition).

USEPA (1998). Environmental Regulations and Technology. Method 1682: *Salmonella spp.* In *Biosolids by Enrichment, Selection and Biochemical Characterization*, Office of Water, Washington, DC.

USEPA (1999). Environmental Regulations and Technology. *Control of Pathogens and Vector Attraction in Sewage Sludge-Appendix F-EPA/625/R-92/013* – Office of Research and Development.

Potential biogas scrubbing using a high rate pond

G. Mandeno*, R. Craggs**, C. Tanner**, J. Sukias** and J. Webster-Brown*

*School of Geography and Environmental Science, University of Auckland, Private Bag 92019, Auckland, New Zealand (E-Mail: geoffmandeno@hotmail.com)
**National Institute of Water and Atmospheric Research, PO Box 11-115, Hamilton, New Zealand

Abstract The potential to scrub biogas in a high rate pond (HRP) was evaluated using apparatus designed to maximize gas–liquid contact. Experiments compared the removal of carbon dioxide from synthetic biogas by an "in-pond angled gutter" to that by a simulated "counter-current pit." Results showed that the counter current pit has potential for use in biogas scrubbing, with synthetic biogas carbon dioxide composition consistently reduced from 40% to <5%. The in-pond angled gutter was less effective due to bubble coalescence which reduced the total bubble surface area available for gas transfer. Measurement of oxygen levels in the scrubbed biogas showed that despite supersaturation of oxygen in the HRP water, there was little transfer to the biogas, so that explosive methane/oxygen mixtures would not be formed. Theoretical calculations indicated that the amount of biogas likely to be formed during anaerobic treatment of municipal wastewater could be scrubbed in the HRP of the same advanced pond system with little influence on HRP pH, algal growth and treatment performance. These encouraging results justify further research on this method of biogas purification.
Keywords Biogas scrubbing; carbon dioxide; greenhouse gas; hydrogen sulphide; methane; pH; high rate pond

Introduction

Advanced pond systems include initial anaerobic treatment typically in either an anaerobic pond or advanced facultative pond, followed by aerobic treatment in a high rate pond (HRP) and subsequent pond stages. The biogas produced in the anaerobic stage could potentially be collected for power generation (e.g. Green et al., 1995). However, the biogas may require scrubbing to prevent hydrogen sulphide corrosion of engines, pipelines and biogas storage structures, and to improve engine efficiency by reducing the carbon dioxide concentration (e.g. Metcalf and Eddy, 1991; Huang and Crookes, 1998).

Biogas is usually scrubbed using chemical methods, but scrubbing biogas with HRP water could avoid the use and disposal of expensive chemicals. In this study we investigated biogas scrubbing using two simple apparatus to improve gas–liquid contact and thus gas transfer: an "in-pond angled gutter" (Figure 1) and a "counter-current pit" (Figure 2). Because algal photosynthesis in HRPs can result in dissolved oxygen levels exceeding 300% saturation, oxygen levels in the final purified gas were measured, to assess the transfer of oxygen into the biogas and evaluate the apparent potential for explosive methane/oxygen mixtures to form.

HRP have high daytime pH from assimilation of dissolved carbon dioxide by algae during photosynthesis. Elevated pH may have a negative feedback by reducing algal assimilation of carbon dioxide and growth (e.g. Azov, 1982; Azov et al., 1982). However, high pH augments a number of treatment processes in the HRP, including nutrient removal (by phosphorus precipitation and ammonia volatilisation) and disinfection (Nurdogan and Oswald, 1995; Davies-Colley et al., 2003). Thus, significant lowering of HRP pH associated with carbon dioxide addition from biogas may adversely affect these processes, but could improve algal growth and nutrient assimilation. Theoretical calculations were made to examine the potential effects of biogas carbon dioxide addition on HRP pH.

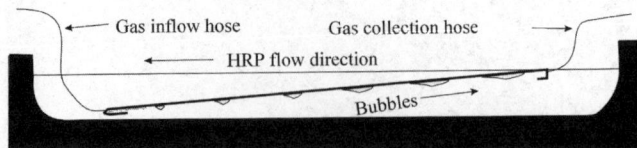

Figure 1 Cross-section of a HRP with an in-pond angled gutter

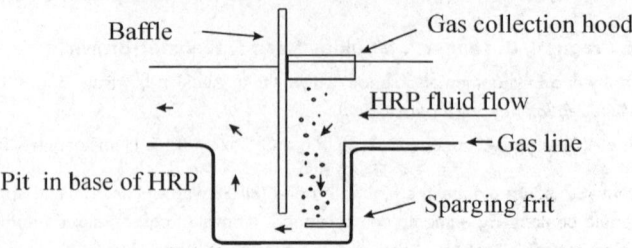

Figure 2 Schematic drawing of the counter current pit concept

Methods

Practical experiments

Two angled gutters (PVC, 3.75% slope, 5 m and 8 m long) were placed within a pilot-scale HRP at the Ruakura Research Facility, Hamilton, New Zealand. A 1 m deep counter-current pit was simulated using a 1 m vertical column (50 mm polycarbonate and PVC pipe) down which HRP water was pumped at a fluid velocity of 15 cm/sec (the horizontal velocity of water in a typical HRP). Both apparatus were evaluated for their ability to remove carbon dioxide from synthetic biogas (40% carbon dioxide, 60% nitrogen). Gas was sparged into each apparatus using an aquarium sparging stone, with gas flow rates fixed at various levels between 250 and 2000 mL/min using a calibrated rotameter. Bubble behaviour was observed and exit gases were collected in 600 mL Tedlar gas bags (Alltech Ltd.) for analysis of carbon dioxide (Draeger tube method CH20301, 5–60% CO_2) and a 250 mL floating gas collector for oxygen measurement (TPS WP-82Y meter with YSI 5739 probe). All gas measurements were carried out in triplicate. The pH, temperature and DO of the HRP water was measured during all experiments, which were carried out during the afternoon (when HRP pH is high) in March 2003 (NZ autumn).

Theoretical calculations

Estimation of biogas production. Assuming a typical volatile suspended solids (VSS) concentration in raw wastewater of 180 g/m^3 (75% of total suspended solids (TSS) 240 g/m^3) and multiplying by values of typical biogas production from anaerobic treatment, (0.5–1.5 m^3 biogas/kg VSS) (e.g. Hobson *et al.*, 1981; Metcalf and Eddy, 1991) gives 0.09–0.27 m^3 biogas/m^3 wastewater. Assuming HRP of the APS system has an 8-day residence time, the relationship of daily potential biogas production (in the upstream anaerobic pond) per m^3 of volume in the HRP is calculated by dividing these values by 8 which gives 0.011–0.034 m^3 biogas/m^3 HRP volume.

Prediction of pH change. The PHREEQC computer chemical equilibrium model was used to predict changes in pH due to biogas carbon dioxide addition. Major ion concentrations in the HRP water measured on several samples using HPIC and bicarbonate titrations and average NH_4-N and DRP concentrations from routine monitoring data measured using *Standard Methods* (APHA, 1998) were used as the model inputs. Starting pH for the modelled HRP water was 9.5 (typical of afternoon pH levels measured in the Ruakura HRPs). The model was set to maintain ionic equilibrium using pH. Carbonate concentrations in the modelled HRP water were increased by amounts equivalent to the number of moles of carbon dioxide contained in the daily biogas production, assuming biogas carbon dioxide concentrations of both 20% and 40%.

Results and discussion
Practical experiments

Upright column: bubble behaviour and carbon dioxide removal. Gas flow rates below 750 mL/min in the upright column resulted in small (2–6 mm diameter) bubbles that were carried away with the downward flow of the pumped HRP water. In practice this would represent loss of valuable methane. Higher gas flow rates resulted in larger (6–10 mm diameter) bubbles that had sufficient buoyancy to slowly rise up the column for collection. Results from exit gas carbon dioxide analysis for gas flow rates between 750 and 2000 mL/min (Table 1) showed that this apparatus has potential for use in biogas purification. Concentrations of carbon dioxide in the purified gas were consistently reduced to less than 5% (the detection limit of the method used) even with gas flow rates up to 1500 mL/min.

Angled gutter: bubble behaviour and carbon dioxide removal. Bubble coalescence in the angled gutters seemed to reduce the extent and consistency of carbon dioxide absorption. The amount of coalescence increased as the gas flow rate was increased. Concentrations of carbon dioxide in the exit gas from the 5m gutter rose from 10% to 24% with increasing gas flow (Table 1). Coalescence of the bubbles (reducing bubble surface area per unit gas volume) as they rose along the gutter reduced the extent of carbon dioxide absorption occurring in the latter length of the gutter. Thus, the longer 8 m gutter (data not shown) did not perform significantly better than the 5 m gutter. Gas flow rates above 1000 mL/min led to large bubbles spilling out from underneath the apparatus.

Oxygen levels in exit gas. Mixtures of methane and pure oxygen are explosive at between 5% and 60% methane content (e.g. Strahle, 1993). The results for carbon dioxide absorption and oxygen desorption shown in Table 1 suggest that exit gas composition would be in excess of 60% methane, and therefore too rich a mixture to combust without the provision of additional oxygen.

Theoretical calculations

Changes in HRP pH due to biogas carbon dioxide addition. Results for PHREEQC modelling of the decrease in HRP water pH with carbon dioxide addition are shown in Table 2. At the maximum amount of biogas carbon dioxide added, HRP pH decreased from 9.5 to 8.7. Because PHREEQC modelling does not take into account the complex and interactive nature of HRP wastewater treatment processes (e.g. Moutin *et al.*, 1992;

Table 1 HRP fluid DO, pH and temperature and exit gas CO_2 and O_2 concentrations at various gas flow rates for 1 m upright column and 5 m gutter using synthetic gas (40% CO_2, 60% N_2)

Apparatus	Gas flow (mL/min)	Temp (°C)	pH	DO (% sat.)	CO_2 (%)	O_2 (%)
1 m upright	750	22	9.38	177	<5	5.4
	1000	22	9.42	185	<5	6
	1500	22	9.42	186	<5	5.2
	2000	22	9.41	185	6	5.9
5 m gutter	250	24	10	245	10	5.3
	500	24	9.6	288	17	3.6
	750	24	9.6	284	18	3.5
	1000	24	9.6	280	24	2.4

Table 2 PHREEQC model outputs of changes to pH after addition of biogas CO_2 to the HRP

m³ biogas / m³ HRP	pH at 20% CO_2	pH at 40% CO_2
0.03	9.2	8.7
0.01	9.4	9.3
0.00	9.5	9.5

Mesple *et al.*, 1996), no attempt is made here to interpret PHREEQC output data in terms of wastewater treatment. However, the relatively low decreases in pH predicted by these calculations lessen concerns over potential reductions of HRP wastewater treatment due to decreased pH, and justify further research into biogas scrubbing in a HRP. The small decline in HRP pH also suggests that any increase in algal biomass concentration as a result of biogas carbon dioxide addition is likely to be minor.

Conclusions

These experiments indicate that the counter-current pit apparatus has the potential to be used for in-pond biogas scrubbing and should be explored further. The angled gutter apparatus was ineffective due to bubble coalescence. Oxygen mass transfer from the HRP water into the gas phase did not lead to the formation of explosive methane/oxygen mixtures under the conditions tested. Chemical equilibrium modelling indicated that the amount of biogas likely to be formed during anaerobic treatment of municipal wastewater could be scrubbed in a HRP with minor effects on HRP pH and algal growth, and thus HRP performance. The encouraging results of this study justify further research into this method of biogas purification.

References

APHA (1998). *Standard Methods for the Examination of Water and Wastewater*, American Public Health Association, Washington.

Azov, Y. (1982). Effect of pH on inorganic carbon uptake in algal cultures. *Applied and Environmental Microbiology*, **43**(4), 1300–1306.

Azov, Y., Shelef, G. and Moraine, R. (1982). Carbon limitation of biomass production in high-rate oxidation ponds. *Biotechnology and Bioengineering*, **24**(1), 579–594.

Davies-Colley, R.J., Craggs, R.J. and Nagels, J.W. (2003). Disinfection in a pilot scale 'advanced' pond system (APS) for domestic sewage treatment in New Zealand. *Water Science and Technology*, **48**(2), 81–87.

Green, F.B., Bernstone, L., Lundquist, T.J., Muir, J., Tresan, R.B. and Oswald, W.J. (1995). Methane fermentation, submerged gas collection and the fate of carbon in advanced integrated wastewater pond systems. *Water Science and Technology*, **31**(12), 55–65.

Hobson, P.N., Bousfield, S. and Summers, R. (1981). *Methane production from agricultural and domestic wastes*, Applied Science Publishers Ltd, Barking, Essex, England.

Huang, J. and Crookes, R.J. (1998). Assessment of simulated biogas as a fuel for the spark ignition engine. *Fuel*, **77**(15), 1793–1801.

Mesple, F., Casellas, C., Troussellier, M. and Bontoux, J. (1996). Modelling orthophosphate evolution in a high rate algal pond. *Ecological Modelling*, **89**(1), 13–21.

Metcalf and Eddy (ed.) (1991). *Wastewater Engineering: Treatment, disposal and reuse* McGraw Hill Inc, New York.

Moutin, T., Gal, J.Y., Halouani, H.E., Picot, B. and Bontoux, J. (1992). Decrease of phosphate concentration in a High Rate Pond by precipitation of calcium phosphate: Theoretical and experimental results. *Water Research*, **26**(11), 1445–1450.

Nurdogan, Y. and Oswald, W.J. (1995). Enhanced nutrient removal in High Rate Ponds. *Water Science and Technology*, **31**(12), 33–43.

Strahle, W.C. (1993). *An Introduction to Combustion*, Gordon and Breach Science Publishers, Amsterdam.

Treatment of wastewater containing high phenol concentrations using stabilisation ponds enriched with activated sludge

M.S. Ramos*, J.L. Dávila*, F. Esparza**, F. Thalasso**, J. Alba**, A.L. Guerrero*** and F.J. Avelar*

*Departamento de Fisiología y Farmacología, ***Departamento de Morfología; Universidad Autónoma de Aguascalientes, Av. Universidad 940, CP 20100, Aguascalientes, Ags., México
(E-mail: fjavelar@correo.uaa.mx)

**Departamento de Biotecnología y Bioingeniería, CINVESTAV-IPN, Av. IPN 2508, San Pedro Zacatenco, CP 07000, México DF, México (E-mail: fesparza@mail.cinvestav.mx; thalasso@mail.cinvestav.mx)

Abstract Treatment of wastewater containing high phenol concentrations (up to 4,000 mg/l, 1,600 kg/ha.d) in laboratory-scale stabilisation ponds enriched with activated sludge was studied. Phenol was biodegraded efficiently, even when fed as the sole carbon source. At influent concentrations of 1,000, 1,300, 1,600, 1,900, 2,500, 3,000 and 4,000 mg/l of phenol (loading rates of 400, 520, 640, 760, 1,000, 1,200 and 1,600 kg phenol/ha.d), the phenol removal efficiencies were 92, 89, 81, 81, 76, 65 and 22%, respectively. At 4,000 mg/l of phenol, the enriched ponds were significantly inhibited. The maximum phenol removal rate observed was 780 kg/ha.d, which is 7.7 times higher than the maximum value reported for attached-growth waste stabilisation ponds. All along the experiments, the enriched ponds showed removal rates 1.8–20.5 times higher than the values observed in control pond (not enriched). The results suggest that enrichment is an effective method to increase xenobiotic removal rates of stabilisation ponds.
Keywords Aerobes; algae; anaerobes; bioflocs; enriched ponds; phenol

Introduction

Waste stabilisation ponds are a cost-efficient method for the treatment of municipal and non-toxic industrial effluents. They are widely used in both developed and developing countries (Pearson, 1996; Zhao and Wang, 1996). This technology is characterised by a low active biomass concentration and low biodegradation rates, which imply large land requirements (Polprasert and Sookhanich, 1995). Several methods based on biofilm growth have been proposed to increase biomass concentration (Shin and Polprasert, 1987; Zhao and Wang, 1996). Recently, an enrichment procedure using activated sludge from a municipal wastewater treatment plant has been proposed to increase the biomass concentration and activity in stabilisation ponds (Avelar et al., 2001; Avelar et al., 2003). This paper presents the results obtained in enriched stabilisation ponds treating high phenol loading rates (up to 1,600 kg phenol/ha.d, 4,000 mg/l). The effects of conventional carbon source and biomass acclimatisation on phenol removal were also studied.

Methods

Four laboratory-scale ponds (83 l volume, 0.8 m long, 0.37 m wide and 0.28 m deep) were built according to the original Eckenfelder design (Eckenfelder and Ford, 1970). Three of them were enriched using 21 l of activated sludge (8 g VSS/l) obtained from the secondary settling tank of a municipal wastewater treatment plant. The working volume of the ponds was completed with synthetic wastewater (Avelar et al., 2003) to obtain a final biomass concentration of 1.8 g VSS/l. After enrichment, biomass was progressively acclimatised to high organic loading rates (1,000 kg COD/ha.d) using gelatine peptone and meat extract as the conventional carbon source (Avelar et al., 2001). The fourth pond (control pond) was not enriched but was inoculated with 21 l of the effluent of a municipal stabilisation pond. Except for the enrichment procedure, the control and the enriched ponds were operated identically. Throughout, the experiments, the hydraulic retention time was kept at 7.5 days and each pond was lighted by 690 lux during 12 h/day. After the acclimatisation, the conventional carbon source was progressively substituted by phenol. Throughout, the experiments, samples of the influent, effluent as well as from the top, medium depth and bottom of the ponds were taken. These samples were used to determine total and soluble COD (open reflux method), phenol concentration (4-aminoantipirine, chloroform extraction method), volatile suspended solids (VSS), heterotrophic aerobes, facultative anaerobes plate count and chlorophyll "a" concentration. Temperature, dissolved oxygen, pH, redox potential and conductivity were also frequently monitored. All measurements were made according to the *Standard Methods for the Examination of Water and Wastewater* (1995).

Results and discussion

During the first experiment, the total organic loading rate was kept constant (1,000 kg COD/ha.d, 2,500 mg COD/l), and the conventional carbon source was gradually substituted by phenol. Each substitution step was maintained during 30 days, and the influent phenol concentrations tested were 20, 150, 375, 660 and 1,005 mg/l. Table 1 shows the results obtained. The phenol removal rates observed in the enriched ponds were 1.8–2.6 times higher than the respective rates observed in the control pond.

During a second experiment, phenol as the sole carbon source was evaluated. With that purpose, in one experimental pond, the conventional carbon source was abruptly removed from the influent. In two more experimental ponds, the conventional carbon source was progressively reduced from 12 to 0.0% of the total organic loading rate, through five reduction steps of 30 days each. Figure 1 shows the results obtained. It was observed that after abrupt elimination of the conventional carbon source, the phenol removal efficiency showed an important decrease, from 65% to 48%. It was also observed that, on the contrary, when the conventional carbon source was progressively removed from the influent, the phenol removal rate increased and reached a removal efficiency of 92%. When phenol was the sole carbon source applied (1,000 mg/l), the phenol removal rate reached 370 kg/ha.d, which is 1.7 and 4.8 times higher than the values previously reported for enriched ponds (Avelar *et al.*, 2003) and attached-growth waste stabilisation ponds (Polprasert and Sookhanich, 1995), respectively.

During a third experiment, phenol as the sole carbon source was fed at concentrations higher than 1,000 mg/l (400 kg/ha.d). Phenol influent concentration was progressively increased through six steps of 30 days each. The corresponding phenol loading rates applied were 520, 640, 760, 1,000, 1,200 and 1,600 kg/ha.d (1,300, 1,600, 1,900, 2,500, 3,000 and 4,000 mg/l, respectively). Figure 2 shows the results obtained. Up to an influent

Table 1 Phenol removal observed in the enriched ponds and in the control pond (not enriched), during the gradual substitution of the conventional carbon source by phenol

Step	Total loading rate (kg COD/ha.d)	Conventional carbon source loading rate (kg COD/ha.d)	Phenol loading rate (kg COD/ha.d)	Phenol removal efficiency (%)	
				Enriched ponds	Control pond
1	983 ± 50.4	964 ± 49.2	19 ± 1.4	75 ± 1.2	5 ± 5.1
2	1000 ± 40.7	857 ± 30.9	143 ± 9.5	61 ± 4.0	34 ± 3.0
3	1026 ± 25.4	669 ± 16.8	357 ± 9.5	62 ± 4.9	24 ± 2.5
4	1004 ± 29.2	378 ± 28.9	626 ± 28.6	70 ± 4.3	30 ± 4.3
5	1093 ± 18.8	136 ± 18.9	957 ± 19.0	65 ± 5.5	27 ± 1.4

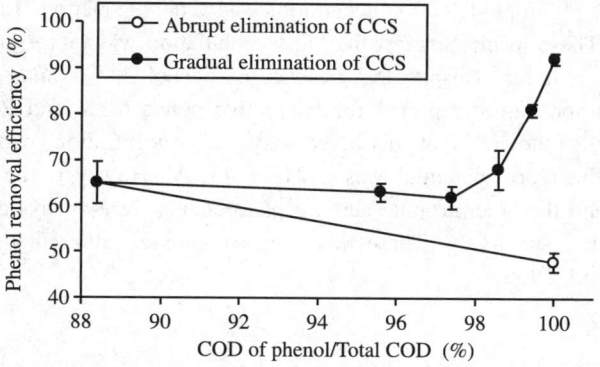

Figure 1 Effect of conventional carbon source (CCS) on phenol removal in enriched ponds. Influent phenol concentration was constant (1,000 mg/l, 400 kg phenol/ha.d)

phenol concentration of 3,000 mg/l (1,200 kg/ha.d), the enriched ponds did not show evidence of significant inhibition. This was not the case when 4,000 mg/l (1,600 kg/ha.d) of phenol was fed, since a clear inhibition was observed. By contrast, the control pond was almost completely inhibited at concentrations superior to 1,600 mg/l (640 kg/ha.d). The maximum phenol removal rate observed in the enriched ponds (780 kg/ha.d) was 7.7 times higher than the maximum value reported for attached-growth waste stabilisation ponds (Polprasert and Sookhanich, 1995). During this experiment, soluble COD analyses (data not shown) confirmed the phenol removal rates observed and suggest that phenol was completely mineralised. Control experiments done with sterilised biomass obtained from the enriched ponds, showed that the abiotic removal of phenol was insignificant (less than 2%). The formation of an active settled biomass bed could explain the results observed in the enriched ponds (Avelar *et al.*, 2001). According to Avelar *et al.* (2003), this settled biomass contained more than 90% of the total microbial population and was about 100 times more active than the suspended biomass. The settled biomass was also significantly more resistant to the phenol inhibitory effects than the suspended biomass (Avelar *et al.*, 2003).

The main biomass indicators were measured before and after phenol was added to the ponds (before experiment one and after experiment three). Before phenol feeding, the heterotrophic aerobes and facultative anaerobes plate counts were 1.1×10^6 and 2.1×10^6 UFC/ml, respectively. After phenol feeding, these indicators were 2.8×10^6 and 1.7×10^7 UFC/ml, for heterotrophic aerobes and facultative anaerobes, respectively.

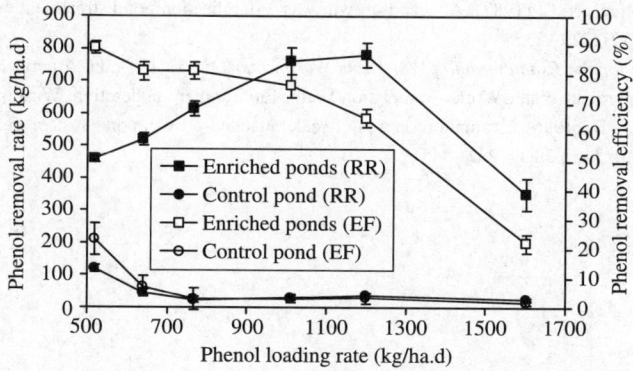

Figure 2 Phenol removal rate (RR) and efficiency (EF) observed at high loading rates in enriched ponds and in the control pond (not enriched). Phenol fed as sole carbon source

By contrast, the chlorophyll "a" concentration showed a reduction, from 5.1 ± 0.6 to 3.1 ± 0.8 mg/l. These results suggest that algae population was more sensitive to phenol than bacterial population. Despite the presence of phenol, the biomass indicators were superior to common values reported for facultative ponds (Shin and Polprasert, 1987). Additional results showed that dissolved oxygen concentration was always below 0.01 mg/l, and the redox potential was −247 ± 43 mV. However, the chlorophyll "a" concentrations and the heterotrophic aerobes plate counts were always above 3 mg/l and 1×10^6 UFC/ml, respectively. These data suggest anoxic, rather than strict anaerobic conditions (Pescod, 1996).

Conclusions

Stabilisation ponds enriched with activated sludge and progressively adapted to the pollutant to be treated were significantly more efficient than conventional ponds. The progressive substitution of conventional carbon source allowed the degradation of phenol as the sole carbon source at removal rates of up to 780 kg/ha.d (1,860 kg COD/ha.d). Moreover, it was surprisingly observed that the absence of peptone and meat extract, previously used as model conventional carbon source, favoured the phenol degradation process. The enrichment procedure could be therefore a potential suitable technique to improve ponds efficiency in the treatment of high loading rates of toxic compounds.

Acknowledgements

This study was supported by a SHIGO-CONACYT grant (19980206029).

References

Avelar, F.J., Martínez-Pereda, P., Thalasso, F., Rodríguez-Vázquez, R. and Esparza-García, F.J. (2001). Upgrading of facultative waste stabilisation ponds under high organic load. *Biotechnol. Lett.*, **23**, 1115–1118.

Avelar, F.J., Martínez-Pereda, P., Thalasso, F., Rodríguez-Vázquez, R., Alba, J. and Esparza-García, F.J. (2003). Phenol removal in upgraded facultative waste stabilisation ponds. *Environ. Technol.*, **24**, 465–470.

Eckenfelder, W.W. and Ford, D.L. (1970). *Water Pollution Control: Experimental Procedures for Process Design*, Jenkins Publishing Co., The Pemberton Press, Austin, Texas, USA.

Pearson, H.W. (1996). Expanding the horizons of pond technology and application in an environmentally conscious world. *Wat. Sci. Tech.*, **33**(7), 1–9.

Pescod, M.B. (1996). The role and limitations of anaerobic pond systems. *Wat. Sci. Tech.*, **33**(7), 11–21.

Polprasert, C. and Sookhanich, S. (1995). Upgrading of facultative ponds to treat a toxic organic wastewater. *Wat. Sci. Tech.*, **31**(12), 201–210.

Shin, H.K. and Polprasert, C. (1987). Attached-growth waste stabilisation ponds treatment evaluation. *Wat. Sci. Tech.*, **19**(12), 229–235.

Standard Methods for the Examination of Water and Wastewater (1995). 19th edn, American Public Health Association/American Water Works Association/Water Environment Federation, Washington DC, USA.

Zhao, Q. and Wang, B. (1996). Evaluation on a pilot-scale attached-growth pond system treating domestic wastewater. *Wat. Res.*, **30**(1), 242–245.

Photosynthetically oxygenated acetonitrile biodegradation by an algal-bacterial microcosm: a pilot-scale study

R. Muñoz, C. Rolvering, B. Guieysse and B. Mattiasson

Department of Biotechnology, Center for Chemistry and Chemical Engineering, Lund University, P.O. Box 124, SE-22100 Lund, Sweden (E-mail: *Raul.Munoz@biotek.lu.se*; *catherinerolvering@web.de*; *Benoit.guieysse@biotek.lu.se*; *Bo.Mattiasson@biotek.lu.se*)

Abstract A 43-L column photobioreactor was tested for the treatment of acetonitrile using a symbiotic consortium consisting of a *Chlorella sorokiniana* strain and a *Comamonas* strain. Complete biodegradation of 1 g acetonitrile/l was achieved in 79 hours under continuous illumination at 500 $\mu E/m^2 s$ and 26 °C. When the photobioreactor was operated at 26 °C under a 14/10 hours light/dark illumination regime at 500 $\mu E/m^2 s$, complete mineralization of 1 g acetonitrile/l was achieved in 111 hours. However, when acetonitrile was supplied at 2 g/l, the biodegradation process was severely inhibited by the increase of pH and NH_4^+ concentration during cultivation. In addition to saving energy for aeration, the microalgae assimilated 33% of the NH_4^+ released during acetonitrile biodegradation, which significantly reduces the need for subsequent nitrogen removal.

Keywords Acetonitrile; algal-bacterial consortium; ammonium removal; biodegradation; photobioreactor

Introduction

By producing O_2 during photosynthesis, microalgae can support the aerobic degradation of many toxic pollutants (Guieysse *et al.*, 2001). In addition of being cheaper compared to mechanical aeration, this is safer (since there is no risk of pollutant volatilization) and reduces the release of CO_2 (produced from the pollutant mineralization) as greenhouse gas in the atmosphere (Oswald, 1988). Algal-bacterial systems are also especially advantageous for the degradation of N-containing organic compounds such are acrylonitrile or acetonitrile because microalgae can assimilate a significant fraction of the NH_4^+ released during the degradation of organonitriles, which reduces the need for subsequent nitrogen removal (Muñoz *et al.*, 2004). These organonitriles, which are commonly found in effluents emanating from acrylonitrile production plants, polymers or metal plating industries, are a priority target due to their high toxicity and sometimes carcinogenic and mutagenic effects on aquatic life (Nawaz *et al.*, 1989).

Traditionally, algal-bacterial associations have been employed in High Rate Algal Ponds (HRAPs). These photobioreactors, although based on photosynthetic oxygenation, still require some kind of mechanical agitation to avoid the formation of O_2 and nutrient gradients and to maintain the microalgae in suspension, which can lead to the stripping out of volatile contaminants (Oswald, 1988). In this case, closed photobioreactors should be prefered.

In this context, a consortium consisting of a *Chlorella sorokiniana* strain in symbiosis with a CH_3CN degrading *Comamonas* strain was tested for acetonitrile biodegradation under various illumination regimes in a 43-L column photobioreactor.

Materials and methods
Microorganisms and culture conditions
Chlorella sorokiniana strain 211/8k was obtained from the Culture Center of Algae and Protozoa (Cambridge, UK). A *Comamonas* strain (GenBank accession number

AY566233; Manolov et al., 2004) was used for CH_3CN degradation. An NO_3-free mineral salt medium (MSM) previously described by Muñoz et al. (2004) was used for microbial cultivation. CH_3CN (99.93%) was obtained from Sigma-Aldrich.

Photobioreactor set-up

The photobioreactor consisted of a Plexiglas column (1500 mm of height × 200 mm of external diameter × 5 mm of wall thickness and 43 L of working volume) illuminated by 8 fluorescent lamps in a circular configuration (GRO-LUX F58W/GRO-T8, Sylvania, Germany). In order to homogenise the reactor broth and to keep the microorganisms in suspension, the liquid medium was continuously recirculated from the top of the reactor to its basement at 1.6 m^3/h, using a centrifugal pump (UPE-25-80 180, GRUNDFOS, Denmark). The temperature in the photobioreactor was maintained constant by recirculating the reactor broth through a compact plate heat exchanger (B5x10, SWEP, Sweden) connected to a thermostatic water bath (CB8-30E, Heto Lab Equipment, Denmark).

The photobioreactor was first supplied with CH_3CN at 1 or 2 g/l in MSM and inoculated with a *C. sorokiniana* strain and a *Comamonas* strain at 30 ± 3 and 10 ± 0 mg/l, respectively. The photobioreactor was operated in batch mode, under continuous illumination at 500 μE/m^2 s and at 26 °C. To study the influence of the illumination regime, the photobioreactor was then supplied with 1 g CH_3CN/l in MSM, inoculated as described above and operated at 26 °C in batch mode under a 14/10 light/dark illumination regime at 500 μE/m^2 s. A sterile control experiment was performed at 8 g acetonitrile/l, pH 9 and 30 °C under continuous illumination at 500 μE/m^2 s for 4 days in order to check for abiotic acetonitrile disappearance.

The dissolved oxygen concentration (DOC), temperature, and illumination were monitored periodically. A liquid sample of 60 ml was periodically withdrawn from the top of the reactor to determine the pH, the culture absorbency and the CH_3CN, CH_3COOH, NH_4^+, total chlorophyll, total inorganic carbon and dry weight concentrations.

Analytical procedures

All the analyses above mentioned were performed as described by Muñoz et al. (2004). CH_3CN and CH_3COOH were analysed by GC-FID and HLPC-UV, respectively. NH_4^+ was determined colorimetrically using a FIAStar® 5000 analyser. The total chlorophyll concentration was determined according to Porra and Grimme (1974). Dry weight was determined according to *Standard Methods* (1998). The total inorganic carbon concentration (CO_2 /HCO_3^-) was determined by titration using a CO_2 test kit HI 3818 (Hanna Instrument, UK).

Results and discussion

CH_3CN biodegradation was carried out under photosynthetic oxygenation since it was neither absorbed nor photodegraded under the experimental conditions tested (data not shown). The green microalga *C. sorokiniana* was not capable of assimilating or biotransforming CH_3CN during photoautotrophic growth (Muñoz et al., 2004), which confirms that CH_3CN disappearance in the culture medium was exclusively due to the *Comamonas* activity. When CH_3CN was supplied at 1 g/l, complete degradation was achieved in 79 hours at a maximum degradation rate of approx. 0.44 ± 0.02 g/l day (Figure 1a). CH_3COOH was detected from the 16th until the 55th hour of cultivation. NH_4^+ concentration constantly increased to reach a final concentration of 0.24 g/l at the end of the cultivation (Figure 1a). The DOC remained below 0.1 mg/l during the entire experimentation and increased up to 14 mg/l after CH_3CN was completely depleted. The pH steadily increased from 6.8 to 8.8.

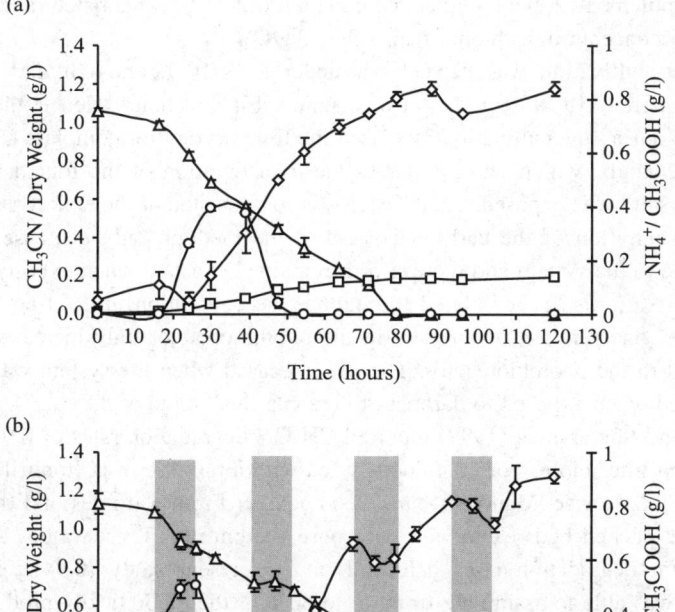

Figure 1 Time course of CH$_3$CN biodegradation (△), CH$_3$COOH production (○), NH$_4^+$ production (□) and dry weight (◇) in the experiments performed at 1 g/l of initial CH$_3$CN concentration under continuous illumination (a) or under a 14/10 hours light/dark illumination regime (b). Shaded areas indicate the periods when the photobioreactor was not illuminated. Vertical bars represent the standard deviation of the presented data

From these results, it can be suggested that CH$_3$CN was hydrolysed by the bacteria into CH$_3$COOH and NH$_4^+$ (as reported by Nawaz et al., 1989). Both the *Comamonas* strain and the microalga strain used in this study have been shown to be able to use CH$_3$COOH as a carbon and energy source (Lee, 2001; Manolov et al., 2004) and it is therefore difficult to state if CH$_3$COOH was consumed by the bacteria, the microalgae or both. However, CH$_3$COOH was ultimately biodegraded into CO$_2$, which was rapidly taken up by *C. sorokiniana* to produce the O$_2$ necessary for further mineralization of CH$_3$COOH. The concentration of dissolved inorganic carbon decreased from 300 mg/l to zero in 64 hours due to photoautotrophic assimilation by microalgae.

When supplied at 2 g/l, CH$_3$CN was rapidly degraded during the first 48 hours of cultivation at a maximum degradation rate of approx. 0.94 ± 0.05 g/l day. After 48 hours of cultivation the degradation rate decreased to 0.04 ± 0.00 g/l day and at 160 hours of cultivation 0.5 g/l of CH$_3$CN still remained in the culture medium. This decrease in the biodegradation efficiency was likely due to microalgae inhibition by the combined effect of high pH and NH$_3$ concentration (Azov and Goldman, 1982). Hence, after 160 hours of cultivation the pH of the culture medium was adjusted to 7 by addition of concentrated H$_2$SO$_4$ in order to decrease the potential toxic effects of NH$_3$ at high pH. In approx. 4 days after the pH adjustment, CH$_3$CN was completely degraded, and the total chlorophyll concentration increased from 6.7 ± 0.3 to 9.4 ± 0.0 nmol/ml. Thus, unless a pH control strategy is implemented, the complete degradation of CH_3CN in a batch mode is not feasible at initial concentrations higher than 1.2–1.3 g/l.

When the cultivation was carried out under a 14/10 hours light/dark illumination regime, complete CH_3CN degradation was achieved in 111 hours. Hence, the presence of dark periods during the cultivation increased the time needed for complete CH_3CN degradation by 32 hours, which was similar to the total duration of the four dark periods to which the system was exposed. CH_3COOH was not detected at the end of the cultivation, but the concentration of the carboxylic acid in the medium only decreased when light was available to the system and increased or remained constant when exposed to darkness (Figure 1b). NH_4^+ also accumulated throughout the cultivation up to a concentration of 0.23 g/l. The concentration of biomass in the photobioreactor only increased when light was supplied to the photobioreactor. The pH increased when the system was illuminated and decreased when exposed to darkness to reach a final value of 8.8.

Dhillon and Shivaraman (1999) reported CH_3CN degradation rates of 0.91 g/l day in a 19-L trickling filter bioreactor continuously fed with a mixture of acetonitrile at acrylonitrile at 0.5 g/l. Likewise, Manolov et al., 2004 achieved removal rates of 1.04 g/l day in a 20-L aerobic packed bed reactor fed with pure acetonitrile. These values are similar to the maximum degradation rates achieved in the present study (0.94 ± 0.05 g/l day). The system was able to assimilate from 46 to 50% of the theoretically produced $N-NH_4^+$, which is in accordance with previously reported data using a similar algal-bacterial consortium (Muñoz et al., 2004). Hence, the microalgae assimilated 33% of the NH_4^+ released during the acetonitrile biodegradation by the Comamonas sp. (the Comamonas strain was not able to nitrify NH_4^+, data not shown).

Conclusions

Photosynthetic oxygenation in closed photobioreactors offers an interesting alternative to conventional aeration since inexpensive O_2 is generated *in-situ* using sunlight and the CO_2 produced from CH_3CN mineralization. Under experimental conditions, CH_3CN biodegradation was totally supported by photosynthetic oxygenation when CH_3CN was supplied at concentrations below 1.2–1.3 g/l. At higher concentration, the combination of high pH and high NH_4^+ concentrations inhibited microalgal activity, leading to process failure.

Acknowledgements

This research was supported by SIDA (The Swedish International Development Cooperation Agency). The Bertil Andersson fund is gratefully acknowledged for financial support in the photobioreactor construction. The Ernhold Lundströms fund is also acknowledged for financial support.

References

Azov, Y. and Goldman, J.C. (1982). Free ammonia inhibition of algal photosynthesis in intensive cultures. *Appl. Environ. Microbiol.*, **43**(4), 735–739.

Dhillon, J.K. and Shivaraman, N. (1999). Biodegradation of organic and alkali cyanide compounds in a trickling filter. *Ind. J. Environ. Protect.*, **19**(11), 805–810.

Guieysse, B., Borde, X., Muñoz, R., Hatti-Kaul, R., Nugier-Chauvin, C., Patin, H. and Mattiasson, B. (2001). Influence of the initial composition of an algal-bacterial microcosm on the biodegradation of salicylate. *Biotech. Lett.*, **24**, 531–538.

Lee, Y.-K. (2001). Microalgal mass culture systems and methods: Their limitation and potential. *J. Appl. Phycol.*, **13**, 307–315.

Manolov, T., Håkansson, K. and Guieysse, B. (2004). Continuous acetonitrile degradation in a packed-bed bioreactor. *Appl. Microbiol. Biot.*, **5**, 567–574.

Muñoz, R., Jacinto, M.A.S., Guieysse, B. and Mattiasson, B. (2004). Acetonitrile biodegradation using algal-bacterial systems. In: *Appl. Microbiol. Biot.*, (in press).

Nawaz, M.S., Chapatwala, K.D. and Wolfram, J.H. (1989). Degradation of acetonitrile by Pseudomonas putida. *Appl. Environ. Microbiol.*, **55**(9), 2267–2274.

Oswald, W.J. (1988). Micro-algae and waste-water treatment. In: *Micro-algal biotechnology*, in Borowitzka, M.A. and Borowitzka, L.J. (eds), Cambridge University Press, Cambridge, UK, pp. 305–328.

Porra, R.J. and Grimme, L.H. (1974). A new procedure for the determination of Chlorophylls a and b and its application to normal and regreening. *Chlorella. Anal. Biochem.*, **57**, 255–267.

Standard Methods for the Examination of Water and Wastewater (1998). 20th edn, American Public Health Association/American Water Works Association/Water Environment Federation, Washington DC, USA.

Microphyte and macrophyte-based lagooning in tropical regions

I.M. K. Noumsi*, J. Nya*, A. Akoa*, R. A. Eteme*, A. Ndikefor*, T. Fonkou** and F. Brissaud***

*Wastewater Research Unit, Faculty of Science, University of Yaounde I, P.O. Box 8404 Yaounde, Cameroon (E-mail: ives_kengne@yahoo.fr)
**Faculty of Science, University of Dschang, P.O. Box 67 Dschang, Cameroon (E-mail: tfonkou@yahoo.fr)
***Hydrosciences, University of Montpellier II, 34095 Montpellier Cedex 05, France
(E-mail: brissaud@msem.univ-montp2.fr)

Abstract A 720 m^2 plant made of 8 ponds in series, set in Yaounde (Cameroon), was successively operated as a macrophyte-based system (type M) from November 1997 to October 98, a microphyte-based system (type m) from October 1999 to September 2000 and a combination of macrophyte and microphyte ponds (type M + m) from May to July 2001. Average applied loads varied over the years; from 420 kg. BOD$_5$ ha^{-1} d^{-1} on the year 1997/98, the loads reached 510 kg BOD$_5$ ha^{-1} d^{-1} in 1999/2000 and 500 in 2001. Though the system became more and more overloaded and sludge accumulated rapidly in the first ponds, it provided average removals of SS, BOD$_5$ and COD that were always higher than 90% whatever the type of lagooning. Performances in the removal of SS, organic matter and the abatement of N-NH$_4^+$ and PO$_4^{3-}$ did not significantly differ according to the type of lagooning and the applied load. Macrophyte lagooning did not show any definitive superiority as to nutrient removal when compared to microphyte lagooning. Microphyte lagooning was the most effective process in faecal indicators removal.
Keywords Cameroon; macrophyte; microphyte; removal; waste stabilisation ponds

Introduction

Natural systems, such as stabilisation ponds and constructed wetlands are nowadays recognised as suitable and cost-effective methods of wastewater treatment in developing countries, particularly those of sub-Saharan Africa which lack capital and qualified manpower (Denny, 1997; Mashauri *et al.*, 2000; Kivaisi, 2001). Indeed, construction and maintenance costs of conventional sewage treatment plants require large amounts of money that countries facing structural and financial adjustment cannot afford. Most of the conventional systems (especially activated sludge systems) present in the Sub-Sahara Africa lack maintenance, are overloaded or simply out of order (SOGREAH, 1993; Agendia, 1995).

Although land may be a limiting factor in densely populated areas, waste stabilisation ponds (WSPs) generally are an effective decentralised wastewater treatment technique for small communities where municipal land is not in short supply. Most significant advantages of WSP technology are: easy construction and operation, low operational costs, and a capability to withstand both organic and hydraulic shock loadings. Furthermore, most developing countries have warm tropical and subtropical climates that favour a year-round high biological activity and productivity, hence better efficiency of natural systems.

Among the natural systems present in Africa, WSPs, based on either microphytes or macrophytes, are gaining importance. In Cameroon for instance, all the towns planning sewerage programmes advocate their use for sewage treatment since almost all the activated sludge plants built after independence are out of function. The topography of the town of Yaounde, with many small watersheds allowing a gravitational flow of wastewater to inhabited lowlands should favour the adoption of decentralised WSP systems (SOGREAH, 1993). However, there is a growing controversy about the efficiency and possible reuse of by-products of the various types of WSP, particularly a confrontation between microphyte and macrophyte-based systems.

Several studies have shown the ability of both types to remove nutrients and pathogens from raw sewage (Reddy and Smith, 1987; Agendia, 1995; Verhoeven and Meuleman, 1999; Amahmid et al., 2002), but very few have compared the treatment efficiency of both systems in the same operation and environmental conditions (Mandi et al., 1993; Ouazzani et al., 1995). Most of the works dealing with this comparison and reported in the literature have been carried out in Europe and America, the inlet wastewater often being primary or secondary effluents with a relatively low pollution concentration (Reddy and Smith, 1987; Vyzamal et al., 1998). To the best of our knowledge, no such comparative investigation has been carried out in Western Africa, probably because of the slow expansion of this eco-technology in the region. Furthermore, the performance of these plants when receiving highly concentrated influents, as is the case in most African cities, has been little documented.

A WSP facility was built in 1986 by local authorities to treat domestic sewage from approximately 650 inhabitants of a residential quarter of Yaounde (Cameroon). The plant received high organic concentration wastewater ($BOD_5 > 450 \, kg \, ha^{-1} d^{-1}$). It offered the opportunity to document the controversy in the local context. This paper presents the results of a comparative analysis of the treatment efficiency of the plant operated in three ways: floating macrophytes (LM), microphytes (Lm) and a combination of macrophyte and microphyte ponds (LM + Lm).

Materials and methods

This work was carried out at the Biyem-Assi pilot plant constructed to treat approximately $45 \, m^3 d^{-1}$ of raw sewage from a quarter of the capital city Yaounde (latitude 3° 52 N, longitude 11° 32 E, altitude 760 m). The total surface of the plant is approximately 0.1 ha (Figure 1). The system is made up of eight ponds dug in lateritic soil and separated with dykes of compacted soil (Table 1). The first pond (B0) is designed as an anaerobic pond. The other ponds (B1–B7) are rectangular with a length to breadth ratio of 5, allowing the hydraulic pattern to resemble a piston-like flow and to minimize the dead zones; with a velocity generally very small in Yaounde the wind hardly contributes to mixing. Wastewater flows by gravity and the hydraulic retention time is between 9 and 16 days.

From November 1997 to October 1998, ponds B1–B7 were planted with floating macrophyte *Pistia stratiotes* (LM), water hyacinth being absent in Cameroon. Previous studies carried out in the same plant showed that *Pistia stratiotes* should be harvested between the 15th and the 25th day of cultivation; this duration takes the rapid turnover of the absorbed nutrients into account (Agendia, 1995). For the current studies, $2/5^{th}$ of each pond were harvested every 25 days to minimize the harvesting stress.

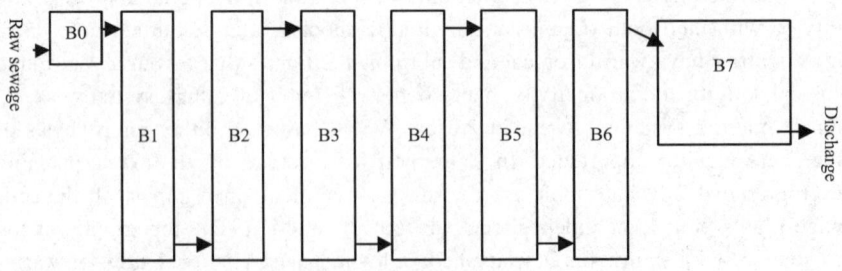

Figure 1 Layout of the Biyem-Assi waste stabilisation ponds (B0: anaerobic settling pond; B1–B7: lagoons)

Table 1 Characteristics of the Biyem-Assi sewage treatment plant

Pond	Length (m)	Breadth (m)	Depth (m)	Surface (m^2)	Volume (m^3)
B0	7.5	3.5	1.8	26.3	47.3
B1	22	4.4	0.7	97.0	67.9
B2	22	4.3	0.8	94.0	75.2
B3	22	4.4	0.9	96.0	86.5
B4	22	4.3	0.8	95.0	76.0
B5	22	4.3	0.9	95.0	85.5
B6	22	4.4	0.9	96.0	86.4
B7	18	6.6	0.5	119.0	59.5

After this first stage, the plant was allowed from October 1999 to September 2000 to function mainly with microphytes (Lm) by clearing ponds B2–B7, where micro algae could develop. Pond B1 was planted with a mixture of *Ipomoea aquatica* and *Enydra fluctuans* since these macrophytes help to reduce the emanation of foul odors.

Lastly, from May to July 2001, macrophyte and microphyte ponds were combined (LM + Lm). Pond B1 was left vegetated with *Enydra* and *Ipomoea*, the following three ponds, B2–B4, were planted with *Pistia stratiotes* and the rest (ponds B5–B7) remained without macrophytes.

Organic load (COD, BOD$_5$, SS), nutrients (N-NH$_4^+$, PO$_4^{3-}$) and bacterial indicators of fecal contamination (fecal coliforms and fecal streptococci) were monitored during the experiments two to five times per month. Water was sampled at the exit of each pond except for fecal indicators which were collected only at the entry and exit of the plant during the first stage. Analyses were performed using a Hach manometric apparatus for BOD and Hach DR 2010 for the rest of physico-chemical parameters (Hach, 1992). Fecal coliforms and fecal bacteria determination were done by MPN or membrane filtration technique using appropriate media (APHA, 1992). Rainfall and mean daily temperature are given in Figure 2.

Results

Despite some variations, the average quality of inlet wastewater did not change dramatically over the investigation period (Table 2). However, high standard deviation values showed a significant variability in the short term. As often observed in developing countries where access to water is limited either by its cost or by a short supply, suspended solids and organic matter concentrations were two to three times higher than those generally observed in European domestic wastewater. Studies generally assume that approximately 50–70 L are rejected per inhabitant per day against 160–200 L in developed countries (Nduka Okafor, 1985; Eckenfelder, 1982). High pollution content could also be explained by womens' extra activities such as traditional cassava processing and meal cooking. Furthermore, it also suggests poor maintenance of the wastewater collection network.

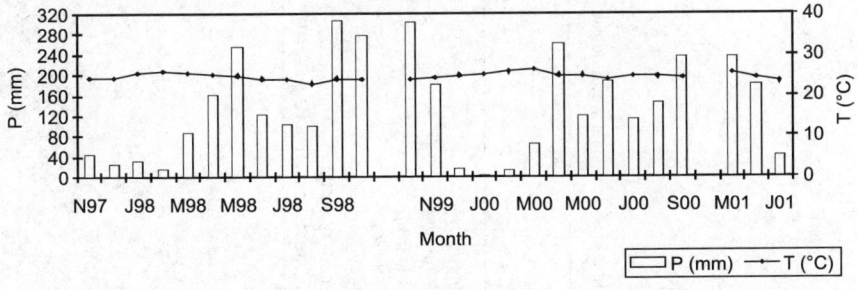

Figure 2 Rainfall and temperature in Yaounde

Table 2 Inlet and outlet water quality

	LM		Lm		LM + Lm	
	Inlet	Outlet	Inlet	Outlet	Inlet	Outlet
SS (mg/L)	835 ± 572	30 ± 19	880 ± 389	87 ± 32	865 ± 325	145 ± 90
BOD_5 (mg/L)	687 ± 138	103 ± 68	756 ± 136	148 ± 76	803 ± 200	128 ± 63
COD (mg/L)	1630 ± 726	140 ± 72	2367 ± 914	229 ± 131	1946 ± 812	250 ± 117
PO_4^{3-} (mg/L)	32 ± 39	12 ± 8	52 ± 20	16 ± 7	46 ± 10	13 ± 4
$N-NH_4^+$ (mg/L)	121 ± 45	52 ± 24	144 ± 55	52 ± 16	114 ± 34	39 ± 20
Fecal coliforms (nb/100 ml) × 10^6	39 ± 50	0.19 ± 0.14	85	0.015	79 ± 81	0.46 ± 0.3
Fecal streptococci (nb/100 ml) × 10^6	30 ± 39	0.067 ± 0.12	13	0.006	40 ± 78	0.03 ± 0.04

Average applied loads varied over the years, mainly because of inflow variation. From respectively 520, 420, 1000 and 19 kg ha^{-1}d^{-1} SS, BOD$_5$, COD and PO$_4^{3-}$ on the year 1997/98, the loads reached 590, 510, 1600 and 35 kg ha^{-1}d^{-1} SS, BOD$_5$, COD and PO$_4^{3-}$ in 1999/2000 and then decreased to 540, 500, 1200, and 29 kg ha^{-1}d^{-1} SS, BOD$_5$, COD and PO$_4^{3-}$ when the ponds were operated as a combined macrophyte–microphyte system (Table 3). Mean ammonia load was about 75 kg N-NH4$^+$ha^{-1}d^{-1} not to mention organic nitrogen which could not be analyzed. Fecal indicator concentrations in the influent were generally above 10^7 bacteria/100 ml. Compared to the European standards of microphyte ponds, the Biyem-Assi facility would be appraised as being 4 to 5 times smaller than required. Fortunately, the climate of Yaounde is tropical with small seasonal variations. Mean water temperatures were between 22.9 and 25.9 °C, with a range of temperature less than 1.5 °C. However, the organic load, higher than 420 kg BOD$_5$ ha^{-1}d^{-1}, is almost two times higher that the maximum generally recommended for these plants (Mara, 1976). Despite a real under sizing, all types of WSP provided high pollutant removal.

Pollution patterns along the plant showed fair similarity whatever the type of WSP (Figure 3). A great reduction of the pollutants was observed mainly in the first three ponds (B0–B2), followed by a relatively slight decrease of pollutant contents in the following ponds (B3–B7). Reduction of SS and organic matter in the first ponds accounted for approximately 60–70% of the total pollution removed. As total phosphorus and Kjeldahl nitrogen could not be measured, assessing nutrient removal requires some caution; nevertheless, as organic P and N are likely to be mineralized in the first ponds, it is not inappropriate to think that the first three ponds also play a very important part in the removal of N and P. Though concentrations decreased slowly in ponds B3–B7, their role must not be underestimated. Indeed, thanks to evaporation and possible infiltration, mean water flow rate decreased steadily from 44, 48 and 45 m^{3}d^{-1} at the inlet of the plant to respectively 15, 25 and 17 m^{3}d^{-1} at the outlet of pond B7. Therefore, mass flow diminished more than is shown through concentration values.

The rapid decrease of pollution in the first three ponds (B0–B2) highlights the predominance of physical processes in the removal of pollutants as reported by Kadlec (1997). The partitioning of the plant into a series of ponds led to sedimentation and important sludge accumulation in the first ponds. A 25 cm/year deposition of sludge was noted in pond B1, causing a rapid filling of the ponds. Therefore water residence time was smaller during Lm and Lm + LM experiments in this pond, leading to a lack of efficiency in BOD and P abatement (Figure 3). Analysing pond malfunction in France, Racault (1993) suggested an increase of the surface and the depth of the first ponds to limit the frequencies of cleaning-out.

Despite the high concentration of the influent, the treatment efficiency of the plant remained relatively high. Some little differences were nevertheless observed amongst the

Table 3 Applied loads and pollution removal

	LM		Lm		LM + Lm	
	Load (kgha^{-1}d^{-1})	Removal (% or log units)	Load (kgha^{-1}d^{-1})	Removal (% or log units)	Load (kgha^{-1}d^{-1})	Removal (% or log units)
SS	513	94 ± 4	592	94 ± 3	542	92 ± 7
BOD$_5$	422	95 ± 3	508	89 ± 7	503	93 ± 5
COD	1020	97 ± 2	1590	95 ± 4	1220	94 ± 4
PO$_4^{3-}$	19	81 ± 15	35	81 ± 4	29	89 ± 4
N-NH$_4^+$	76	84 ± 9	97	79 ± 7	72	87 ± 8
Fecal coli.		2.0		3.8		2.8
Fecal strept.		3.0		3.3		2.2

Removal is a percentage of the applied load

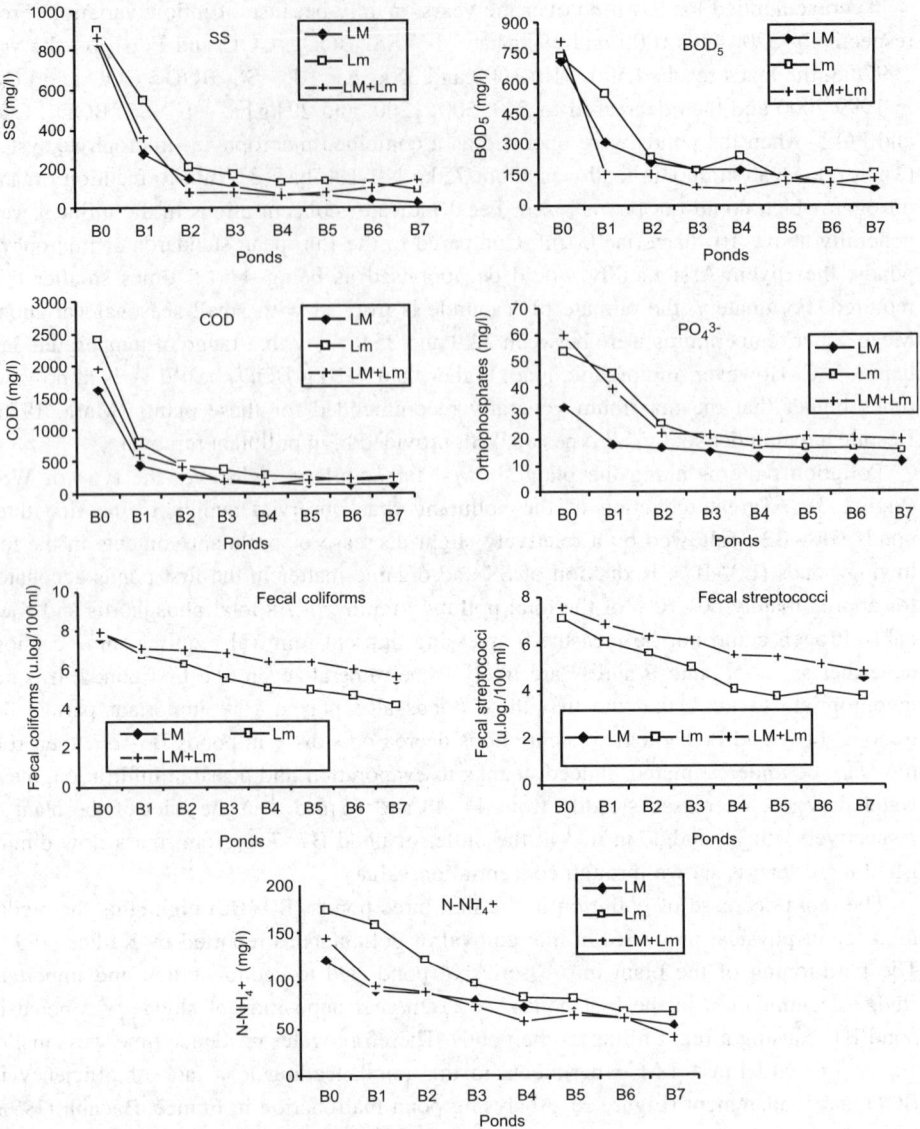

Figure 3 Pollutant patterns along the plant (LM: ponds with macrophytes; Lm: ponds with microphytes; LM + Lm: ponds with macophytes and microphytes)

types of treatment (Table 3). The macrophyte-based treatment system (LM) appeared to be the best system in the removal of organic load (mean COD removal: 97%, mean BOD_5 removal: 95%), followed by the combined LM + Lm system (mean COD removal: 94%, mean BOD_5 removal: 93%) and the microphyte-based one (Lm) with a rough average removal of 95% and 89% for respectively, COD and BOD_5. Removal of SS was very effective in LM, with a very low outlet content of $30\,\text{mgL}^{-1}$. Though the outlet content was higher in Lm, mean SS removal was 94% in both cases due to differences in hydraulic flow rates. The relative high efficiency of the macrophyte based-system to remove organic load and suspended matters could be due the presence of roots which filter the water by trapping the organic particles and enhance their sedimentation speed by reducing the flow velocity of the water (Brix, 1997). Furthermore, there was a greater development of algae in the last ponds when they were operated without macrophytes.

Regarding the nutrients, 87% of $N-NH_4^+$ and 89% of PO_4^{3-} were removed by the combination of macrophytes and microphytes (LM + Lm) against 79% and 81% respectively for Lm and 84% and 81% respectively for LM. These figures do not take organic N and P into account, therefore actual removals of N and P were likely to be higher than shown in Table 3. For the same reason, discussing and comparing P and N removals is not that easy. It was difficult to explain the best abatement of nutrients in LM + Lm. $N-NH_4^+$ removal looked slightly higher in LM than in Lm. Removal of PO_4^{3-} was about the same. Equivalent nutrient removal efficiency of the plant with and without macrophytes certainly clashes with the assertion presenting aquatic macrophytes as a key factor for the removal of nutrients in natural systems. Ouazzani et al. (1995) found similar results when comparing the efficiency of waste stabilisation ponds with and without plants in arid conditions. Indeed, they found that 60% and 62% of NH_4^+ and PO_4^{3-} respectively were removed in microphyte-based waste stabilisation ponds against 50% and 26% respectively in water hyacinth ponds in Marrakesh (Morocco). The increase of phosphorus precipitation in response to pH fluctuations and a high volatilisation of ammonia or nitrification/denitrification in free water may certainly explain these observations.

The reduction of faecal indicators was considerably higher in Lm than Lm and LM + Lm. Indeed, fecal coliform removal was 3.8 log. units when the plant was vegetated with microphytes against 2 and 2.2 respectively for LM and LM + Lm. Fecal streptococci removal was approximately the same for LM and Lm (3.3 and 3 log. units respectively), but slightly higher than that of LM + Lm (2.8 log units). The higher fecal coliforms numbers in the effluent in the presence of floating macrophytes (LM) confirm their negative impact on die-off. The removal efficiencies obtained were in accordance with those mentioned in others studies (Mandi et al., 1993; Ouazzani et al., 1995; George et al., 2002). The key factors generally pointed out are bactericidal effects of sunlight and high pH values (Mezrioui, 1987; Davies-Colley et al., 1999).

Conclusions

Despite the low dimensioning of the plant with respect to the organic load, the high reduction of pollutants confirms the flexibility of the natural systems. Indeed, the pond surface of Biyem-Assi WSP is less than $1\,m^2$ per p.e, which is very low compared to the $2-5\,m^2$ p.e^{-1} generally recommended in hot climates (Mara, 1976). The abatement is higher than 2 log. units for the fecal indicator organisms, more than 90% for the organic load and more than 80% for the nutrients. This system, when properly maintained, particularly with respect to desludging, would provide higher quality effluent whatever the type of WSP. However, with respect to European standards, pollution concentration remains relatively high in the effluent. If required by water uses in the receiving body, this situation might be improved by adding supplementary ponds to the Biyem-Assi facility. These data reinforce the place that such natural systems could take in the battle against pollution of the environment and for the improvement of the sanitary quality of water in the African context.

Slight differences were observed in the performance of the different types of WSP tested. The macrophyte-based treatment showed a small advantage over other WSP types in its ability to remove organic load, while the microphyte-based treatment was more able to reduce the fecal indicators and the combination of macro and microphytes to remove nutrients. These results revealed the high complexity of a cut-and-dried choice of a natural system to treat wastewater. This choice should therefore be done in accordance to the priority of the wastewater treatment in the area of concern. In Cameroon, as in many other African countries, the protection of populations against waterborne diseases is of primary concern. The microphyte-based treatment better responds to this requirement and should therefore be promoted. The respective management constraints must be considered; among which the regular harvesting of floating macrophites is the most exacting, not to mention the disposal and fate of the harvest. Moreover, environmental impacts must be taken into account, particularly mosquito breeding (Kengne *et al.*, 2005). However further investigations for a better dimensioning, cost-benefit analysis and environmental impact of the system need to be done.

Acknowledgements

We are indebted to IFS (International Foundation for Science, Grant no 1580/3F) and the French Ministry of Foreign Affairs (Programme CAMPUS Fédérateur de Recherche sur l'Assainissement des Eaux Usées en Afrique Subsaharienne) who contributed financially to this work.

References

APHA (1992). *Standards methods for the examination of water and wastewater*, 18th edn, American Public Health Association, Washington DC Part 9000: Microbial examination.

Agendia, P.L. (1995). *Treatment of Sewage using Aquatic Plants: Case of the Biyeme Assi Domestic sewage (Yaounde)* Doctorat d'Etat thesis, University of Yaounde I.

Amahmid, O., Asmama, S. and Bouhoum, K. (2002). Urban wastewater treatment in stabilization ponds: occurrence and removal of pathogens. *Urban Water*, **4**, 255–262.

Brix, H. (1997). Do macrophytes play a role in constructed wetlands? *Wat. Sci. Tech.*, **35**(5), 11–17.

Davies-Colley, R.J., Donnison, A.M., Speed, D.J., Ross, C.M. and Nagels, J.W. (1999). Inactivation of faecal indicator microorganisms in waste stabilization ponds: interactions of environmental factors with sunlight. *Wat. Sci. Tech.*, **33**(5), 1220–1230.

Denny, P. (1997). Implementation of constructed wetlands in developing countries. *Wat. Sci. Tech.*, **35**(5), 27–34.

Eckenfelder, W.W. (1982). *Gestion des eaux usées urbaines industrielles: Caractérisation, Techniques d'épuration, Aspects économiques*, Lavoisier, Paris.

George, I., Crop, P. and Servais, P. (2002). Fecal coliform removal in wastewater treatment plants studied by plate counts and enzymatic methods. *Wat. Res.*, **36**, 2607–2617.

Hach, (1992). *Water Analysis Handbook*, 2nd edition, Colorado.

Kadlec, R.H. (1997). Overview: surface flow constructed wetlands. *Wat. Sci. Tech.*, **32**(3), 1–12.

Kengne Noumsi, I.M., Akoa, A., Atangana Eteme, R., Nya, J., Ngniado, P., Fonkou, T. and Brissaud, F. (2005). Mosquito development and biological control in a macrophyte-based wastewater treatment plant. *Wat. Sci. Tech.*, **51**(12), 201–204.

Kivaisi, A.K. (2001). The potential for constructed wetland for wastewater treatment and reuse in developing countries. A review. *Ecol. Eng.*, **16**, 545–560.

Mandi, L., Ouazzani, K., Bouhoum, K. and Boussaid, A. (1993). Wastewater treatment by stabilisation ponds with and without macrophytes under arid climate. *Wat. Sci. Tech.*, **28**(10), 177–181.

Mara, D. (1976). *Sewage Treatment in Hot Climates*, ELBS and Wiley, Chichester, UK.

Mashauri, D.A., Mulungu, D.M.M. and Abdulhussein, B.S. (2000). Constructed wetland at the University of Dar es Salam. *Wat. Res.*, **34**(4), 1135–1144.

Mezrioui, N. (1987). *Etude expérimentale des effets du pH, du rayonnement et de la température sur la disparition des bactéries d'intérêt sanitaire et évaluation de la résistance aux antibiotiques de Escherichia coli lors de l'épuration des eaux*. Thèse de Doctorat, Université des Sciences et Techniques du Languedoc, Montpellier II.

Nduka Okafor, (1985). *Aquatic and waste microbiology*, F.D.P., Nigeria, 169 P.

Ouazzani, K., Bouhoum, K., Mandi, L., Bouarab, L., Habbari, K.H., Rafiq, F., Picot, B., Bontoux, J. and Schwartzbrod, J. (1995). Wastewater treatment by stabilization ponds: Marrakesh experiment. *Wat. Sci. Tech.*, **31**(12), 75–80.

Racault, Y. (1993). Pond malfunction: case study of three plants in the South-West of France. *Wat. Sci. Tech.*, **28**(10), 183–192.

Reddy, K.R. and Smith, W.H. (1987). *Aquatic Plants for Wastewater Treatment and Resource Recovery*, Reddy, K.R. and Smith, W.H. (eds), Magnolia Publishing Inc., Orlando, Florida.

SOGREAH (1993). *Etude du Plan Directeur d'Assainissement de la ville de Yaoundé. Schéma Directeur*, Mémoire.

Verhoeven, J.T.A. and Meuleman, A.F.M. (1999). Wetlands for wastewater treatment: opportunities and limitations. *Ecol. Eng.*, **12**, 5–12.

Vyzamal, J., Brix, H., Cooper, P.F., Harberl, R., Perfler, R. and Laber, J. (1998). Removal mechanisms and types of constructed wetlands. In *Constructed wetlands for wastewater treatment in Europe*, Vyzamal, J., *et al* (eds), Backuys Publishers, Leiden, The Netherlands, pp. 17–66.

The effect of water hyacinths for wastewater treatment under Cuban climatic conditions

C. Rodriguez* and P.D. Jenssen**

*Hydraulic Research Center, Higher Polytechnical Institute "José Antonio Echeverría", Havana City, Cuba
(E-mail: celia@cih.cujae.edu.cu)
**Department of Mathematical Sciences and Technology, Norwegian University of Life Sciences, 1432 Aas, Norway (E-mail: petter.jenssen@umb.no)

Abstract The purification capacity of systems using floating aquatic plants depend on the climatic conditions under which they are used. This study from Cuban conditions evaluate the effects of the organic loading rate, hydraulic loading rate and water depth on the purification capacity of water hyacinths, as well as the effect of some climatic variables on the kinetics of the treatment processes. The experimental system consisted of two consecutive tanks simulating a system of ponds in series. The water depths used were 0.5 m and 1.12 m. In the shallower system with shorter retention times and greater superficial organic loading higher removal efficiencies are obtained. With the data obtained, empirical relations were sought. From these correlations it is possible to determine the values for some parameters used in the design of aquatic treatment systems with water hyacinths. The results revealed a relationship between the purification capacity of the water hyacinth and its velocity of growth. The specific velocity of growth varied with the months of the year and was associated with the temperature and the solar radiation. A multiple correlation equation describing these relations was obtained.
Keywords Aquatic systems; design criteria; water hyacinth; wastewater treatment; tropical climate

Introduction

Systems using floating aquatic plants are low cost and effective wastewater treatment systems (Tchobanoglous, 1993). Water hyacinth is the most common plant used in these systems and functions as a biological filter for removal of organic as well as inorganic contaminants, toxic substances and pathogenic microorganisms (DeBusk and Reddy, 1991; Sipauba-Tavares *et al.*, 2002; Yedla *et al.*, 2002; Kim *et al.*, 2003; So *et al.*, 2003). The purifying capacity of water hyacinth depends on its growth. The growth of the plant varies with the climatic conditions, as well as the composition of nutrients of the water and the methods of harvesting (DeBusk and Reddy, 1987; Aoyama and Nishizaki, 1993; Costa *et al.*, 2003). In this paper an experimental study is presented which is carried out at pilot scale. The objective was to evaluate the effects of the organic loading rate, hydraulic loading rate and water depth on the purification capacity of water hyacinth ponds, as well as the effect of some climatic variables on the kinetics of the processes.

Materials and methods

The experimental system consisted of two consecutive tanks simulating a system of ponds in series. In one system water hyacinths were grown, while the other was without plants and functioned as a control. The water depths used were 0.5 m and 1.12 m.

The system with 0.5 m water depth was made using four fibrocement tanks with the following dimensions: 1.50 m (length); 0.56 m (depth); 0.74 m (top width) and 0.51 m (bottom width), the surface area and the effective volume being $1.11\,m^2$ and $0.47\,m^3$ respectively. The 1.12 m deep pilot system was similar to the previous one, but the surface area of each pond was $1.05\,m^2$ and the effective volume was $1.14\,m^3$. Domestic wastewater was supplied by continuous flow at a constant loading rate. The samples of the water were taken at the inlet and outlet of each pond in the series. The water samples were analyzed according to standard methods *Standard Methods for Examination of Water and Wastewater* (1995). During the study the density of plants were kept at $5\,\text{kg}\,\text{m}^{-2}$ by harvesting every week. The statistical analysis of the data was performed using the computation program package "Statgraphics".

Results and discussion

The analysis of the effect the organic and hydraulic loading rate for the experimental systems with plants and without plants and depth of 0.5 m was carried out in three phases. In the first phase, the system worked with a retention time of three days for each pond in the series. In the second phase the retention time was one and a half days in each pond. In the third phase the retention time was one day for each pond. The results obtained in the three phases are presented in Table 1.

Table 1 shows that the removal achieved for the different contaminants in the three phases studied was always higher in the ponds with plants. Most of the removal of contaminants occurred in the first pond, both in the vegetated and non-vegetated systems. The quality of the effluent from the pond with plants (with organic loading up to 33.8 g $\text{BOD}\,\text{m}^{-2}\,\text{d}^{-1}$ and hydraulic loading of $0.28\,\text{m}^3\,\text{m}^{-2}\,\text{d}^{-1}$ and only 1.5 days of retention time) exceeded the effluent values given in the literature for a system with 4.2 times lower organic loading and approximately 10 times smaller hydraulic loading (Crites and Tchobanoglous, 1998). The water hyacinth ponds removed up to $3.8\,\text{g}\,\text{TKN}\,\text{m}^{-2}\,\text{d}^{-1}$ and $1.3\,\text{g}\,\text{TP}\,\text{m}^{-2}\,\text{d}^{-1}$. These values are 2.6 times above the maximum reported for nitrogen and 3.7 times for phosphorus by Reddy and Turcker (1983) and DeBusk and Reddy (1991). The plant growth values fluctuated between 307 and $594\,\text{g}\,\text{m}^{-2}\,\text{d}^{-1}$ (wet weight). The growth of the plants is influenced by season and the nutrient content in the pond.

The data obtained experimentally were used to establish empirical relationships. For the systems without plants the data of the soluble fraction of the analyzed parameters were used. From these correlations it is possible to determine the values of some parameters for use in the design of such systems (Table 2). The results show that organic loading removed in the systems with plants, were between 1.02 and 1.24 times higher than those obtained in the systems without plants. The organic loading that it is possible to apply to a 0.5 m deep system for the climatic conditions of Cuba, are more than double the loading reported by other authors (Crites and Tchobanoglous, 1998).

The study of the 1.12 m deep systems was also organized in three phases. The first phase had a retention time of 4.3 days for each pond. The second phase had a retention time of 3 days for each pond. The third phase had retention times of 2.5 days in each pond. The results obtained in the three stages are presented in Table 3.

Table 3 shows that the removal achieved for the different contaminants in the three phases studied were higher in the ponds with plants. This is similar to the behavior observed for 0.5 m deep systems. In Table 4 basic parameters and a similar regression analysis as for the 0.5 m deep systems is presented. The results show that for 1.12 m deep systems the removal of organic matter was between 1.02 and 1.15 greater in the vegetated ponds than in ponds without plants. The results also indicate the effect of depth for the removal of the contaminants by these systems. To obtain 80% BOD removal efficiency in a vegetated system with a depth of 1.12 m it is required to apply an organic loading 1.7 smaller than the organic loading applied in a 0.5 m deep system. The retention time necessary to achieve the 99% removal of fecal coliforms in a 1.12 m deep is 1.9 greater than the required retention time for the 0.5 m deep system.

Table 1 Characteristics of the raw and treated wastewaters in the 0.5 m deep experimental pond systems with water hyacinths and without plants

Phases	RT	OL	HL	BOD mg L^{-1} T	BOD mg L^{-1} S	COD mg L^{-1} T	COD mg L^{-1} S	TKN mg L^{-1} T	TKN mg L^{-1} S	N-NH$_4$ mg L^{-1} T	N-NH$_4$ mg L^{-1} S	TP mg L^{-1} T	TP mg L^{-1} S	TSS mg L^{-1} T	VSS mg L^{-1} T	FC	GP
Phase I																	
Influent				141		338		28.8		18.6		4.6		209	93	1.1×10^7	
Pond 1	3	19.9	0.14	48	23	228	99		12.6		9.1		2.6	129	103	2.9×10^5	
Pond 2	3	6.7	0.14	42	19	188	91		10.4		5.6		1.9	132	104	6.2×10^9	
Hyacinth 1	3	19.9	0.14	15		61		8.4		7.4		1.2		35	11	1.3×10^5	461
Hyacinth 2	3	2.1	0.14	11		44		7.7		5.8		0.5		26	6	7.3×10^4	307
Phase II																	
Influent				120		263		24		12.6		4.3		96	42		
Pond 1	1.5	33.8	0.28	64	36	208	94		17.3		9.5		2.2	200	170		
Pond 2	1.5	18.1	0.28	59	25	190	88		14		6.5		2.2	199	177		
Hyacinth 1	1.5	33.8	0.28	16		50		13.7		7.9		1.5		27	11		522
Hyacinth 2	1.5	4.5	0.28	12		44		12.1		6.5		1.3		15	8.5		402
Phase III																	
Influent				121		201		24.5		15.1		5.5		92	56	1.1×10^7	
Pond 1	1	51.2	0.42	57	43	112	72		16.2		11.2		2.9	121	105	2.5×10^6	
Pond 2	1	24.1	0.42	53	43	91	71		15.3		8.4		2.3	119	102	1.1×10^6	
Hyacinth 1	1	51.2	0.42	36		61		15.5		10.9		2.5		32	16	1.1×10^6	594
Hyacinth 2	1	15.3	0.42	32		58		14.3		10.6		2		20	14	3.7×10^5	471

T: Total content (unfiltered sample); S: Soluble content (filtered sample); RT: Retention time (d); OL: Organic loading rate (g m^{-2} d^{-1}); HL: Hydraulic loading rate (m^3 m^{-2} d^{-1}); FC: Fecal coliform (MPN/100 mL); GP: Growth of plant (g m^{-2} d^{-1})

Table 2 Parameters in systems with depth of 0.5 m depth and corresponding regression equations

RT	HL	OLa	OLr H	OLr P	L TKN (a)	L TKN (r)H	L TKN (r)P	L TP (a)	L TP (r)H	L TP (r)P	N/N₀ H	N/N₀ P
3	0.141	19.9	17.8	16.7	2.04	1.49	1.3	0.65	0.48	0.28	1.18×10^2	2.64×10^2
6	0.071	9.97	9.19	8.63				0.33	0.29	0.19	6.64×10^3	5.64×10^3
1.5	0.282	33.8	29.33	23.7	6.77	2.9	1.89	1.21	0.79	0.6		
3	0.141	16.9	15.2	13.4	3.38	1.68	1.41	0.61	0.42	0.3		
1	0.423	51.2	36	33	10.4	3.81	3.51	2.33	1.27	1.1	1.0×10^1	2.27×10^1
2	0.212	25.6	18.8	18.4	5.19	2.16	1.95	1.16	0.74	0.68	3.36×10^2	1.0×10^1

Independent variable	Dependent variable	Regression equation	Correlation coefficient
Vegetated systems (Water hyacinths)			
Ola	OLr	$Y = 0.656x + 3.8$	$R = 0.981$
HL	OLr	$Y = 76.1x + 4.96$	$R = 0.978$
RT	OLr	$Y = 38.9e^{-0.26x}$	$R = 0.945$
RT	N/N₀	$Y = 0.10e^{-0.5x}$	$R = 0.91$
Non vegetated systems			
Ola	OLr	$Y = 0.58x + 3.77$	$R = 0.996$
HL	OLr	$Y = 66.9x + 4.82$	$R = 0.989$
RT	OLr	$Y = 34.2e^{-0.25x}$	$R = 0.947$
RT	N/N₀	$Y = 0.39e^{-0.73x}$	$R = 0.919$

OLa: Organics, loading rate (g m^{-2} d^{-1}); OLr: Organics, removal rate (g m^{-2} d^{-1}); L TKNa: TKN loading applied (g TKN m^{-2} d^{-1}); L TKNr: TKN loading removed (g TKN m^{-2} d^{-1}); L TPa: TP loading applied (g TP m^{-2} d^{-1}); L TPr: TP loading removed (g TP m^{-2} d^{-1}); N/N₀(H): Fraction of fecal coliform in hyacinth pond; N/N₀(P): Fraction of fecal coliform in pond

Table 3 Characteristic of the raw and treated wastewaters in experimental ponds with water hyacinth and without plants (1.12 m of depth)

Phases	RT	OL	HL	BOD mg L^{-1}		COD mg L^{-1}		TKN mg L^{-1}		FC MPN	GP
				T	S	T	S	T	S		
Phase I											
Influent				72		124		21.5		1.1×10^7	
Pond 1	4.3	18.2	0.25	26	18	82			11.8		
Pond 2	4.3	6.6	0.25	22	12	72			8.9	1.2×10^5	
Hyacinth 1	4.3	18.2	0.25	15		35					480
Hyacinth 2	4.3	3.8	0.25	8		20		7.3		1.8×10^5	354
Phase II											
Influent				91		130		18.5		2.4×10^7	
Pond 1	3	33.3	0.37	45	30	74	45		15.5	4.3×10^6	
Pond 2	3	16.5	0.37	38	20	81	47		11.7	4.3×10^5	
Hyacinth 1	3	33.3	0.37	21		47		11.8		2.7×10^6	425
Hyacinth 2	3	7.7	0.37	12		37		9		1.9×10^5	357
Phase III											
Influent				66		129		17.6		1.6×10^7	
Pond 1	2.5	28.9	0.44	37	25	82	48		13.2	3.2×10^6	
Pond 2	2.5	16	0.44	26	15	74	52		10.4	1.2×10^6	
Hyacinth 1	2.5	28.9	0.44	23		45		11.1		5.2×10^6	532
Hyacinth 2	2.5	10.1	0.44	14		31		8.5		1.2×10^6	447

Effect of the temperature and the solar radiation on the growth of the water hyacinths

For this analysis a 10-month study regarding growth velocity of the water hyacinth plants without nutrient limitation was carried out. In addition to the growth velocity, solar radiation and ambient temperature was registered in terms of the monthly mean values. The results obtained are presented in Figure 1.

Figure 1 shows that the specific growth velocity of the water hyacinths varied with the months of the year. It was evident that this variation was associated to the temperature and solar radiation. Through analysis of the data the following multiple correlation equation was obtained (Rodríguez, 1997):

$$\mu = 0.025 \cdot 1.0386^T \cdot 1.0317^{SR}$$

where:
  μ: specific speed of growth (d)
  T: temperature of the air (°C)
  SR: solar radiation (MJ m^{-2})

These results show that for the climatic conditions of Cuba, the growth velocity of the water hyacinths depends on the temperature and the solar radiation.

By combining the results obtained in the study relating growth of water hyacinths to solar radiation and temperature and the removal of the contaminants in ponds with water hyacinths, a mathematical model describing the kinetics of the processes in pond systems with hyacinths was determined. The results obtained experimentally in the vegetated pond systems in series for depths of 0.5 m and 1.12 m indicate a dependence on BOD removal efficiency with the BOD concentration of the influent. For this reason it was decided to apply the model of Grau (Grau *et al.*, 1975):

$$\frac{S_o - S}{\theta \cdot X} = K \cdot \frac{S}{S_o} \quad \text{(Grau equation)}$$

Table 4 Parameters in systems with depth of 1.12 m depth and corresponding regression equations

RT	HL	OLa	OLr H	COr P	L TKN (a)	L TKN (r)H	L TKN (r)P	N/N0 H	N/N0 P
4.3	0.252	18.2	14.4	13.6	5.43	2.98	2.45		
8.6	0.126	9.09	8.08	7.57	2.71	1.79	1.59	1.64×10^2	1.09×10^2
3	0.366	33.3	25.6	22.3	6.77	2.45	1.1	1.13×10^1	1.79×10^1
6	0.183	16.7	14.5	13.02	3.39	1.74	1.25	1.46×10^2	1.79×10^2
2.5	0.437	28.9	18.8	17.9	7.69	2.84	1.92	3.25×10^1	2.0×10^1
5	0.219	14.4	11.4	11.2	3.85	1.99	1.57	7.5×10^2	7.5×10^2

Independent variable	Dependent variable	Regression equation	Correlation coefficient
Vegetated systems (Water hyacinths)			
COa	COr	$Y = 0.65x + 2.4$	$R = 0.974$
CH	COr	$Y = 42.7x + 4.2$	$R = 0.810$
TRH	COr	$Y = 31.4e^{-0.16x}$	$R = 0.880$
TRH	N/N_0	$Y = 0.64e^{-0.48x}$	$R = 0.890$
Non vegetated systems			
COa	COr	$Y = 0.56x + 3.1$	$R = 0.989$
CH	COr	$Y = 38.6x + 4.09$	$R = 0.867$
TRH	COr	$Y = 28.9e^{-0.16x}$	$R = 0.921$
TRH	N/N_0	$Y = 0.75e^{-0.52x}$	$R = 0.961$

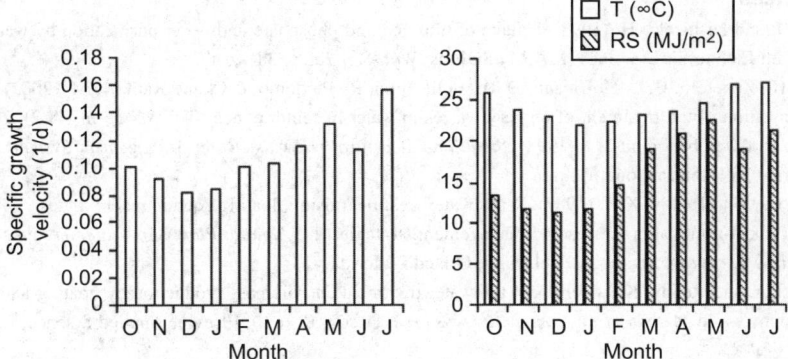

Figure 1 Monthly growth velocity of water hyacinths and monthly mean temperatures (T) and solar radiation (RS)

where:
- S_o: concentration of BOD influent (mg L^{-1})
- S: concentration of BOD effluent (mg L^{-1})
- θ: hydraulic retention time (d)
- X: concentration of microorganisms in the reactor (mg L^{-1})
- K: constant of reaction velocity (d^{-1})

Fitting the model of Grau to the results obtained experimentally for the two pond depths (0.5 and 1.12 m) the following equation for calculation of the removal rate coefficient K for BOD was obtained (Rodríguez, 1997):

$$K = 0.0108 \cdot (0.988)^{Bv} \cdot (0.8157)^{T} \cdot (1.0485)^{BODi} \cdot (1.091)^{SR}$$

where:
- Bv: volumetric organic loading rate (g BOD m^{-3} d^{-1})
- T: ambient temperature (°C)
- $BODi$: BOD influent (mg L^{-1})
- SR: solar radiation (MJ m^{-2})

Conclusions

It has been demonstrated that in systems with water hyacinths higher removal was always achieved and most removal occurred in the first of the tanks in series. The water hyacinth ponds removed up to 3.8 g TKN m^{-2} d^{-1} and 1.3 g TP m^{-2} d^{-1}. The plant growth values fluctuated between 307 and 594 g m^{-2} d^{-1} (wet weight) and it was influenced by the month of the year, nutrient concentration, temperature and solar radiation. For the climatic conditions of Cuba the removal of contaminants were higher than for climatic conditions less favorable to growth water hyacinths.

The results obtained at 1.12 m depth was similar to those observed for the depth of 0.5 m but in the shallower system with shorter retention times and greater superficial organic loading higher removal efficiencies are obtained.

The results revealed a relationship between the purification capacity of the water hyacinth and its velocity of growth.

A mathematical model for the design of water hyacinth systems including climatic variables such as temperature and solar radiation as well as the effect of the depth has been obtained.

References

Aoyama, I. and Nishizaki, H. (1993). Uptake of nitrogen and phosphate and water purification by water hyacinth Eichhornia crassipes (MART) SOLms. *Wat. Sci. Tech.*, **28**(7), 47.

Costa, R.H., Zanotelli, C.T., Hoffmann, D.M., Belli Filho, P., Perdomo, C.C. and Rafikov, M. (2003). Optimization of the treatment of piggery wastes in water hyacinth ponds. *Wat. Sci Tech.*, **48**(2), 283–289.

Crites, R. and Tchobanoglous, G. (1998). *Small and Decentralized Wastewater Management Systems*, WCB/McGraw-Hill, International Edition.

DeBusk, T.A. and Reddy, K.R. (1987). Wastewater treatment using floating aquatic macrophytes: containment removal processes and management strategies. In: *Aquatic Plants for Water Treatment and Resource Recovery*, Magnolia Publishing, Orlando, Florida.

DeBusk, T.A. and Reddy, K.R. (1991). Wastewater treatment and biomass production by floating aquatic macrophytes. In *Methane from community waster*, Isaacson, R. (ed.), Elsevier Applied Science, London, UK.

Grau, P., Dohanyas, M. and Chudoba, J. (1975). Kinetics of multicomponent substrate removal by activated sludge. *Wat. Res.*, **9**, 637.

Kim, Y., Giokas, D.L., Chung, P.G. and Lee, D.R. (2003). Design of water hyacinth ponds for removing algal particles from waste stabilization ponds. *Wat. Sci Tech.*, **48**(11–12), 115–123.

Reddy, K.R. and Turcker, J.C. (1983). Effect of nitrogen source on productivity and nutrient uptake of water hyacinth. *Econ. Bot.*, **37**, 236.

Rodríguez Pérez de Agreda.(1997). *Uso de las plantas acuaticas en la depuracion de las agues residuals domésticas*. PhD thesis, Centro de Investigaciones Hidráulicas. Instituto Superior Politécnico José Antonio Echeverría, Ciudad de La Habana, Cuba.

Sipauba-Tavares, L.H., Favero, E.G. and Braga, F.M. (2002). Utilization of macrophyte biofilter in effluent from aquaculture: I Floating plant. *Braz. J. Biol.*, **62**(4A), 713–723.

So, L.M., Chu, L.M. and Wong, P.K. (2003). Microbial enhancement of Cu^{2+} removal capacity of Eichhornia crassipes (Mart.). *Chemosphere*, **52**(9), 1499–1503.

Standard Methods for Examination of Water and Wastewater (1995) 19th edn, American Public Health Association/American Water Work Association/Water Environmental Federation, Washington DC, USA.

Tchobanoglous, G. (1993). Constructed wetland and aquatic plant system. In *Constructed Wetland for Water Quality Improvement*, Moshiri, G. (ed.), Lewis Publishers, Pensacola, Florida.

Yedla, S., Mitra, A. and Bandyopadhyay, M. (2002). Purification of pulp and paper mill effluent using Eichornia crassipes. *Environ. Tech.*, **23**(4), 453–465.

Contribution of floating macrophytes (*Lemna* sp.) to pond modelization

H. Jupsin, H. Richard and J.L. Vasel

Sciences and Environmental Management Department, University of Liège, Arlon, Belgium
(E-mail: *jl.vasel@ugl.ac.be*)

Abstract The objective of the present study was to develop a methodology for the quantification of the growth rate of Lemnaceae biomass by digital image analysis. The effect of biomass surface coverage on the oxygen transfer coefficient (Kla) was also quantified. Contribution of Lemnaceae to oxygen balance was evaluated by closed respirometry. Monod-like equations could be derived from growth rate coefficients in various experimental conditions. This opens the way to a deterministic model of Lemnaceae ponds where uptake of nitrogen and phosphorus (even heavy metals) can be calculated.
Keywords Floating macrophytes; duckweeds; growth kinetics; mathematical model; oxygen balance; uptake

Introduction

Various types of floating macrophytes such as Lemnaceae, water lettuces (*Pistia stratiotes*) and water hyacinths (*Eichhornia crassipes*) have been used in constructed wetlands and ponds. The contribution of macrophytes to treatment plants has, however, mostly been regarded in the literature as a black or grey box model, as global first-order kinetic coefficients are for example compared with similar coefficients for microphyte waste stabilization ponds. Yet the contribution of macrophytes to processes such as biochemical oxygen demand (BOD) and chemical oxygen demand (COD) removal, N (nitrogen) and P (phosphorus) uptake are directly related to the biomass production even if the precise driving mechanisms are still unknown (in the case of the effect of plants on BOD removal for example).

In this paper we describe how the growth rate of Lemnaceae biomass can be monitored by digital image analysis. The growth rate of the Lemnaceae biomass can then be modelized, enabling the evaluation of its influence on the pond system.

Material and methods

Several methods have been used to quantify the lemna biomass in those systems but most are very tedious. Moreover they do not enable the evaluation of the percentage of surface covered by the biomass. The authors therefore aimed to develop a method using digital image analysis.

Oxygen transfer coefficient

Oxygen transfer coefficients were measured by the standard method (ASCE, 1984) in identical tanks with and without biomass. Oxygen transfer coefficients starting from zero dissolved oxygen concentration (after N_2 injection) to reach saturation have been quantified for various percentages of surface coverage by biomass.

Digital image analysis

Camera: hp photosmart 715 (3.3 MP resolution: 2048 × 1536 pixels) Software: UTHSCSA Image Tools (IT Version 2.0) software. More sophisticated software is available but the authors first wanted to check the feasibility of the procedure.

Closed respirometry

A closed respirometer has been designed to measure O_2 and CO_2 transfer rates between the gas phase and the liquid phase during an experiment. The "transparent" respirometer has known volumes of liquid and gas. *Lemna* can grow and cover the water surface while gas and liquid phases can be sampled without opening the respirometer. From those measurements mass balances on oxygen and CO_2 can be quantified to evaluate the fluxes between those phases (Figure 1).

Biomass and growth rate measurements

Various steps are needed: the conversion of the image to a grey/black image, the definition of a threshold value to conserve only black and white pixels, and each object on the figure has to be numbered. An example is given in Figure 2.

Depending on the type of camera, various tests were made with a view to optimizing the results, such as changing the distance between the camera and the water surface, and trials with the zoom or artificial light.

An object of known dimensions was also placed on the water surface in order to facilitate the conversion of pixels into metric units. From image analysis the geometric properties of the *Lemna* "objects" could be characterized.

If the minimum size could be clearly defined the maximum size is more difficult to characterize. We observed that for one individual the main parameters (surface, perimeter, large axis, small axis) have Gaussian distribution. We quantified the same properties for objects composed of more than one frond. Most of those properties (except the one in grey) provided in Table 1.

It is difficult to distinguish individual *Lemna* when they are in contact. This means that although the total surface of *Lemna* can be calculated, it is much harder to give the exact number of individuals.

We also checked that when we gathered the same number of individuals on a smaller surface we obtained the same total surface coverage (±2%) which validated the method. Some results indicated that the physiological status of the biomass could also be obtained

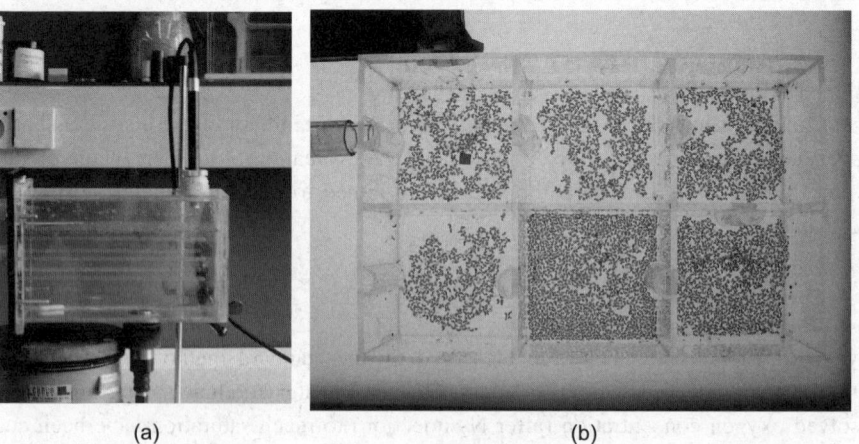

(a)   (b)

Figure 1 (a) Closed respirometer with known volumes of gas phase and liquid phase (b) Reactor for growth kinetic and gas transfer measurements

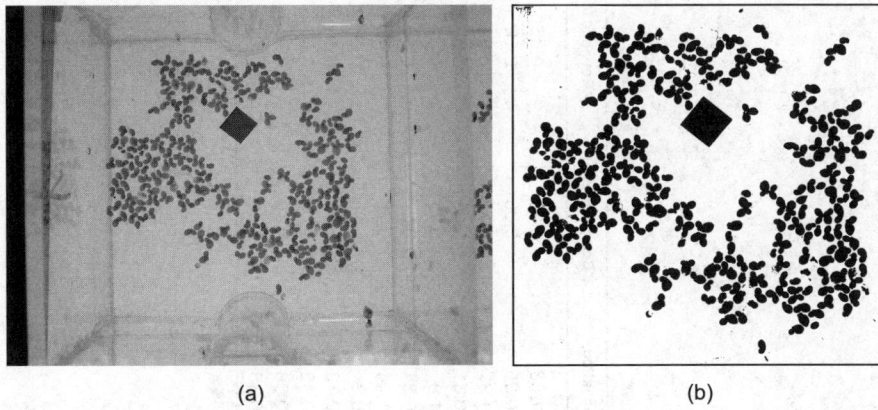

Figure 2 (a) Digital image analysis with calibrated object (b) Processed black and white image (object count and surface coverage)

from (color) image analysis. Elements were numbered and finally the percentage of water area covered by macrophytes was calculated.

Correlation with biomass measurements

Figure 3 illustrates the correlation between biomass (dry weight) and surface area, demonstrating the feasibility of the method for the quantification of the Lemnaceae biomass. Conducting various experiments we found that the method was valid when the area coverage percentage was <65%, otherwise overlapping individuals produced measurement errors (in this case a "dilution" of the surface covered is needed).

More precise image analysis was carried out, indicating that bud identification was also possible, though results will not be presented here.

As can be observed in Figure 3, there was a strong positive correlation between area coverage (<60%) and dry biomass (and even with fresh biomass if the sampling procedure was carefully defined).

Growth kinetic measurement

Using the methodology developed to measure the biomass, the following experimental conditions were assayed. Light intensity: $18.5 \, W/m^2$ (dark period = light period = 12 hours). Temperature = 25 °C. Initial nutrients conditions: 0.031 mg $P\text{-}PO_4$/l and 2.42 mg $N\text{-}NH_4$/l, total alkalinity: 43 meq/l. Table 2 compares our results with the literature.

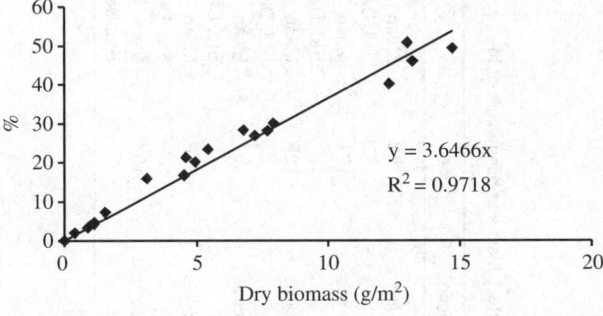

Figure 3 Relationship between dry weight (g m^{-2}) and percentage of area covered (%) by *Lemna minor*

Table 1 Main size parameters of the *Lemna* sp

Category	N		Area (mm²)	Perimeter (mm)	Major axis (mm)	Minus axis (mm)	Elongation	Compactness	Roundness
1 frond	50	Mean	3.67	8.67	2.79	1.77	1.63	0.76	0.65
		Std. Dev.	1.71	3.05	0.63	0.52	0.35	0.06	0.18
2 fronds	171	Mean	5.81	10.97	3.78	1.94	1.97	0.72	0.61
		Std. Dev.	1.62	1.98	0.68	0.31	0.31	0.05	0.09
3 fronds	125	Mean	8.28	14.71	4.58	2.62	1.8	0.71	0.49
		Std. Dev.	1.74	2.31	0.53	0.58	0.34	0.05	0.34
4 fronds	51	Mean	10.07	17.33	4.9	3.54	1.44	0.73	0.43
		Std. Dev.	2.05	2.8	0.66	0.77	0.3	0.05	0.08
5 fronds	6	Mean	11.4	19.26	5.37	3.9	0.73	0.71	0.4
		Std. Dev.	2.31	3.67	0.65	0.96	0.42	0.06	0.6
All objects	403	Mean	6.91	12.87	4.07	2.36	0.58	0.72	0.54
		Std. Dev.	2.73	3.82	0.96	0.78	0.38	0.06	0.13

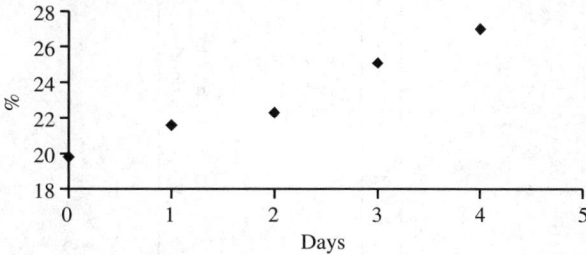

Figure 4 Example of growth kinetic measurements (*Lemna minor*)

Mortality could also be evaluated thanks to the color modification of Lemnaceae (see Figure 2a). The first growth kinetic experiments yielded a mortality coefficient of 0.0068 g dry g^{-1}d^{-1}.

Following this methodology therefore enabled the evaluation of the growth of biomass species such as Lemnaceae, as well as kinetic measurements (cf Figure 4) in various conditions from which Monod kinetic coefficients could be deduced.

Based on those experiments and on data from the literature it was possible to modelize the growth kinetics by the following formula:

$$\mu = f(I)f(T)\hat{\mu} \cdot \left(\frac{1}{1 + \frac{K_{S,N}}{N-NH_4^+} + \frac{N-NH_4^+}{K_{I,N}}} \right) \left(\frac{P - PO_4^{3-}}{P - PO_4^{3-} + K_{S,P}} \right) \quad (1)$$

with $f(T)$ = temperature effect function and $f(I)$ = light effect function.

As can be seen, the effect of nitrogen on growth rate includes an inhibition function (also observed by other authors (Caicedo *et al.*, 2000; Oron *et al.*, 1986)).

With

$$f(T) = A \exp\left(\frac{-(T - t_{opt})^2}{dti^2} \right) \quad (2)$$

where
    A = Maximum activity at optimal temperature
    T = Temperature (°C)
    t_{opt} = Optimal temperature (°C)
    dti = Temperature sensitivity (°C)

$$f(I) = A_I \frac{I}{I_M} \exp\left(1 - \frac{I}{I_M} \right) \quad (3)$$

    I = average light intensity
    I_M = optimal light intensity (for studied species)
    A_I = parameter accounting for the differences between the solar and artificial wavelength spectra (= 1 for artificial illumination).

Experiments to better fit those functions are still in progress.

Effect of surface coverage on gas exchange with the atmosphere

Consequently the percentage of surface covered by the biomass indeed also appears to have a drastic influence on factors such as extinction coefficients in the liquid phase under the surface or oxygen transfer coefficients, which were evaluated (Figure 5).

Table 2 Growth kinetic measurements and bibliographic values (RGR = relative growth rate)

Authors	Species		Data		Photo irradiance
This study	*Lemna minor*	RGR	0.05	gDW (gDW)$^{-1}$ d^{-1}	18.5 W/m^2
Oron et al., 1986	Duckweed	RGR	0.10–0.35	gDW (gDW)$^{-1}$ d^{-1}	Outdoor
	Duckweed	Doubling times	2.3–7.3	Days	Outdoor
Reddy and Debusk, 1987	Duckweed	Annaul productivity	6.0–26.0	tDW ha^{-1} y^{-1}	?
	Duckweed	Growth rate	16–71	gDW m^{-2} d^{-1}	?
Cedergreen and Madsen, 2002	*Lemna minor*	RGR	0.04–0.31	gDW (gDW)$^{-1}$ $^{-1}$	68.5 W/m^2
Zimmo et al., 2002	*Lemna gibba*	Production of duckweed	7.5–12.3	gDW (gDW)$^{-1}$ d^{-1}	Outdoor

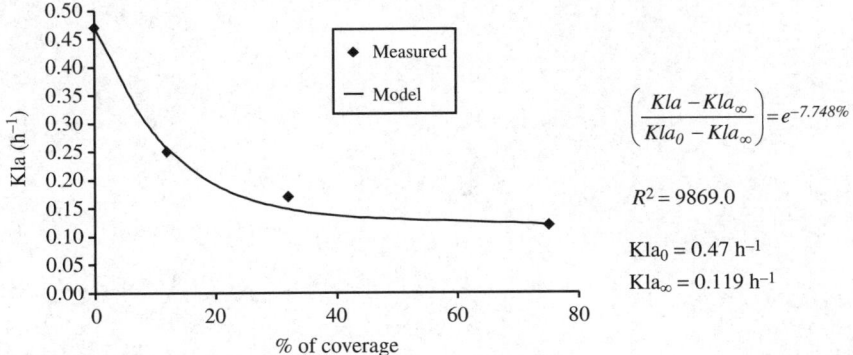

Figure 5 Effect of surface coverage on oxygen transfer coefficient

Effect of biomass activity on oxygen and CO_2 balances in the liquid phase

In another part of the project a special closed respirometer was developed and installed in an experimental "phytotron" to measure oxygen and CO_2 fluxes between the liquid phase and the gas phase. From those results we concluded that for *Lemna minor* less than 10% of the oxygen (CO_2) produced is released (consumed) in the liquid phase. This means that this species will restrict the oxygen transfer with atmosphere and will not produce oxygen in the liquid phase. Similarly CO_2 is exchanged directly with the gas phase and the effect on alkalinity can be neglected in equation (1).

Conclusion

Digital image analysis provides a promising method of quantification of growth kinetics (and population statistics) of floating macrophytes. Other characteristics such as rate of division or decaying should also be possible to obtain from further developments. The coverage of the water surface by floating macrophytes has a large effect on gas exchange rates with the atmosphere. In the case of oxygen the relationship between the oxygen transfer coefficient (*Kla*) and oxygen levels is exponential. In the case of Lemnaceae, the contribution of macrophytes to O_2 and CO_2 balances can be quantified in a close respirometer. Exchanges happen mostly directly within the gas phase. As growth rates are easier to measure, Monod-type growth kinetic models (such as equation 1) can be made to fit experimental data. The Lemnaceae model is now being combined with a microphyte pond model (Jupsin *et al.*, 2003).

References

ASCE (1984). *ASCE Standard for Measurement of oxygen transfer in clean water*, American Society Civil Engineers.
Caicedo, J.R., Van Der Steen, N.P., Arce, O. and Gijzen, H.J. (2000). Effect of total ammonia nitrogen concentration and pH on growth rates of duckweed (*Spirodela polyrrhiza*). *Water Research*, **34**(15), 3829–3835.
Cedergreen, N. and Madsen, T.V. (2002). Nitrogen uptake by the floating macrophyte *Lemna minor*. *New Phytologist*, **155**, 285–292.
Jupsin, H., Praet, E. and Vasel, J.-L. (2003). Dynamic mathematical model of High Rate Algal Pond (HRAP). *Water Science and Technology*, **48**(2), 197–204.
Oron, G., Porath, D. and Wildschut, L.R. (1986). Wastewater treatment and renovation by different duckweed species. *Journal of Environmental Engineering*, **112**(2).
Reddy, K. and Debusk, T.A. (1987). State of the art utilization of aquatic plants in water pollution control. *Water Science Technology.*, **19**(10), 61–79.
Zimmo, O.R., Al Sa'ed, R.M., Van Der Steen, N.P. and Gijzen, H.J. (2002). Process performance asessment of algae-based and duckweed-based wastewater treatment systems. *Water Science and Technology* **45**(1), 91–101.

Comparison of nutrient cycling in a surface-flow constructed wetland and in a facultative pond treating secondary effluent

A. Sajn Slak, T.G. Bulc and D. Vrhovsek

Limnos, Water Ecology Group, Podlimbarskega 31, 1000 Ljubljana, Slovenia
(E-mail: *alenka.sajn@cgsplus.si*; *tjasa@limnos.si*; *dani@limnos.si*)

Abstract There is a growing interest in the possibilities offered by combinations of waste stabilisation ponds (WSP) and constructed wetlands (CW). The purpose of our study was to compare treatment performances and nutrient cycling in a surface-flow wetland (SFW) and in a WSP treating secondary effluent. In the period between 2000 and 2003, a pilot SFW and a pilot WSP were constructed at the outlet of the wastewater treatment plant and their performance monitored while both were active under the same conditions. The SFW was planted with *Phragmites australis* and *Eichhornia crassipes*, while in the WSP development of algae was spontaneous. Performance efficiency was monitored by means of evaluation of physical and chemical parameters in water, by measurement of plant productivity and by analysis of N and P contents in biomass. The SFW with macrophytes proved more efficient in decreasing the suspended solids (64.6%), settleable solids (91.8%), organic N (59.3%), total N (38%), COD (67.2%) and BOD_5 (72.1%) than the WSP. The WSP with algae was more efficient in treatment of ammonia nitrogen (48.9%) and ortho-phosphate (43.9%). The results of this study provide data that are of help in optimising combinations of SFW and WSP.

Keywords Surface-flow wetland; facultative pond; nutrients

Introduction

The mechanisms that control short- and long-term nutrient removal in WSPs and SFWs are not completely understood. However, it is known that nutrient cycling by plants (both algae and macrophytes) is an important process that regulates net retention (Richardson and Craft, 1993; Vymazal, 1995; Kadlec and Knight, 1996; Mara and Pearson, 1998; Nungesser and Chimney, 2000).

WSP are designed for BOD and pathogen removal but very little data are available on nitrogen and phosphorus removal. According to Mara and Pearson (1998) in anaerobic ponds organic nitrogen is hydrolysed to ammonia, so ammonia concentrations in anaerobic pond effluents are generally higher than in the raw wastewater. In facultative and maturation ponds, ammonia is incorporated into new algal biomass. Eventually the algae become moribund and settle to the bottom of the pond; around 20% of the algal cell mass is non-biodegradable and the nitrogen associated with this fraction remains immobilised in the pond sediment. That associated with the biodegradable fraction eventually diffuses back into the pond liquid and is recycled back into algal cells to start the process again. At high pH, some of the ammonia will leave the pond by volatilization. There is little evidence for nitrification and hence denitrification, unless the wastewater is high in nitrates, in WSP. Phosphorus removal mechanisms are sedimentation of organic P in the algal biomass and precipitation as inorganic P (principally as hydroxyapatite at pH levels above 9.5). As with nitrogen, the phosphorus associated with the non-biodegradable fraction of the algal cells remains in the sediment. Thus the best way of increasing phosphorus removal in WSP is to increase the number of maturation ponds, so that progressively more and more phosphorus becomes immobilized in sediments.

In contrast, the major removal mechanisms of organic N in SFW are the sequential processes of ammonification, nitrification and denitrification. Ammonia is oxidized to nitrate by nitrifying bacteria in aerobic zones. Organic N is mineralized to ammonia by hydrolysis and bacterial degradation. Nitrates are converted to dinitrogen gas (N_2) and nitrous oxide (N_2O) by denitrifying bacteria in anoxic and anaerobic zones. The oxygen required for nitrification is supplied by diffusion from the atmosphere and leakage from macrophyte roots. Nitrogen is also taken up by plants, incorporated into the biomass and released back as organic nitrogen after decomposition. Other removal mechanisms include volatilisation and adsorption (Kadlec et al., 2000). The sustainable P removal processes involve the accretion of new wetland sediments. Uptake by small organisms, including bacteria, algae and duckweed, forms a rapid-action, partly reversible removal mechanism. Macrophytes, such as cattails and bulrushes, follow a similar cycle but on a slower time scale of months or years. The detrital residual from the macrophyte cycle also contributes to long-term storage of P in accreted solids. Direct settling and trapping of particulate P can contribute to the accretion process. There can also be biological enhancement of mineralogical processes, such as iron and aluminium uptake and subsequent P binding in detritus, and the algae-driven precipitation of P with calcium (Kadlec et al., 2000).

SFWs have some properties in common with facultative lagoons and also have some important structural and functional differences. Water column processes in deeper zones within treatment wetlands are nearly identical to ponds with surface autotrophic zones dominated by planktonic or filamentous algae, or by floating or submerged aquatic macrophytes. Deeper zones tend to be dominated by anaerobic microbial processes in the absence of light. However, shallow emergent macrophyte zones in treatment wetlands and aerobic lagoons can be quite dissimilar. Emergent wetland plants tend to cool and shade the water. Net carbon production in vegetated wetlands tends to be higher than that in facultative ponds because of high gross primary production in the form of structural carbon, accompanied by resistance to degradation and low rates of decomposition of organic carbon in the oxygen-deficient water column. This high availability of carbon and the short diffusional gradients in shallow vegetated wetlands result in differences in biogeochemical cycling compared with ponds and lagoons (Kadlec et al., 2000).

There is a growing interest in the possibilities offered by combinations of WSP and CW for treatment improvement and better environmental disposal. To find optimal combinations basic comparative studies are needed. The purpose of our 3-year study was to compare treatment performances and to test above mentioned nutrient pathways in a pilot SFW and in a pilot WSP treating secondary effluent.

Methods
Pilot system

The study was carried out at a conventional treatment plant (CTP) in Ajdovscina, SW Slovenia, treating municipal and technological wastewater. Pilot SFW and WSP of identical design were constructed (5 m × 9 m × 0.6 m). Both basins were lined with high-density polyethylene and covered with a 200 mm soil layer (soil:sand = 6:1). SFW was planted with *Phragmites australis* (15 m^2) and *Eichhornia crassipes* (22.5 m^2) a year before monitoring started (Figure 1).

Secondary treated effluent from the CTP entered each basin separately at the surface at a controlled flow rate. The characteristics of the inflowing wastewater are shown in Table 1. The hydraulic loading rate of both basins was 0.13 m/d and the retention times were 4 days. Outlets were situated at the water surface. The SFW and WSP operated

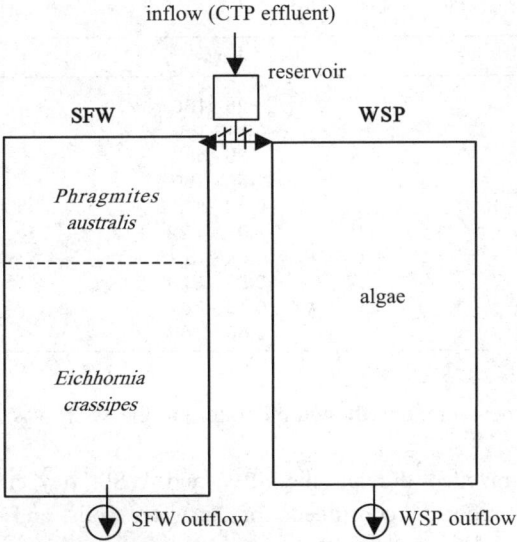

Figure 1 Design of the SFW and WSP pilot system

under the same environmental and hydraulic conditions and therefore comparison was appropriate.

Monitoring

The treatment performance of the SFW and WSP was determined by measuring their influent and effluent water quality on a monthly basis in 2001. Influent and effluent samples (24-h discrete samples at 1 h intervals) were taken with an autosampler (SIST ISO 5667/10) with delay of the retention time. Retention time was determined empirically with conductivity. Samples were kept chilled and in the dark prior to analysis at the Ajdovscina CTP chemical laboratory within 24 hours of collection. The water quality variables were analysed according to *Standard Methods* (APHA, 1998). At the sampling sites temperature, pH and dissolved oxygen were measured with a WTW MultiLine F/Set.

Samples for chlorophyll-a analysis were filtered through a Whatman GF/C filter glass microfibre (Whatman Ltd., Maidstone, England) and extracted with hot methanol. Chlorophyll-a was determined according to Vollenweider (1974). Qualitative 25 μm mesh net samples were taken as a vertical profile, preserved in 5% formaldehyde and analysed for phytoplankton species community composition. Species composition was analysed using a light microscope.

Macrophyte dry weight was monitored on a monthly basis by harvesting of aboveground shoots of *Phragmites australis* and the whole plant of *Eichhornia crassipes* from a surface of $0.25\,m^2$ (SIST ISO 11465). TP and TN were determined in plant biomass according to ISO standards (SIST ISO 5516:1995, SIST ISO 6491:1999 and modified SIST ISO 11261).

Results and discussion

T, dissolved oxygen (DO), pH

Inflow temperatures ranged from 11.7 to 20.8 °C, with an average of 17.1 °C. Diurnal and seasonal oscillations were larger at the WSP than at the SFW which exhibited 'buffer' capacity. The average outflow temperature at the SFW was 14.0 °C, with a range of

Table 1 The characteristics of the influent

	Range	Mean
SS, mg/l	26–136	88
Sett.s., mg/l	<0.1–1.6	–
COD, mg/l	115–582	391
BOD_5, mg/l	40–350	219
Ammon.N, mg/l	0.18–12.78	3.28
Nitrate, mg/l	0.03–0.22	0.07
Org.N, mg/l	3.64–13.15	10.11
Total N mg/l	3.86–26.0	13.51
Ortho-P, mg/l	0.15–2.67	1.27
Total P, mg/l	0.59–2.68	2.07

3.4–23.0 °C, while the average outflow temperature at the WSP was 14.5 °C, with a range of 3.1–24.4 °C.

As wastewater travelled through the SFW and WSP, oxygen was consumed by carbonaceous and nitrogenous constituents of the water, soils, and detritus (Kadlec and Knight, 1996; Bulc and Sajn Slak, 2003). The average inflow concentration of DO was 1.82 mg/l, decreasing to 0.7 mg/l in the SFW and to 1.1 mg/l in the WSP. The WSP exhibited higher DO concentrations during the whole year.

SFW also exhibited 'buffer' capacity with respect to pH. The average pH value of the inflow was 6.9, the outflow was 7.3 at the SFW and 7.5 at the WSP. High diurnal oscillations of pH value were measured in the WSP in spring and summer because of algal photosynthetic activity.

Suspended and settleable solids

The mean input mass loading rate for suspended solids (SS) was 11.8 g/m^2/d. The mean annual mass removal efficiency was higher in the SFW (64.6%) than in the WSP (31.6%). The specific mass removal rate was 7.6 g/m^2/d in the SFW and 3.7 g/m^2/d in the WSP. The higher mass removal rate in the SFW occurred because of sorption on the biofilm which developed on the surface of macrophytes and detritus. Macrophytes also shaded the water surface and thus slowed down development of algae. On the other hand, resuspension caused by wind and algal development were higher in the WSP which resulted in a lower treatment efficiency compared to the SFW (Figure 2).

The inflowing concentrations of settleable solids ranged from <0.1 to 1.6 mg/l. Mean reduction efficiency was 91.8% in the SFW and 66.9% in the WSP.

COD and BOD

The mean input mass loading rate for COD was 52.6 and for BOD_5 29.5 g/m^2/d. The mean COD concentration at the outflow of the SFW was 144 mg/l and the mean BOD_5 concentration 69 mg/l. The SFW in Ajdovscina showed a higher treatment efficiency for BOD_5 than NADB treatment wetlands for secondary treatment (EPA, 2000). The mean annual mass removal efficiency was 67.2% for COD and 72.1% for BOD_5. A higher treatment efficiency was noticed for both parameters in the warmer periods of the year. The temperature effect was not eliminated because Ajdovscina SFW was not a highly productive wetland. The BOD_5 rate constant was k = 0.16 m/d at a background concentration of 5.5 mg/l and a hydraulic loading rate of 0.13 m/d. The BOD_5 rate constant is comparable with FWS wetlands in the USA (Kadlec et al., 2000). Loading rates for BOD_5 were still too high to reach the Slovenian permitted concentration limit of 30 mg/l (OG RS 35/96). The mean COD concentration at the outflow of the WSP was 197 mg/l

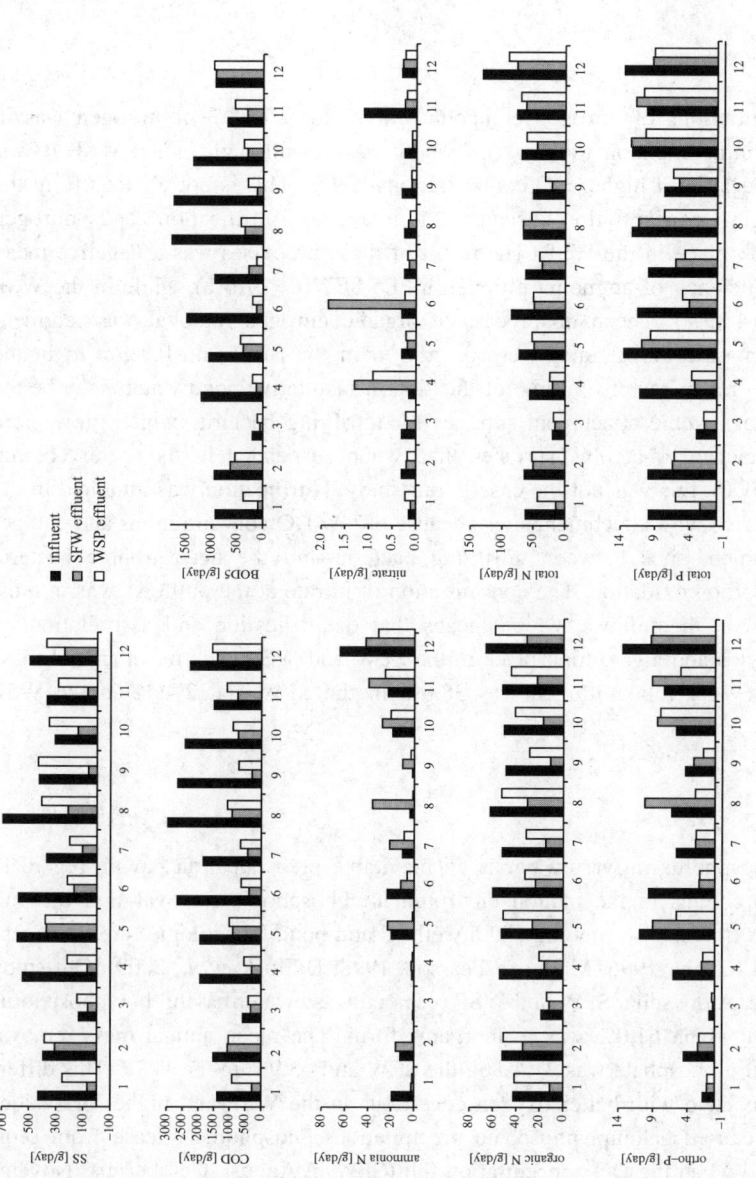

Figure 2 Mass input and output rates at Ajdovscina SFW and WSP at single samplings in 2001 (from January to December)

and the mean BOD$_5$ concentration 86 mg/l. The mean annual mass removal efficiency was 55.5% for COD and 65.2% for BOD$_5$. Treatment efficiencies were lower than in the SFW because of the presence of algae in the effluent (samples were not filtered). The same seasonal dynamics were noticed in the WSP as in the SFW (Figure 2). The BOD$_5$ mass loading rate was too high according to Mara and Pearson (1998) to reach facultative conditions in the WSP during the whole year. The WSP was anaerobic in the colder months of the year.

Nitrogen

Inflowing concentrations of nitrogen compounds were low. 74.8% of nitrogen was in organic form. Ammonification took place in both basins, although in the WSP it was more intensive because of higher DO concentrations and T. This is not obvious from the ammonia nitrogen concentration because simultaneous nitrification and nitrogen volatilization took place in the WSP. The result of these processes was a negative mean mass removal efficiency of ammonia nitrogen in the SFW (-17.6%), while in the WSP it was positive (48.9%). The main process of organic nitrogen removal was sedimentation. Specific mass removal rates were 0.8 g/m^2/d in the SFW and 0.3 g/m^2/d in the WSP. Although a major factor limiting nitrification in facultative pond waters may be the lack of permanent aerobic attachment surfaces for nitrifying bacteria, which grow more readily when attached to aerobic surfaces than when suspended in the water column (Craggs et al., 2000), this was not the case in our study. Nitrification was inhibited in the SFW which was rich with attachment sites because of low DO concentrations and competition for attachment sites between nitrifying bacteria and the heterotrophic bacteria responsible for carbon oxidation. The concentration of nitrate at the outflows was in most cases lower than at the inflows, which means that denitrification and assimilation by plants (macrophytes and algae) took place in the SFW and WSP. The mean annual mass removal efficiency for total nitrogen was 38.0% in the SFW and 27.9% in the WSP (Figure 2).

Phosphorus

Reduction of phosphorus in wastewater is an inevitable preoccupation having regard to eutrophication problems in the natural environment. Phosphorus removal is a difficult task in any water treatment technology, and wetland and pond technologies are no exception (Kadlec and Knight, 1996; Mara and Pearson, 1998; DeBusk et al., 2000). Our study showed the same. Ajdovscina SFW and WSP received wastewater having low phosphorus concentrations of which 61.0% was in inorganic form. The mean annual mass removal efficiency for ortho-phosphate was 21.4% in the SFW and 43.9% in the WSP. This difference can be associated with better oxygen conditions in the WSP and higher pH values that sometimes caused calcium phosphate precipitation. Phosphorus release from sediment was noticed when the DO concentration fell (SFW in August, October and November and in the WSP in October). The mean annual mass removal efficiency for total phosphorus was about equal in the SFW and WSP (31.3% and 34.6%). Organic phosphorus content increased from 39% at the inlet point to 47.2% at the outlet as a result of the presence of algae in the WSP effluent. We can conclude that the WSP was more efficient in inorganic phosphorus removal while the SFW was more efficient in organic phosphorus removal (Figure 2).

The role of macrophytes and algae

Shoots of *Phragmites australis* in the SFW reached a maximum average height of 187 cm and density of 101 shoots/m^2. We recorded a maximum aboveground biomass dry weight of 2.09 kg/m^2 at the end of November. Nitrogen content was 46 g/m^2 and phosphorus 2.95 g/m^2. Maximum nitrogen and phosphorus uptake was recorded at the end of June and the start of July (0.79 g/m^2/d of total nitrogen and 0.05 g/m^2/d of total phosphorus). *Eichhornia crassipes* (water hyacinth) increased its biomass until the end of November, reaching 0.642 kg/m^2 before it was harvested. The fastest growing rate was recorded in the second part of August and in the first part of September (9.27 g/m^2/d). Nutrient concentration was highest in November: 3.2% of total nitrogen and 4.1 mg/g of total phosphorus. Nutrient contents were 20.2 g/m^2 of total nitrogen and 2.6 g/m^2 of total phosphorus at that time. Nutrient uptake by water hyacinth reached its maximum in the second part of August and in the first part of September (0.23 g/m^2/d of total nitrogen and 0.03 g/m^2/d of total phosphorus).

It was found that DO concentrations in the SFW were lower than those in the WSP due to vegetation, which affected nutrient cycling. Macrophytes shaded the water surface and thus slowed down production of algae in the SFW. Additionally, wetland vegetation dampened temperature oscillations. According to a rough estimate, 7.5% of yearly ammonia nitrogen input and 2.5% of yearly ortho-phosphate input was removed by harvesting of water hyacinth at the Ajdovscina SFW.

The yearly chl-a mean value at the Ajdovscina SFW was 48 μg/l while at the WSP the mean value was 234 μg/l. The WSP was facultative in the summer and at the start of the autumn. Otherwise, it was anaerobic. Phytoplankton species diversity was similar to that usually found in facultative ponds. There was not much difference between samples from the SFW and WSP. The most frequent species was *Synechococcus sp*. For the first time in Slovenia, *Uva casinoensis* was recorded in September in a WSP sample. Measurements showed that algae from the WSP caused a lower treatment efficiency of suspended and settleable solids, organic and total nitrogen, organic phosphorus, COD and BOD$_5$ if compared with the SFW. In addition, algae also had some indirect impacts: a higher pH in the WSP than in the SFW due to high photosynthetic activity. Consequently, ammonia volatilisation took place from time to time and ortho-phosphate precipitation was better. Furthermore, algae with photosynthetic oxygen production contributed to better oxygen conditions in the WSP compared to those in SFW in combination with atmosphere aeration. All this resulted in more efficient retention of ortho-phosphates in the WSP, more intensive nitrification and better ammonia nitrogen removal efficiency as compared to the SFW.

Conclusions

Macrophytes in the SFW enhanced sedimentation and sorption on biofilm. Therefore the SFW was more efficient in decreasing suspended solids, settleable solids, organic N, total N, COD and BOD$_5$ than was the case with the WSP. Macrophytes shaded the water surface and so dampened temperature oscillations, algal development and gas exchange with the atmosphere. Uptake of nutrients by macrophytes had a minor role.

Algae in the WSP with photosynthetic activity caused a higher pH (and consequently ammonia volatilisation and ortho-phosphate precipitation) and a higher DO concentration (consequently higher ortho-phosphate retention and more intensive nitrification) compared to the SFW. On the other hand, algae in the effluent caused a lower treatment efficiency of suspended solids and BOD$_5$.

The study showed that there are differences in treatment efficiencies between the SFW and WSP for particular parameters and that these facts should be considered in planning of ecoremediational biotechnologies such as SFWs and WSPs and their combinations.

References

APHA (1998). *Standard Methods for the Examination of Water and Wastewater*. 20th edn., American Public Health Association/American Water Works Association/Water Environment Federation, Washington DC, USA.

Bulc, T. and Sajn Slak, A. (2003). Performance of constructed wetland for highway runoff treatment. *Wat. Sci. Tech.*, **48**(2), 315–322.

Craggs, R.J., Tanner, C.C., Sukias, J.P.S. and Davies-Colley, P.R.J. (2000). Nitrification potential of attached biofilms in dairy farm waste stabilisation ponds. *Wat. Sci. Tech.*, **42**(10-11), 195–202.

DeBusk, T.A., Forrest, E.D., and Reddy, K.R. (2001). The use of macrophyte-based systems for phosphorus removal: an overview of 25 years of research and operational results in Florida. *Wat. Sci. Tech.*, **44**(11-12), 39–46.

EPA, U.S. Environmental Protection Agency (2000). *Constructed wetlands treatment of municipal wastewaters*. Manual EPA/625/R-99/010, Office of Research and Development, Cincinnati, Ohio, USA.

Kadlec, R.H. and Knight, R.L. (1996). *Treatment Wetlands*, Lewis Publishers, CRC Press, Inc., Boca Raton, Florida, USA.

Kadlec, R.H., Knight, R.L., Vymazal, J., Brix, H., Cooper, P. and Haberl, R. (2000). *Constructed wetlands for pollution control*, Scientific and technical report No. 8. IWA Publishing, London, UK.

Mara, D.D. and Pearson, H.W. (1998). *Design Manual for Waste Stabilization Ponds in Mediterranean Countries*, Lagoon Technology International Ltd, Leeds, UK.

Nungesser, M.K. and Chimney, M.J. (2000). Evaluation of phosphorus retention in a south Florida treatment wetland. In *7th International Conference on Wetland Systems for Water Pollution Control*, Lake Buena Vista, Florida, USA, pp. 179–186.

Richardson, C.J. and Craft, C.B. (1993). Effective phosphorus retention in wetlands: Fact or fiction. In *Constructed Wetlands for Water Quality Improvement*, Hoshiri, G.A. (ed.), Lewis Publishers, Boca Raton, FL, pp. 271–282.

Vollenweider, R.A. (1974). *Primary Production in Aquatic Environments*, Blackwell Science, Oxford, UK.

Vymazal, J. (1995). *Algal and Element Cycling in Wetlands*, Lewis Publishers, Boca Raton, Florida, USA.

Integrated natural treatment systems for developing communities: low-tech N-removal through the fluctuating microbial pathways

O. Shipin, T. Koottatep, N.T.T. Khanh and C. Polprasert

Environmental Engineering and Management Program, School of Environment, Resources and Management, Asian Institute of Technology, P.O. Box 4, Pathumthani, 12120, Thailand (E-mail: *oshipin@ait.ac.th*)

Abstract Integration of natural treatment systems (NTS) (WSP, wetlands *etc.*) with each other as well as with advanced unit processes (biofiltration) offers a second lease of life to NTS. Long-term full and pilot scale experience in South Africa and Thailand have shown that contrary to a common view, a low tech N-removal from municipal and light industrial wastewater is a reality for a developing community The high treatment efficiency is ascribed to interplay of N-related processes complementing each other. The present FISH-based (Fluorescence *In Situ* Hybridization) approach to microbial community structure is a pioneering effort in the field of NTS. It establishes interrelationships between major N-removing groups (aerobic and anaerobic ammonia oxidizers (ANAMMOX), denitrifiers) within integrated systems and links them to the high treatment performance. Seasonally fluctuating presence of the ANAMMOX bacteria (0–2.5% of total bacterial numbers) in the NTS (free surface flow wetland) is reported for the first time. Their numbers correlate with metabolically dependent ammonia-oxidizers (2.0–3.0%) but not with stable overall *Planctomycetes* population (4.5–5.1%). As a result of the flexible microbial structure the robust low cost removal down to TN < 10 mg/L is routinely feasible at the loading rates ranging from 0.005 to 0.08 TN kg/m^3/day.

Keywords FISH technique; integrated natural treatment systems; ponds; wetlands N-removal

Introduction

Integration of various types of natural systems with each other as well as with more advanced treatment systems is a major way to provide natural treatment technologies (NTS) with a second lease of life in the new millennium (Polprasert and Koottatep, 2005; Meiring and Shipin, 2005). Ironically these systems are much more extensively used by the developed communities than by developing communities of the warmer and more conducive climates. The systems are arguably characterized by a greater biological complexity hence higher robustness and operational stability, the very qualities required in the context of developing communities.

The aim of the present study is to gain understanding of the microbiological mechanisms underlying both a unit process *per se* and an integrated whole. Specifically it deals with a still *terra incognita* of interrelationships between N-removing microbial consortia within the warm climate natural treatment systems. Performance data for several demonstration and full-scale integrated facilities are related to microbial community analysis based on the Fluorescence *In Situ* Hybridization (FISH) technique. Culture-independent techniques (FISH and PCR-based detection) developed over the last two decades have revolutionized our understanding of microbial processes *in vivo*. However these powerful techniques are yet to be fully utilized by the researchers investigating natural treatment systems. In reality these are being extensively used for studies of far more microbiologically uniform and straightforward processes (e.g. activated sludge, anaerobic digestion). One of the very few, if not the only, attempts to employ the techniques was undertaken by Baptista *et al.* (2003) and PCR-generated profiles were correlated with the C-removing performance of the vegetated/non-vegetated subsurface flow constructed wetlands (CW). The present paper aims at further filling the gap.

Materials and methods
Description of wastewater treatment (WWT) facilities

WSP-based plants. Detailed description of the South African WWT facilities (A-C) particularly, patented PETRO® process has been presented elsewhere (Shipin et al., 1999). A. Ponds/Trickling Filter (TF) Newcastle PETRO® plant (integrated WWT system A, Table 1, I) is characterized by average dry weather water flow (ADWF) 19,000 m^3/day, average winter T_{water} = 18 °C and summer T_{water} = 22 °C. It consists of 2 primary ponds and 3 secondary ponds upstream of 4 parallel vertical rock (slag) trickling filters (VTF) with TN-loading rate: 0.06 TN kg/m^3 day. B. Ponds/TF/Free Surface Flow wetland Letlhabile PETRO® system plant (integrated WWT system B, Table 1, IV and Table 6) consists of 2 primary ponds and 4 secondary ponds upstream of 4 vertical rock (granite) trickling filters followed by a FSF CW (Free Surface Flow Constructed Wetland) and is characterized by a similar climate and ADWF 5,000 m^3/day. Cattail (*Typha sp.*)-based FSF CW is characterized by the N-loading rate 0.02 TN kg/m^3 day, HRT 1.6 d, V = 10,000 m^3. C. Ponds/TF Bloemhof plant in a non-PETRO® mode (integrated WWT system C, Table 1, II) consists of a primary pond upstream of 4 parallel vertical rock (granite) trickling filters. ADWF 5,000 m^3/day, average winter T_{water} = 18 °C and summer T_{water} = 22 °C. Rock VTF N-loading rate: 0.08 TN kg/m^3/day. D. Wetland system (Rayong Industrial estate, Thailand, integrated WWT system D, Table 1, III and Tables 2–3; full description of the system is presented elsewhere, Koottatep *et al.*, 2002). It is characterized by average T_{water} = 27 °C. Each pilot-scale HDPE sheet-lined subsurface vertical flow CW unit: 18 × 35 × 0.4–0.5 m was planted with several emergent plant species such as cattail (*Typha latifolia*), canna (*Canna speciosa*), bulrush (*Scirpus* spp.) and bird of paradise flower (*Heliconia* spp.) on the sand-gravel bed. V = 1,000 m^3 (2 sequential units), surface area of each CW unit 630 m^2, HRT 1–4.0 days; N-loading rate: 0.005 kg TN/m^3 day. Data in Table 3 averages 68 sample analyses for each parameter (Sept. 2001–Feb. 2003). Industrial Effluent Standard is quoted as issued by the Thai Department of Industrial Works, 1992. E. Asian Institute of Technology WWT plant (Pathumthani, Thailand): FWS constructed wetland system integrated with 4 upstream ponds (integrated WWT system E). It is characterized by ADWF 750 m^3/day, average T_{water} = 28 °C, CW N-loading rate: 0.052 kg TN/m^3 day; HRT 2.4 days; dimensions: 60 × 12 × 0.5 m; V = 360 m^3. All analyses were performed in accordance with the *Standard Methods* (APHA, 1995).

Table 1 High N-removing performance of the integrated system (ponds, wetlands) and biofiltration. Only performance of the main N-removing process units is presented: Vertical Trickling Filter, VTF; Vertical Subsurface (VSS) Constructed Wetland, VF CW; Free Water Surface Constructed Wetland, FWS CW. Data averaged over 3.5 years (over 50 values), 2000–2002. Standard deviations did not exceed 15%

Type of the integrated system Parameter	Unit process influent			Final effluent			
	TKN	NH$_4$-N	NO$_3$-N	TKN	NH$_4$-N	NO$_3$-N	N$_{total}$
I. VTF downstream of (anaerobic and facultative) ponds, Tables 4–5	35.6	28.0	0.02	6.7	2.5	3.5	9.5
II. VTF downstream of an anaerobic pond	40.2	31.3	0	7.0	4.4	1.0	8.4
III. VSS CW downstream of ponds (Tables 2–3)	30.0	14.0	0.1	4.0	3.0	6.5	10.5
IV. FWS CW downstream of ponds/VTF, Table 6	18.0	10.0	15.1	8.4	1.6	10.5	18.9

Table 2 Operating conditions of the pilot-scale vertical subsurface flow constructed wetland units (integrated WWT system D). Standard deviations did not exceed 15% (over 50 values)

Run	Period	HLR (L/m²/day)		OLR (g BOD/m² day)		HRT (day)	
		CW1	CW2	CW1	CW2	CW1	CW2
1	September 2001–January 2002	150.5	100.0	12.7	2.4	2.4	3.6
2	February 2002–February 2003	340.0	180.5	32.7	2.1	1.1	2.0

Table 3 Overall treatment performance of Vertical Subsurface Filter Constructed Wetland units 1 and 2 (integrated WWT system D). N.A. Not available. 18 months averages; * Increased due to nitrification. Units: mg/L

Parameters	Industrial standard	Influent	Unit 1 effluent	Unit 2 effluent	Overall removal (%)
BOD	≤ 20.0	96.6	12.7	3.0	97.7
SS	≤ 50.0	100.0	12.2	5.2	95.1
TKN	≤ 100.0	30.1	16.4	4.1	86.0
NH_3-N	N.A.	18.3	12.1	3.0	84.6
NO_3-N	N.A.	0.1	1.2	6.5	*
Total P	N.A.	6.8	3.1	2.2	87.5

Study of the microbial community structure

The FISH technique was performed according to the protocols used by Neef et al., 1996. The well-recognised and tested specific 16S and 23S rRNA-targeted oligonucleotide probes were used: EUB338 (all *Eubacteria*), NON338, Alf1b (all α-*proteobacteria*, Bet24a (all β-*proteobacteria*), GAM42a (all γ-*proteobacteria*), HGC69a (high GC gram-positive bacteria), CF319a + b (*Cytophaga-Flavobacterium* cluster), Ps (γ-proteobacterial pseudomonads), NIT3 (α-proteobacterial nitrite-oxidisers: *Nitrobacter* spp.), Nso1225/NEU (aerobic Ammonia-Oxidizing Bacteria, AOB), PAR651 (*Paracoccus* spp.), NSR447 (nitrite-oxidisers: *Nitrospira* spp.), Amx820 (ANNAMOX bacteria), PLA886 (all plancto-mycetes) with the corresponding applied stringency, formamide concentration (Loy et al., 2003). The probes were labelled with the CY3, CY5, FITC fluorophors. Specific probe counts, i.e. relative abundance, %, represent fractions of total EUB338-positive counts. Probe NON338 was used as a negative control throughout the study.

The difficulty of AOB counting was overcome by use of combined *Nso1225/NEU* probes (Konuma et al., 2000). Axioplan epifluorescence microscope equipped with appropriate filters (Zeiss, Germany) and to a lesser extent a conventional microscope with reflected light fluorescence attachment (Olympus, Japan) were used. General counting methodology reported by Neef et al. (1996) was followed.

Results and discussion
Comprehensive N-removal

Long-term full scale and pilot scale experience in the Mediterranean and subtropical climate (South Africa) as well as in the tropical climate (Thailand) have convincingly demonstrated that contrary to a common view, a low tech low cost N-removal from municipal and light industrial wastewater is a reality for a developing community. Table 1 shows that it is feasible to obtain routine comprehensive N-removal (TKN, NH_4^+, NO_3^-) in an integrated natural treatment system and, most importantly, in a developing community-friendly way. Levels of TN < 10 mg/L at loading rates ranging from 005 to 0.08 TN kg/m³/day could be reliably achieved by treatment facilities of up to at least 19,000 m³/day. Such facilities comprise upfront anaerobic treatment in pond(s) followed by low cost vertical biofiltration in either conventional rock trickling filter(s) (VTF), vertical submerged flow constructed wetland(s) (VF CW) or free water surface (FWS) CW (both vegetated). However in many cases in South East Asia (Thailand, Vietnam, etc.) due to low (COD 150 mg/L) raw sewage strength, for which anaerobic on-site pre-treatment is responsible, wetlands alone would suffice. The strength of typical domestic and light industrial wastewater in South Africa precludes it from being treated in wetlands only and necessitates anaerobic pre-treatment in waste stabilization ponds. Typical example of the existing situation is the Thai case (integrated WWT system E). TKN loading rates for CW units 1 and 2 were maintained at 10 and 3 g/m^2/day (0.005 kg TN/m^3 day), comparable to the suggested loading of 6 g/m^2/day (Polprasert and Koottatep, 2005). Excellent low-tech N-removal was observed over the 18 months run of the 2 sequential CW units (Tables 2–3; Figure 1). Similar to a rock trickling filter, the nitrification-denitrification occured in vegetated vertical flow CW units, fed by organics from wastewater and oxygen provided through plant roots. It was observed that while the effluent DO concentrations reached up to 2.0 mg/L, the influent DO was only 0.5 mg/L. These conditions had important microbiological implications, which are discussed in further sections. Main suggested N-removal mechanisms in CW units are microbial N-uptake, sedimentation of organic N in solid fractions and denitrification. Moreover, the regular harvesting could stimulate N-uptake by the wetland plants.

Plant cultivation and preliminary cost appraisal. It has been observed during 18-months investigation, that cattails, canna, bulrush and golden torch could grow faster than golden ginger and pandanus palm. Since Thailand is located in the tropics (average year round T = 25–35 °C), the biomass yields ranged between 2.0–6.7 kg wet weight/m^2. A unit cost per m^3/day for operation of CW systems amounts to only 12 US$/year, whereas the cost for conventional systems range from 33 to 98 US$/year (payback period 1–2 years). Moreover the plant harvesting can provide further economic benefits, e.g. golden torch and bird of paradise flowers can be sold as a decorative material at the price of approx. 0.2 US$/flower.

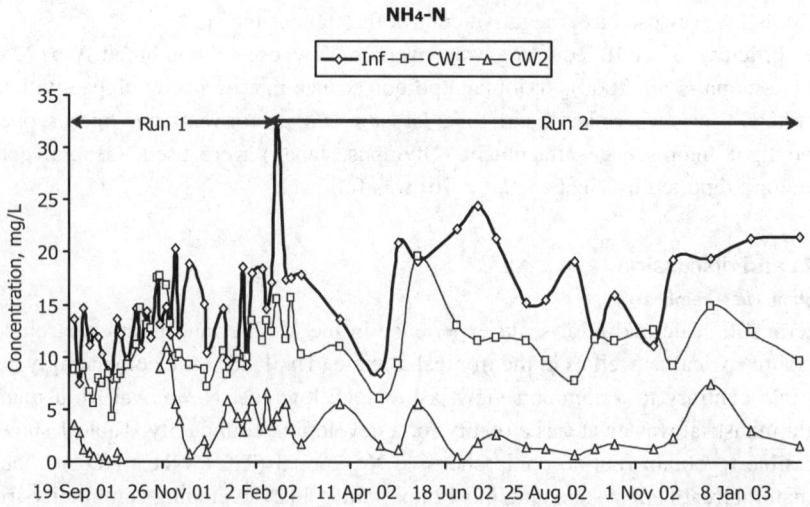

Figure 1 Dynamics of the NH$_4$-N-removal (similar NO$_3$-N removal not shown) in the pilot scale Vertical Subsurface Flow CW units. Inf.: Pretreated influent; CW1: Constructed wetland unit 1 effluent; CW2: Unit 2 effluent (integrated WWT system D)

Table 4 Relative abundance (%) of the key N-removing bacteria in waste stabilization ponds. WWT facility A (Table 1, I). Cross-section: surface 1–10 cm (depth 1 m in brackets)

Target microorganism (FISH probe)*	Primary pond**		Secondary (facultative) pond	
Denitrifying bacteria	Summer, DO 1.0	Winter, DO 0	Summer, DO 4.0	Winter, DO 1.5
Paracoccus spp. (PAR651)	10.5 (2.5)	0.5 (0.1)	5.0 (1.5)	0.1 (0.1)
Pseudomonads (Ps)	1.0 (2.0)	15.0 (10.5)	0.5 (0.5)	3.5 (2.0)

*Annamox bacteria (Amx820) were not detected; **an anaerobic pond with an oxygenating inter-pond recycle from a secondary (facultative) pond with high algal presence

Microbial mechanisms underlying high N-removing performance

In order to elucidate the rationale behind the high N-removing capacity of the microbial consortia in the various integrated NTS and vertical rock biofilters, relative abundance of the main N-related groups was evaluated. A variety of the specific FISH probes were used in order to quantify *in situ* principal β-proteobacterial aerobic ammonia-oxidizers (*Nitrosomonas*, *Nitrosospira*), α-proteobacterial nitrite-oxidizers *Nitrobacter* spp., unrelated nitrite-oxidizers *Nitrospira* spp., anaerobic ammonia oxidizers (ANNAMOX bacteria) and denitrifiers (α-proteobacterial *Paracoccus* spp. and γ-proteobacterial pseudomonads). Principal groups such as α-,β-,γ-proteobacteria, *Planctomycetes*, important hydrolytic and fermentative bacteria (*Cytophaga-Flavobacterium* cluster) as well as *Actinomycetes*-related organisms (high GC-group) were also profiled. The results of the FISH-based analysis are presented in Tables 4–7.

FISH-based microbial structure analysis of various unit processes of the integrated systems indicates a season- and N-load-dependent interplay of the conventionally recognized phenomena (SND, nitrification, anoxic denitrification) with those which only recently became a focus of attention: aerobic denitrification and ANNAMOX. *Paracoccus*-like bacteria represent one of the dominant groups in all the integrated unit processes under study (Tables 4–6).

These known denitrifiers were found to occur under DO concentrations of up to 4.0 mg/L which is corroborated by the literature reports linking *Paracoccus* spp. to the yet poorly understood phenomenon of aerobic denitrification (Neef *et al.*, 1996). Pseudomonads, specific to the 23S rRNA-targeted probe Ps, were selected to represent

Table 5 Relative abundance (%) of the key N-removing bacteria in the vertical TF (Table 1, I). Cross-section: depths 10, 100 and 250 cm in summer (winter in brackets). N.A. Not available

Target microorganism (FISH probe)	Depth of VTF rock strata		
	10 cm	100 cm	250 cm
α-*Proteobacteria* (Alf1b)	N.A.	N.A.	35.5 (27.1)
α-*proteobacterial* denitrifiers:			
Paracoccus spp. (PAR651)	13.0 (6.0)	12.5 (4.0)	6.5 (1.5)
α-*proteobacterial* nitrite-oxidisers:			
Nitrobacter spp. (NIT3)	0 (0.5)	1.5 (0.1)	2.9 (0.1)
β-*Proteobacteria* (*Bet24a*)	N.A.	N.A.	20.4 (24.0)
β-*proteobacterial* aerobic ammonia-oxidizers (Nso1225/NEU)	1.5 (2.0)	3.5 (2.0)	4.5 (1.8)
γ-*Proteobacteria* (GAM42a)	N.A.	N.A.	12.3 (10.6)
γ-*proteobacterial* denitrifiers:			
Pseudomonads (Ps)	3.5 (6.5)	7.0 (7.5)	1.5 (3.5)
Nitrite-oxidisers:			
Nitrospira spp. (NSR447)	0 (1.0)	0.5 (0.1)	1.5 (0.1)
Anaerobic ammonia-oxidizers:			
ANNAMOX bacteria (Amx820)	0 (0)	0 (1.5)	0.5 (1.5)
Cytophaga-Flavobacterium (CF319a + b)	N.A.	N.A.	8.0 (20.9)
High GC gram-positive bacteria (HGC69a)	N.A.	N.A.	14.4 (17.0)

Table 6 Relative abundance (%) of the key N-removing bacteria in the FWS Constructed Wetland downstream of ponds and TFs (integrated WWT system B, Table 1, IV). Values for CW inlet end (outlet end in brackets). N.A. Not available

Target microorganism (FISH probe)	CW with ponds and preceding VTF		CW with preceding ponds	
	Summer DO 1.0	Winter DO 0	Summer DO 4.0	Winter DO 0.5
α-proteobacterial denitrifiers: *Paracoccus spp.* (PAR651)	2.0 (0.5)	0.1 (0.1)	4.0 (3.5)	2.0 (1.5)
Nitrifiers: *Nitrobacter spp.* (NIT3)	0.5 (0.5)	0 (0)	2.0 (N.A.)	2.0 (N.A.)
γ-proteobacterial denitrifiers: Pseudomonads (Ps)	4.5 (4.0)	3.5 (2.5)	2.5 (N.A.)	7.0 (N.A.)
β-proteobacterial aerobic ammonia-oxidizers: (Nso1225/NEU)	2.5 (2.0)	2.5 (2.0)	3.0 (N.A.)	2.5 (N.A.)
Nitrite-oxidizers: *Nitrospira spp.* (NSR447)	0 (0)	0.5 (0.5)	1.5 (0.5)	1.0 (0.1)
Anammox bacteria (Amx820)	1.5 (2.5)	2.0 (2.5)	0 (0)	0 (1.0)
Planctomycetes (PLA886)	(5.1)	N.A.	(4.5)	N.A.

conventional anoxic denitrifiers, a metabolically and taxonomically diverse bacterial group. Seasonally fluctuating numbers of autotrophic ANAMMOX bacteria were found in the low organic load sections of the integrated treatment systems. Size of the ANAMMOX consortium appears to be an inverse function of DO (mainly due to microalgae originating from ponds) and organic load. Furthermore, populations of aerobic and anaerobic ammonia oxidizers appear to be interdependent.

The numbers of the ANNAMOX bacteria (part of *Planctomycetes*) correlate with those of metabolically dependent ammonia-oxidizers (2.0–3.0%) but not with the stable overall planctomycete population (Tables 5–6). Dissimilatory reduction of nitrate to nitrite, nitric acid or nitrous oxide known to be performed by *Nitrobacter* spp. in the anoxic strata is the major source of nitrite for anammox bacteria in the anoxic zones of the system, as well as anoxic denitrification by *Nitrosomonas* spp., both described by the generalized equation (Schmidt et al., 2002):

$$NH_4^+ + NO_2^- \rightleftharpoons N_2 + 2H_2O$$

Character of the documented microbial interrelationships in the post-pond VTF (Table 5) and Free Surface Flow CW (Table 6) is remarkably the same. Activity of β-proteobacterial ammonia-oxidizers follows that of ANAMMOX bacteria rather than that of aerobic nitrite-oxidizing *Nitrobacter* spp. (*NIT* probe). The trend is particularly distinct in winter while microalgae-related oxygenation is minimal. It is relevant to note that over the last decade former obligatory aerobic nitrifiers, AOB, were widely recovered from permanently anoxic natural habitats (Schmidt et al., 2002). Though the microbial

Table 7 Relative abundance of the key N-removing bacteria as related to the pond/wetland integrated system performance (Free Water Surface CW). Further research work is in progress (integrated WWT system E)

Target microorganism (FISH probe)	Relative abundance (%)	N-related parameters	Influent mg/L	Effluent mg/L
All α-*Proteobacteria*	36.3	NO_3-N	6.6	9.4
Nitrite-oxidizers: *Nitrobacter spp.* (NIT3)	4.3	NH_4-N	13.3	3.1
All β-*Proteobacteria*	18.0			
Aerobic ammonia-oxidizers (Nso1225/NEU)	1.7	Organic N	5.1	6.0

community profile of the Vertical Flow CW has not yet been investigated (Table 1, III, Tables 2–3), a very similar microbiological scenario is feasible and requires further investigation. Due to a high effluent DO (up to $2\,\mathrm{mg\,L^{-1}}$) presumably provided through plant roots, a process responsible for the NH_3 to N_2 conversion may well not be anoxic denitrification, but rather the aerobic denitrification effected by *Paracoccus* spp., organisms abundant elsewhere (Neef *et al.*, 1996). Most importantly, it was also shown that in the integrated natural systems comprising anaerobic and facultative pond(s) as a primary stage and either a gravel-based VF CW (two in-series units) or a conceptually similar rock VTF (as a secondary stage) offered a superior N-removing performance in comparison to a horizontal FWS CW and a horizontal TF.

Physico-chemical and molecular microbiological investigation of the 4 unit free surface flow wetland downstream of a 2 pond series (integrated WWT system E) provided insights into its microbial phenomena (Table 7). Though, due to an improper management, the cattail plant biomass is not being removed leading to a significant anaerobiosis due to protein decomposition and sulfate reduction (high sulfide concentration, 12 mg/L, in the final CW effluent), a reasonable (under the circumstances) N-removal takes place with overall 6.5 mg N/L removed from the influent (approx. $0.014\,\mathrm{kg}$ TN/m^3/day). The prominent role of β-proteobacterial aerobic (possibly) aerobic ammonia-oxidizers and α-proteobacterial nitrite-oxidizers testifies to a substantial level of nitrification occurring on plant decaying material. It is counterbalanced by ammonia oxidation. Whether it is aerobic or anaerobic remains to be seen and will be the subject of further investigation.

Conclusions

The work presents a pioneering effort in molecular microbiology of the natural treatment systems (NTS). The microbial mechanisms are elucidated with a view to understanding high N-removing efficiency of the integrated NTS. FISH is a technique used as a tool in the present study. Though employed extensively in the activated sludge and anaerobic digestion research, it is still to receive its dues in the field of NTS. This paper strives to address this imbalance. To the best of our knowledge this is the first report of the significant, though seasonally fluctuating, presence of the ANAMMOX bacteria in NTS (free surface flow constructed wetland). The fact that these bacteria were also found in a secondary rock TF corroborates their previously reported wide distribution in WWT processes. Their numbers appear to be dependent on the particular characteristics of the integrated unit process. Their numbers correlate with those of metabolically dependent ammonia-oxidizers (2.0–3.0%) but not with stable overall planctomycete population (4.5–5.1%). Though anaerobic ammonia oxidation plays a role, the *Paracoccus*-like bacteria and pseudomonads were confirmed to be prominent denitrifiers in NTS such as ponds and constructed wetlands. Overall the high N-removing capacity of the integrated natural treatment systems appears to be dependent on fluctuating interplay of aerobic and anaerobic denitrification running along with aerobic and anaerobic ammonia oxidation. As a result of the flexible microbial structure the robust low cost removal down to TN $<$ 10 mg/L is routinely feasible at the loading rates ranging from 0.005 to 0.08 TN kg/m^3/day. It was shown that concomitant to disposal of wastes products of value can be generated, thereby offsetting the cost of treatment.

References

Baptista, J.D.C., Donnelly, T., Rayne, D. and Davenport, R.J. (2003). Microbial mechanisms of carbon removal in subsurface flow wetlands. *Wat. Sci. Tech.*, **48**(5), 127–134.

Daims, H., Ramsing, N.B., Schleifer, K.-H. and Wagner, M. (2001). Cultivation-independent, semiautomatic determination of absolute bacterial cell numbers in environmental samples by FISH. *Appl. Environ. Microbiol.*, **67**(12), 5810–5818.

Konuma, S., Satoh, H., Mino, T. and Matsuo, T. (2000). Comparison of enumeration methods for ammonia-oxidizing bacteria. *Wat. Sci. Tech.*, **43**(1), 107–114.

Koottatep, T., Polprasert, C. and Surinkul, N. (2002). *Water purification and reuse through constructed wetlands at the Rayong Industrial Estate*. Final Report submitted to the National Centre for Genetic Engineering and Biotechnology/Hemaraj Land Development Co., Ltd, Thailand.

Loy, A., Horn, M. and Wagner, M. (2003). ProbeBase – an online of the rRNA-targeted oligonucleotide probes. *Nucleic Acids Res.*, **31**, 514–516.

Meiring, P.G.J. and Shipin, O.V. (2005). Ponds integrated with biofilters/activated sludge. In: *Pond treatment technology*, Chapter 14, in Shilton, A. (ed.) International Water Association, London (in press).

Neef, A., Zaglauer, A., Meier, H., Amann, R., Lemmer, H. and Schleifer, K.-H. (1996). Population analysis in a denitrifying sand filter: conventional and *in situ* identification of *Paracoccus* spp. in methanol-fed biofilms. *Appl. Environ. Microbiol.*, **67**(12), 5810–5818.

Polprasert, C. and Koottatep, T. (2005). Integrated pond/wetland and pond/aquaculture systems. In: *Pond treatment technology*. Shilton, A. (ed.) International Water Association, LondonIntegrated Environmental Technology series (in press).

Schmidt, I., Hermelink, C., Pas-Schoonen, K., Strous, M., Camp, H.J., Kuenen, J.G. and Jetten, M.S.M. (2002). Anaerobic ammonia oxidation in the presence of nitrogen oxides (NO_x) by two different lithotrophs. *Appl. Environ. Microbiol.*, **68**(11), 5351–5357.

Shipin, O.V., Meiring, P.G.J. and Rose, P.D. (1999). Microbial processes underlying the PETRO® concept (trickling filter variant). *Wat. Res.*, **33**(7), 1645–1651.

Standard Methods for the Examination of Water and Wastewater (1995). 19th edn. American Public Health Association/American Water Works Association/Water Environment Federation, Washington DC, USA.

Comparison of maturation ponds and constructed wetlands as the final stage of an advanced pond system

C.C. Tanner, R.J. Craggs, J.P.S. Sukias and J.B.K. Park

National Institute of Water and Atmospheric Research, PO Box 11-115, Hamilton, New Zealand
(E-mail: c.tanner@niwa.co.nz)

Abstract The treatment performance of a maturation pond (MP), the typical final polishing stage of an Advanced Pond System (APS), is compared with that of a surface-flow constructed wetland (CW) over 19 months. Both received ~67 mm d^{-1} of wastewater after passage through upstream stages of the APS. The MP, with greater sunlight exposure, had higher algal biomass (and associated suspended solids) than the CW, showed higher dissolved oxygen (DO) concentrations and greater diurnal variation in DO and pH. Neither polishing stages reduced nutrients markedly, with the CW exporting slightly more NH_3-N and DRP, and less NO_3-N than the MP. Disinfection was more efficient in the MP (geometric mean 1 log load removal, 12 MPN (100 ml)$^{-1}$) compared to the CW (0.47 log load removal, 53 MPN (100 ml)$^{-1}$). Incorporation of a final rock filter (28% of area) reduced median solids levels to <10 g m^{-3} in both the MP and CW. A hybrid between MPs and CWs with alternating zones of open-water (for enhanced disinfection and zooplankton grazing of algal solids) and wetland vegetation (promoting sedimentation and denitrification, and providing refugia for zooplankton) may provide more consistent effluent quality that either stage alone.

Keywords Disinfection; tertiary treatment; treatment wetland; waste stabilisation pond; wastewater

Introduction

Wetlands have been widely adopted as an option for polishing the effluent from conventional wastewater stabilisation ponds (Kadlec, 2003; Tanner and Sukias, 2003). However, there has been little research on their incorporation into 'improved' pond systems such as advanced pond systems (Craggs *et al.*, 2003; Oswald, 1991). APS comprise an integrated series of treatment stages based around high rate algal ponds (HRP). HRP are shallow, paddlewheel-mixed recirculating ponds that promote algal photosynthetic oxygen production to stimulate aerobic bacterial breakdown of organic matter. Nutrients are removed by assimilation into algal biomass, and may be volatilised (NH_3) or co-precipitated (P) at elevated pH (in response to CO_2 uptake during algal photosynthesis). Daytime oxygen supersaturation and elevated pH also enhance disinfection by solar UV (Davies-Colley *et al.*, 2003). After subsequent settling of algal biomass and associated nutrients in algal settling ponds (ASP), final polishing is normally undertaken in maturation ponds (MP). MPs are designed to promote zooplankton grazing of remaining algal solids, and further disinfection via exposure to solar radiation, sedimentation and protozoan grazing.

Surface-flow constructed wetlands (CW) are a possible alterative to MPs in APS. Emergent wetland plants in CW, disperse flow, and shade and shelter the water column, limiting algal growth and promoting particulate settling (Kadlec and Knight, 1996). The plants also provide attachment sites and substrates for microbial biofilms, assimilate nutrients; mediate gas transport between the sediments, water column and atmosphere (creating a mosaic of anaerobic, anoxic and aerobic conditions that may enhance degradation and N transformations); and provide habitat for micro- and meso-fauna. This paper compares treatment results from a side-by-side study of a MP and a CW as the final stage of an APS.

Methods

The pilot-scale APS in this study treated human sewage and laboratory wastewaters from Ruakura Research Centre, an agricultural and horticultural research facility in Hamilton, North Island, New Zealand. It consisted of an Anaerobic Digester (AD) followed by twin HRP and two-stage ASP systems (operating in parallel). The flow ($\sim 2\,m^3\,d^{-1}$) was then combined and split approximately equally between the MP and CW, which were set up in identical $15\,m^2$ epoxy-coated concrete tanks (11.4 m long, 1.32 m wide, 0.75 m deep). The MP and CW were divided into two equal cells in series. The CW was planted with four emergent wetland plant species (*Schoenoplectus tabernaemontani*, *S. californicus*, *Eleocharis sphacelata* and *Typha orientalis*) in soil-filled $0.33\,m^2$ surface area ($0.66 \times 0.5 \times 0.4\,m$ deep) rectangular polypropylene containers that were arranged in continuous alternating bands across the wetland in three zones occupying $\sim 66\%$ of the total surface area. The planted zones (0.35 m water depth) were interspersed by open-water (0.75 m deep) that became covered ($\sim 3-12\,g\,DW\,m^{-2}$) with the free-floating duckweed *Spirodela punctata*. In April 2003, the *T. orientalis* planted containers were removed from the CW (reducing the planted area to $\sim 47\%$ of the total) and duckweed harvested from the CW open-water at weekly intervals. Monitoring between January and April 2003 (Southern Hemisphere Summer/Autumn) had shown that daytime dissolved oxygen levels in the CW were very low ($<3\%$ saturation).

Although both the MP and CW occupied the same land area, the CW had less volume due to shallower water depths in the planted zones and the volume occupied by plant shoots ($\sim 8\%$ within planted zone), and therefore only had a theoretical hydraulic retention time (HRT) of 7.6 d compared to 11.3 d in the MP. After the first year's operation, rock filters (RF; $\sim 3.2\,m^3$; 20–60 mm gravel, 49% porosity) were incorporated into the final 28% of both the CW and MP as a final polishing stage, and the first cell of the MP was subdivided into 2 cells, so that the resulting 3 cells in the MP each had an HRT of

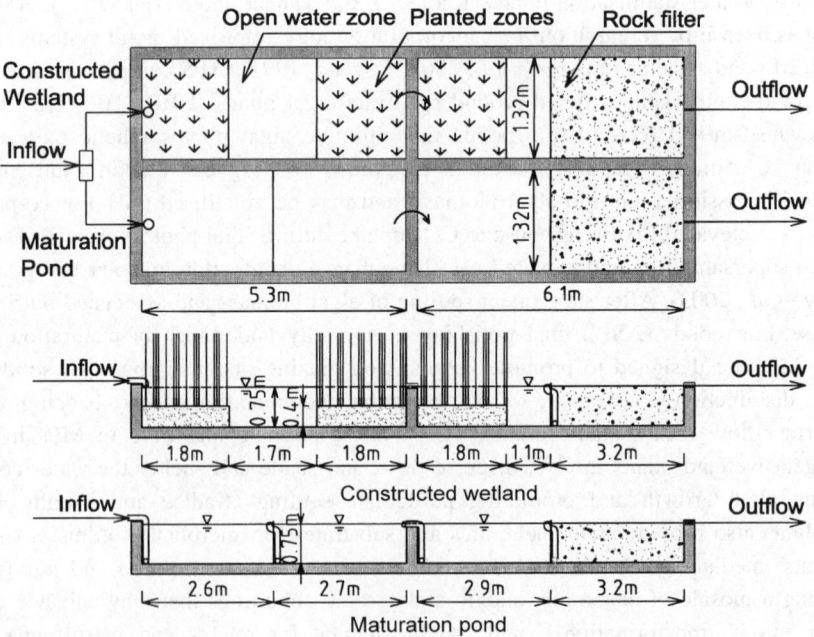

Figure 1 Final layout of MP and CW showing planted zones and rock filter. Plan view (at top) and longitudinal cross-sections (below)

less than 3 days (Figure 1.). The rock filters reduced the theoretical HRT to 6.6 d (and open-water to 25%) in the CW + RF and 9.7 d in the MP + RF.

The inflows and outflows of all stages of the APS, including the CW and MP, were monitored on a monthly basis over 19 months from September 2002. Water temperature, dissolved oxygen (DO) and pH were measured in the field using calibrated field meters and probes (WP models 81 and 82Y, TPS Ltd, Australia). Automated diurnal, hourly composite sampling from directly beside the outflow from July 2003 superseded single grab samples taken (between 10:00 and 10:30 a.m.). Samples were maintained on ice and returned directly to the laboratory for analysis using standard methods as detailed in Craggs et al. (2003). Weekly cumulative flows into and out of the MP and CW were measured with purpose-built stainless steel tipping buckets equipped with mechanical counters. Average monthly flows and contaminant concentrations were multiplied to calculate mass flows in and out of the MP and CW, and compare treatment efficiency.

Diurnal cycles of physico-chemical conditions (water temperature, pH and DO) over weekly periods during January and February 2004 (summer) were measured at 15 minute intervals approximately one-third of the way along the MP and CW using submersible multi-parameter dataloggers (Datasonde versions 3 and 4, Hydrolab Corporation, Austin, TX). The pre-calibrated loggers were positioned in the surface (epilimnion, 0.1–0.2 m depth) and bottom waters (hypolimnion, (0.4–0.5 m depth) of the MP and the open water zones of the CW, and at 0.1–0.2 m depth within the planted zones of the CW. Attenuation of solar photosynthetically active radiation (PAR), UV-A and UV-B through the canopy of emergent and floating-leaved plants was measured at a range of sites under diffuse sunlight with a pair of SD 126QV-Cos submersible light sensors (Macam Photometrics Ltd, Livingston, Scotland, UK) in February 2004.

Results and discussion
Physico-chemical conditions

Outflow water temperatures measured at 10 a.m. were generally 1–2 °C higher in the outflow from the MP than from the CW. Surface waters in the MP showed much greater diurnal variation in temperature than in the CW (Figure 2), becoming thermally stratified on warm sunny days. Shading of sunlight (typically 62–95% reduction of PAR beneath emergent plants and 60–80% under light-moderate duckweed cover, ~3 g DW m^{-2}) markedly reduced diurnal temperature fluctuations (Figure 2) and the incidence and

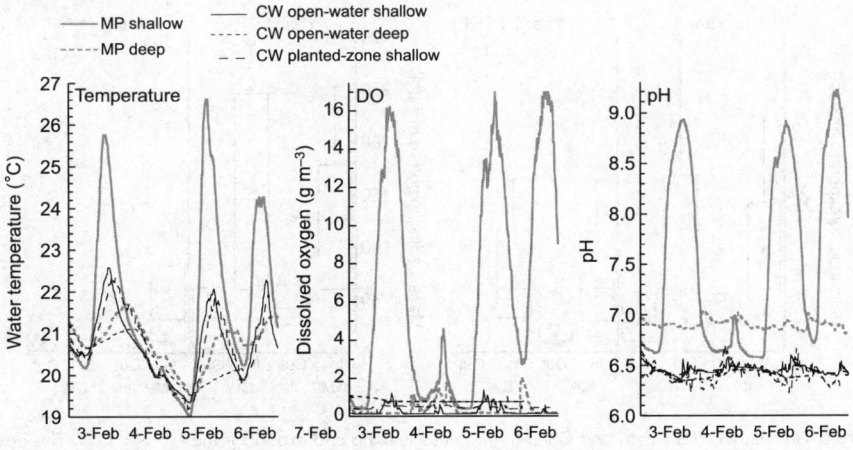

Figure 2 Typical diurnal temperature, pH and DO fluctuations at different depths in the MP and CW over four days during summer 2004. Note effect of dull rainy conditions on 4 February

extent of thermal stratification compared to the MP. The sheltering and shading effects of the vegetation are likely to have been amplified in our relatively small pilot-scale CW. Up-scaling the CW would result in larger open water areas exposed to greater wind fetch, thus limiting the cover of free-floating duckweed and promoting greater wind mixing. We attempted to manage duckweed cover by periodic harvesting from the open-water areas, but redistribution and growth of duckweed harboured amongst the emergent plant zones was rapid.

Marked diurnal fluctuations in pH and DO were recorded in surface waters of the MP (Figure 2) as a result of daytime algal photosynthesis (producing oxygen and consuming dissolved CO_2) in the surface waters. The MP bottom waters showed small increases in pH and DO when turnover occurred during the night, allowing remixing of top and bottom waters. Shallow open-water zones of the CW showed only minor daytime elevations in pH and DO. This is likely to result from reduced algal abundance (e.g. Year 1 outflow Chl a, CW 163 cf. MP 601 mg m^{-3}) and productivity (lower water temperatures and light exposure due to shading by plants) in the CW, and reduced atmospheric gas exchange under the duckweed cover. Unlike algae growing within the water column, the photosynthetic oxygen production and CO_2 uptake of floating and emergent wetland plants is exchanged mainly with the atmosphere.

Treatment performance

In line with previous studies of APS (e.g. Craggs et al., 2003) upstream treatment stages markedly reduced median concentrations of all monitored contaminants before wastewaters entered the MP or CW. Median concentrations (g m^{-3}) for the inflow (and percentage reduction relative to raw wastewater) were: BOD 40 (81%), TSS 80 (75%), TN 25 (55%), TP 5.7 (24%); and E. coli geometric mean 158 ± 794 MPN (100 mls)$^{-1}$ (>4 log reduction). Because the CW received slightly higher inflow rate than the MP during the first year and both were subject to different rates of evapotranspiration, all performance comparisons are made on a mass flow basis. Performance is summarised for the first year of operation without the RF and the final 7 months with the RF (Figures 3–6). The wastewater entering the APS (and hence the MP and CW) was of lower strength during this second period.

VSS comprised ~95% of inflow and ~80% of both MP and CW outflow TSS, suggesting most of the particulates were organic in nature. Lower mass export of TSS

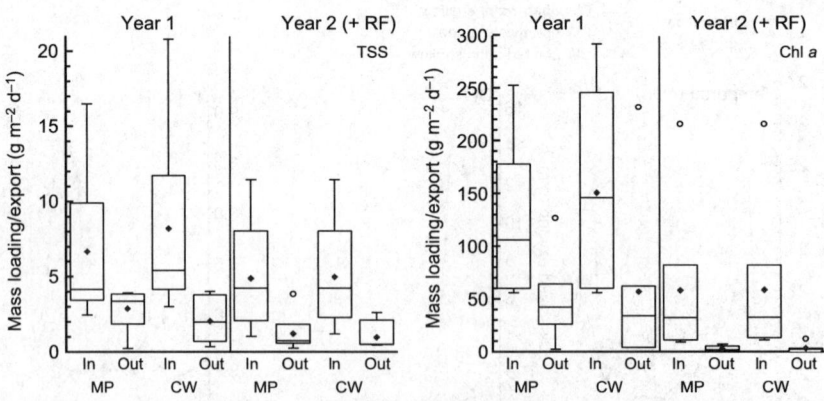

Figure 3 Summary of TSS (left) and Chl a (right) mass loading and export for the MP and CW. The boxes show the bounds of the quartiles above and below the median (dissecting horizontal line). The whiskers show the range of the data within a further 1.5 times the interquartile distance, with data outside this range shown as open circles. The diamonds show mean values

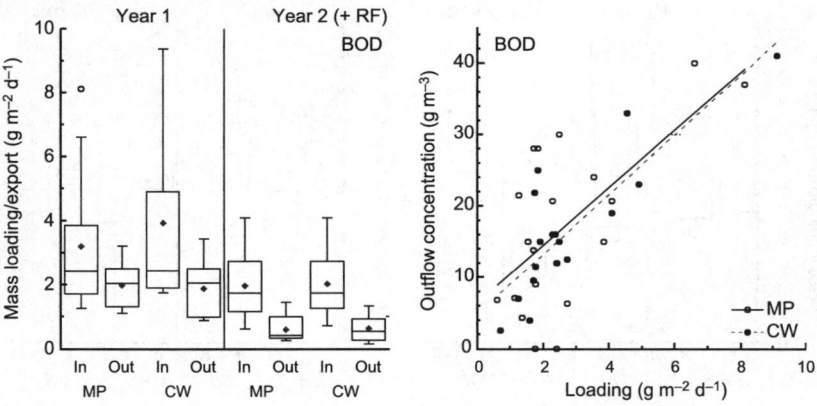

Figure 4 Summary of BOD mass loading and export (left) and relationship of BOD outflow concentration to influent mass loading (right) for the MP and CW. Details for box plots as for Figure 2

and Chl *a* (a measure of algal biomass) was recorded for the CW than the MP in Year 1 (Figure 3; median TSS, 18 and 35 g m^{-3}; Chl *a* 163 and 601 mg m^{-3}, respectively) presumably due to a combination of shading and enhanced settling in the presence of the aquatic plants. Reduction in loading in Year 2 and addition of the RF resulted in improved TSS performance for both the MP and CW with the same median outflow TSS concentrations of 9 g m^{-3} (Figure 3). Sampling before and after the RF showed 45–55% additional TSS removal, with median Chl *a* declining to ~30 mg m^{-3} and VSS concentrations to 6–7 g m^{-3} in the outflows of both the MP + RF and CW + RF.

Overall reductions in BOD mass load, and relationships between mass loading and outflow concentration were very similar for both the MP and CW (Figure 4). Median outflow concentrations of BOD were lower in the CW than the MP during Year 1 (16 and 26 g m^{-3}, respectively), but similar (8–9 g m^{-3}) for Year 2. Improved BOD removal in Year 2 with the RF fitted included reductions in both particulate and dissolved fractions.

TN mass loadings were higher for the CW than the MP in Year 1, but overall relationships between inflow mass loading and outflow concentration were very similar (Figure 5), with similar Org-N export for both systems. In both years the CW exported slightly more NH$_4$-N (median outflow concentrations 9 and 11 g m^{-2} d^{-1}; compared to 6 and 9 g m^{-3} in the MP, respectively). Both the MP and the CW showed elevated outflow

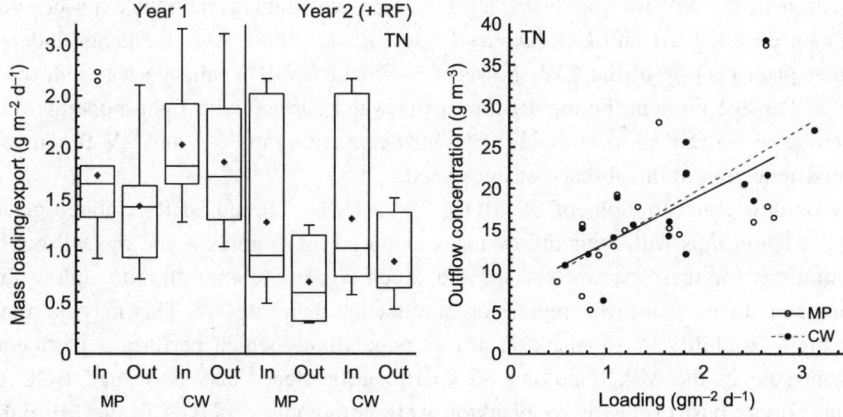

Figure 5 Summary of TN mass loading and export (left) and relationship of TN outflow concentration to influent mass loading (right) for the MP and CW. Details for box plots as for Figure 2

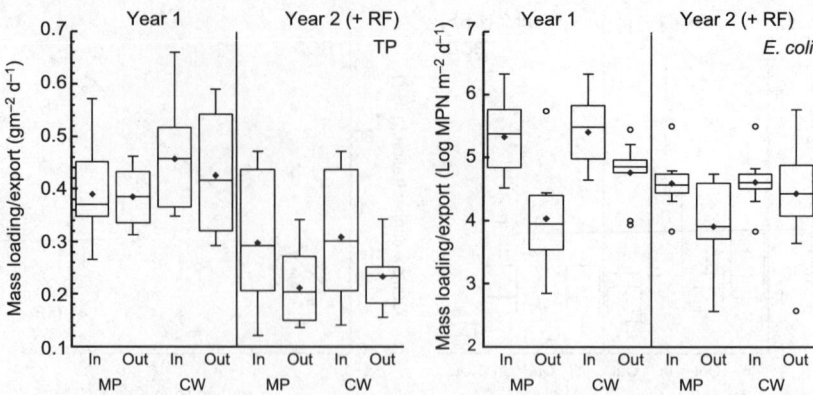

Figure 6 Summary of TP (left) and log *E. coli* (right) loading and export for the MP and CW. Details for box plots as for Figure 2, except that for *E. coli* the diamonds show geometric means

NH_4-N ($\sim 30\,g\,m^{-3}$) during one winter sampling (July 2003) when removal in the upstream HRPs was temporarily reduced. The MP exported slightly more NO_3-N than the CW, but addition of the RF in Year 2 reduced NO_3-N to very low levels in both the MP and CW, probably through promotion of denitrification activity. Little reduction of TP loading was recorded after passage through either the MP or CW in Year 1 (Figure 6), but DRP export increased slightly in both. This difference was balanced by reduced particulate P export in the CW. Approximately 20% mass removal of TP was recorded in both the MP and CW in Year 2, with slight decreases in DRP, but median inflow and outflow concentrations remained very similar.

Disinfection, measured as load reduction of the faecal indicator bacteria *E. coli*, was greater (Figure 6) in the MP (1.34 log load removal) than the CW (0.64 log load removal) during Year 1. During Year 2 influent *E. coli* loads were considerably lower, with a geometric mean influent concentration of 83 MPN $(100\,mls)^{-1}$. Outflow loads were reduced further in the MPs (Figure 6) resulting in geometric mean outflow concentrations of 21 MPN $(100\,mls)^{-1}$ compared to 55 MPN $(100\,mls)^{-1}$ for the CW (in which outflow loads were frequently higher than influent loads). Studies of other CW in New Zealand (Tanner and Sukias, 2003) have noted problems achieving faecal coliforms concentrations below $\sim 300-500$ MPN $(100\,mls)^{1}$. Spot measurements before and after the RF showed little change in *E. coli* concentrations in either the MP or CW. The higher disinfection in the MP was likely to be associated with higher sunlight exposure combined with elevated pH and DO (Davies-Colley *et al.*, 1999). Measurements under the emergent plant canopy in the CW showed UV-A and UV-B irradiance was reduced by 60–99%. Passage through the top 10 mm of the water surface with light-moderate duckweed cover ($\sim 3\,g\,DW\,m^{-2}$) resulted in 80–90% reduction in UV-A and UV-B compared to ~ 60% reduction in the absence of duckweed.

CW outflow concentrations of SS, BOD, TN, NH_4-N, TP and DRP exhibited general positive relationships with their inflow mass loading (e.g. Figures 4 and 5). MP outflow concentrations for these parameters showed much weaker relationships to inflow mass loadings and always had lower regression coefficients than the CW. This may be related to the high variability in zooplankton grazer populations, which perform a particularly important role in the MP, removing SS (particularly algae) and associated BOD and nutrients. Boom-bust cycles of zooplankton were commonly observed in the MP during the trial, associated with alternating periods of clear and turbid conditions. These cycles in zooplankton abundance were not apparent in the CW, either because low DO or other environmental conditions were not favourable, food sources were less abundant and/or the planted zones promoted more stable zooplankton communities. Timms and Moss (1984) showed the importance of macrophytes as refuges for zooplankton grazers in shallow wetland ecosystems, and Lau and Lane (2002) found even relatively low macrophyte densities could provide sufficient refuge to significantly reduce fluctuations in both algal biomass and zooplankton grazers.

Very low oxygen levels ($\sim 1\,\text{g m}^{-2}$) in the CW are likely to have reduced rates of organic matter decomposition and limited development of microbial nitrifier populations. As well as the composition and activity of microbial communities, these conditions are also likely to have affected protozoan, zooplankton and other invertebrate grazer communities active in the wetland (e.g. Stott and Tanner, 2005). Duckweed-based aquatic treatment systems are generally used for secondary treatment and tend to show relatively poor N removal without supplementary aeration (Reed *et al.*, 1995) or intensive harvesting under optimal growth conditions (Korner *et al.*, 2003). To achieve the high levels of solar disinfection and, reduction in SS, BOD, N (particularly NH_4-N) and P desired for final polishing of APS effluents, we consider emergent and free-floating plant coverage of the water surface needs to be restricted below that tested in the present study. Emergent vegetation can be limited by deep-water ($>1.5\,\text{m}$), and duckweed by wind exposure and surface-skimming outlets.

A hybrid between MPs and CWs, where wastewaters pass through alternating zones of wetland vegetation (~ 20–30% of area) interspersed with extensive open-water areas, may benefit overall aquatic community (and associated treatment) stability. High sunlight exposure in the open-water zones would maintain good disinfection and promote aerobic conditions, enhancing decomposition and nitrification during subsequent passage through biofilm-coated plant shoots (Eriksson and Andersson, 1999; Howard-Williams, 1985). Planted zones would provide permeable baffles, reducing short-circuiting and enhancing TSS removal, and (in the final stages) help buffer pH levels in the final discharge. Integration of a wetland component into APS may also provide ancillary benefits by enhancing ecological diversity and providing wildlife habitat (Knight, 1997), although this may compromise final effluent microbiological quality. Incorporation of wetland treatment components may also improve public perception of treatment facilities; for instance in New Zealand, use of wetland and soil contact can improve cultural and spiritual acceptability of wastewater discharges to Maori (indigenous inhabitants). Various configurations need to be compared at realistic scales to optimise designs and determine long-term treatment performance.

Conclusions

MPs and CWs both have potential as polishing stages within APS. The MP provided excellent final disinfection due to high exposure to sunlight and produced an aerobic effluent due to the maintenance of low levels of photosynthetic algae biomass. The shading and enhanced settling action of plants in the CW reduced algal levels and associated TSS, and produced effluents with stable pH, but achieved poorer disinfection and had lower DO levels and slightly higher NH_4-N concentrations. Addition of a final RF stage provided a consistent low TSS final effluent from both MPs and CWs.

Acknowledgements

This project was funded by the NZ Foundation for Science, Research and Technology. NIWA Laboratories undertook water quality analyses. We thank Geoff Mandeno, Don Tindale and John Nagels for technical assistance, and Dr Rob Davies-Colley for reviewing the manuscript.

References

Craggs, R.J., Davies-Colley, R.J., Tanner, C.C. and Sukias, J.P.S. (2003). Advanced pond system: performance with high rate ponds of different depths and areas. *Wat. Sci. Tech.*, **48**(2), 259–268.

Davies-Colley, R.J., Craggs, R.J. and Nagels, J.W. (2003). Disinfection in a pilot-scale "advanced" pond system (APS) for domestic sewage treatment in New Zealand. *Wat. Sci. Tech.*, **48**(2), 81–87.

Davies-Colley, R.J., Donnison, A.M., Speed, D.J., Ross, C.M. and Nagels, J.W. (1999). Inactivation of faecal indicator micro-organisms in waste stabilisation ponds: Interactions of environmental factors with sunlight. *Wat. Res.*, **33**, 1220–1230.

Eriksson, P.G. and Andersson, J.L. (1999). Potential nitrification and cation exchange on litter of emergent, freshwater macrophytes. *Freshwat. Biol.*, **42**, 479–486.

Howard-Williams, C. (1985). Cycling and retention of nitrogen and phosphorus in wetlands: a theoretical and applied perspective. *Freshwat. Biol.*, **15**, 391–431.

Kadlec, R.H. and Knight, R.L. (1996). *Treatment Wetlands.*, CRC Press, Boca Raton, FL.

Kadlec, R.H. (2003). Pond and wetland treatment. *Wat. Sci. Tech.*, **48**(8), 1–8.

Knight, R.L. (1997). Wildlife habitat and public use benefits of treatment wetlands. *Wat. Sci. Tech.*, **35**(5), 35–43.

Korner, S., Vermaat, J.E. and Veenstra, S. (2003). The capacity of duckweed to treat wastewater: ecological considerations for a sound design. *J. Environ. Qual.*, **32**, 1583–1590.

Lau, S.S.S. and Lane, S.N. (2002). Nutrient and grazing factors in relation to phytoplankton level in a eutrophic shallow lake: the effect of low macrophyte abundance. *Wat. Res.*, **36**, 3593–3601.

Oswald, W.J. (1991). Introduction to advanced integrated wastewater ponding. *Wat. Sci. Tech.*, **24**(5), 1–7.

Reed, S.C., Middlebrooks, E.J. and Crites, R.W. (eds.) (1995). *Natural Systems for Waste Management and Treatment*, 2nd Edn McGraw-Hill, New York.

Stott, H.R. and Tanner, C.C. (2005). Influence of biofilm on removal of surrogate faecal microbes in a constructed wetland and maturation pond. *Wat. Sci. Tech.* **51**(9), 315–322.

Tanner, C.C. and Sukias, J.P.S. (2003). Linking pond and wetland treatment: performance of domestic and farm systems in New Zealand. *Wat. Sci. Tech.*, **48**(2), 331–339.

Timms, R.M. and Moss, B. (1984). Prevention of growth of potentially dense phytoplankton polulations by zooplankton, in the presence of zooplanktivorous fish in a shallow wetland ecosystem. *Limnol. Oceanogr.*, **29**, 472–486.

Performance of a combined eco-system of ponds and constructed wetlands for wastewater reclamation and reuse

L. Wang**, J. Peng*, B. Wang* and R. Cao*

*Water Pollution Control Research Center, Harbin Institute of Technology (HIT), 202 Haihe Road, Harbin 150090, China (E-mail: baozhen@public.hr.hl.cn)
**Ocean University of China, 5 Yushan Road, Qingdao 266003, China (E-mail: lwang@mail.ouc.edu.cn)

Abstract An on-site study on the operational performance of a combined eco-system of ponds and SF constructed wetland for municipal wastewater treatment and reclamation/reuse in Donging City, Shandong, China was carried out from January 2001 through October 2003. The removal efficiencies for various main parameters were: TSS 84.8 ± 7.3%, BOD_5 87.2 ± 5.3%, COD_{Cr} 70.2 ± 18.6%, TP 52.3 ± 23.1%, and NH_3-N 54.8 ± 23.9% with effluent concentration of TSS 9.12 ± 5.12 mg/l, BOD_5 6.44 ± 4.58 mg/l, COD_{Cr} 42.8 ± 6.7 mg/l, TP 0.94 ± 0.27 mg/l and NH_3-N 7.95 ± 2.36 mg/l. In addition, the removal efficiencies for faecal coliforms and total bacteria were >99.97% and >99.998% respectively, which well meet Chinese National standards for effluent quality of municipal wastewater treatment plants. The composition of TSS was closely related to COD_{Cr} and BOD_5 variations, and nitrification-denitrification is the major mechanism of nitrogen removal both in ponds and in wetlands. In addition, sedimentation also played an important role in the removal of TSS, nitrogen, phosphorus and BOD_5. The removal efficiencies of various parameters, the number of species and biomass of biological community in the system increased gradually with the ecological maturation.
Keywords BOD_5 transformation; sediments; eco-systems; TSS composition; SF constructed wetlands

Introduction

Both the pond and constructed wetland systems are of low-cost, energy saving, and self-remediation and self-adaptation to surrounding conditions and environment for wastewater treatment with high quality of effluent that can be reclaimed and reused multi-purposely (Wang et al., 1999; Wang et al., 2001). In pond systems, the concentrations of algae TRS that cause oxygen over saturation in daytime and depletion in the night, and TSS increase in receiving waters are usually very high (Steinmann et al., 2003; Gschlößl et al., 1998), therefore the further filtration is needed for reuse or discharge. Constructed reed wetland can be attributed to their role in filtering particulate materials, thereby reducing suspended solid (SS), taking up and storing nitrogen and TP partially (Maynard et al., 1999; Salter et al., 1999). The combined ponds–constructed wetlands system utilizes their characteristics and improves the final effluent to meet the wastewater reuse standards (Wang et al., 2001).

The removal of phosphorus and nitrogen in both pond and wetland systems results from direct uptake by the cultivated plants, the sedimentation of suspended organic substances, volatilization of ammonia at higher pH values and biological nitrification-denitrification (Gross et al., 2000), of which the latter two processes are the important mechanisms for nitrogen removal in pond systems.

The phosphorus removed from the bulk water is mainly by precipitation in sediments and biota uptake, and then the harvest of plants and/or fish. The amount of phosphorus stored in sediments increases with the depth of sediment, and a part of them could release to bulk water again under some environmental conditions (Pettersson et al., 2001; Selig *et al.*, 2002). The adsorption of phosphates by soil particles is faster and regarded as a dominant process in phosphates removal at the initial running period of both the pond and wetland systems. The adsorption capacity is dependent on the presence of Fe^{3+}, Ca^{2+}, Mg^{2+}, Mn^{2+}, Al^{3+} in clay minerals or some humic complex compounds, and will reach saturation capacity soon. The phosphate can also be precipitated with metal ions contained in bulk water at a much slower rate and saturation is hardly reached.

TSS is composed of two parts: inherent SS (iSS), the SS originally contained in raw wastewater and external SS (eSS), the fraction produced in the combined ponds–wetlands system, such as algae, aquatic plant detritus, plankton, sediments re-suspension and fish excreta. The removal of TSS in pond and wetland systems is mainly from physical settlement and filtration. The biological activities do not play any important role for the TSS removal. However, if the most part of TSS is from eSS, biological activities and relevant conditional factors will have a significant effect on the removal of TSS and BOD_5.

The objective of this paper is to study the performance and removal mechanisms of pollutants in combined ponds–constructed wetlands systems and the main factors affecting the performance of the system.

Methods

The combined ponds–wetland eco-system for municipal wastewater treatment and reclamation in Dongying City, Shandong was employed to conduct the on-site investigation study, and the layout of the eco-system is shown in Figure 1.

Raw wastewater → screens → grit chambers → hybrid facultative ponds (HFPs) → aeration ponds (APs) → aerated fish ponds (AFPs) → fish ponds (FPs) → hydrophyte ponds (HPs) → constructed wetland (CWs) → effluent to the reservoir for agricultural uses or to the Guangli River. The design parameters of each of the treatment units are shown in Table 1.

HFPs, APs and AFPs are all facultative ponds with the increase depth of aerobic layer along the flow chart from HFPs, APs to AFPs, especially in AFPs the anaerobic layer at bottom was less than 0.2 m deep in summer 2002. In HPs, the algae and duckweed are the dominant hydrophytes. The density of algae is considerably low and the contribution to DO ranged from 1.3 mg/l to 8.4 mg/l.

The surface flow constructed wetland (SF-CW) is used as the final treatment unit with densely vigorous reeds, which prevented sunlight penetration, suppressed algae photosynthesis and enhanced disintegration of algal cells through decay (Kim and Kim, 2000). Besides, there is a large area of ponding zone in the middle part of the constructed wetland, which was reshaped with operation for the free self-adaptation to flow pattern and ecological optimization (Odum and Odum, 2003) as well as surface re-aeration to increase DO and even re-distribution of water flow.

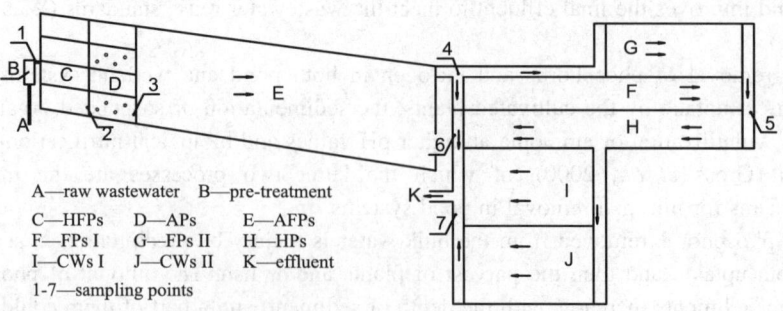

Figure 1 Layout of combined ponds–wetland system in Donying City

Table 1 Main design parameters of treatment unit of the eco-system in Dongying

Pond unit	Surface area (ha)	HRT (d)	Water depth (m)	Remarks
HFPs	3.5	1.5	5	Fermentation pits at the bottom and the biofilm carrier layer 1 m deep and 0.3 m below water surface
APs	3.5	1.3	3.6	16 surface aerators (2.2 kW)
AFPs	29	10.6	3.4	16 surface aerators (2.2 kW) + 8 water fall springs
FPs	12.2	2.4	2.0	Earth dike with grass cover
HPs	7.6	0.8	1.0	Earth dike with grass cover
CWs	35.2	1.8	0.5	Earth dike with grass cover

Materials and analytical methods

The water samples were collected twice or thrice a week from January 2001 through October 2003. Temperature and pH were measured *in situ*. The original water samples were immediately sent to the laboratory where they were passed through a 2 mm sieve to remove large suspended matters like duckweed. Then the subdivided samples were filtered through a Whatman GF/F 0.70-μm filter paper to determine the dissolved components and through the Millipore APFF 0.70-μm filters to determine the particular component. NH_3-N, total nitrogen (TN), NO_3^--N, NO_2^--N and total phosphate (TP), dissolved phosphate (DP), BOD_5, dissolved BOD_5 ($DBOD_5$) were analyzed at once in accordance with *Standard Methods* (1995). The forms of phosphorus were determined on wet samples by extracting phosphate according to Rydin (2000).

Results and discussion

The removal efficiencies of some main water quality parameters of the combined ponds-wetlands system are shown in Table 2, which indicate that their removal efficiency increased gradually from 2001 to 2003 with maturation and increase of species and biomass of the biological community in the system, which is similar to other study (Tanner, 1996). In the initial periods, the stable ecological system was not formed and so the removal efficiencies of various parameters were very low.

Table 2 The performance of the combined system varied with time

Parameter	2001	2002	2003
BOD_5			
Influent (mg/l)	39.7 ± 7.83	49.3 ± 22.5	46.8 ± 9.82
Effluent (mg/l)	10.9 ± 8.77	6.44 ± 4.58	4.67 ± 0.51
Removal (%)	72.54 ± 9.71	89.2 ± 5.3	90.0 ± 9.74
CODcr			
Influent (mg/l)	130.3 ± 15.0	143.9 ± 85.9	172.1 ± 75.1
Effluent (mg/l)	46.4 ± 3.6	42.8 ± 6.7	41.6 ± 6.1
Removal (%)	64.5 ± 10.1	70.2 ± 18.6	75.8 ± 17.2
NH_3-N			
Influent (mg/l)	15.37 ± 4.34	17.6 ± 5.98	21.78 ± 10.03
Effluent (mg/l)	9.38 ± 4.74	7.95 ± 2.36	7.12 ± 3.59
Removal (%)	42.07 ± 24.99	54.83 ± 23.9	67.31 ± 21.7
TP			
Influent (mg/l)	1.73 ± 0.51	1.98 ± 0.45	2.11 ± 0.91
Effluent (mg/l)	0.57 ± 0.47	0.94 ± 0.27	0.86 ± 0.18
Removal (%)	67.05 ± 17.3	52.28 ± 23.14	59.23 ± 22.01
TSS			
Influent (mg/l)	69.9 ± 5.74	59.93 ± 33.8	71.8 ± 26.7
Effluent (mg/l)	18.01 ± 8.71	9.12 ± 5.12	8.53 ± 0.79
Removal (%)	60.33 ± 15.94	84.78 ± 7.30	88.2 ± 12.1

BOD_5 and COD_{Cr} removals

The BOD_5 removal ranged from 75.6% to 90.7% in winter and 85.1% to 93.0% in summer respectively as shown in Figure 2, with effluent BOD_5 of 7.29–16.70 mg/l in winter and 1.50–5.91 mg/l in summer. Both Figure 2 and Table 2 also indicate that 57–92% of BOD_5 was removed in the facultative ponds.

The COD_{Cr} variation with the time is shown in Figure 3, and indicates that 7.5%, 25.5%, 18.0% and 22.3% of removal efficiency were reached in HFPs, AFPs, HPs and CWs, which was higher than that of APs (10.3%) and FPs (3.51%). As shown in Table 1 with the temperature increase and hydrophytes maturation, the COD_{Cr} removal efficiency increased from 64.5% in 2001 to 75.8% in 2003 correspondingly.

From April to September 2003, the removal efficiencies of COD_{Cr} in AFPs and FPs, ranged from 9.3% to -11%, less than in other months. The effluent of $SBOD_5$ in these ponds in the same season was also higher than that in others as shown in Figure 4. The analysis of the correlation of COD_{Cr} and $SBOD_5$ shows that the organic particulates in these two ponds led to the decline of removal efficiencies of the ponds and finally were removed effectively by sedimentation, filtration and the biodegradation in the wetland.

The concentrations of $SBOD_5$ (suspended BOD_5) decreased gradually when wastewater passed through the system, as shown in Figure 4, because of the sedimentation of suspended organic matters. The substantial effect of temperature on the effluent $SBOD_5$ and $DBOD_5$ also shown in Figure 4, which is evident from the variation of $SBOD_5$ and $DBOD_5$ in different months. The effluent $SBOD_5$ of HFPs and APs began to decrease from March, and reached 1.8 mg/l at the lowest point in August, and then it began to increase in December, and the effluent $SBOD_5$ of AFPs and FPs displayed a similar trend, which reached maximum in winter and minimum in summer.

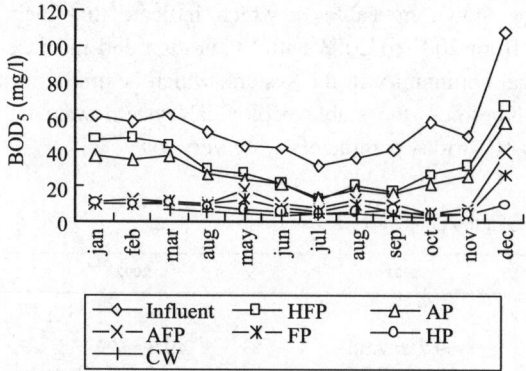

Figure 2 Monthly effluent BOD_5 in different stages

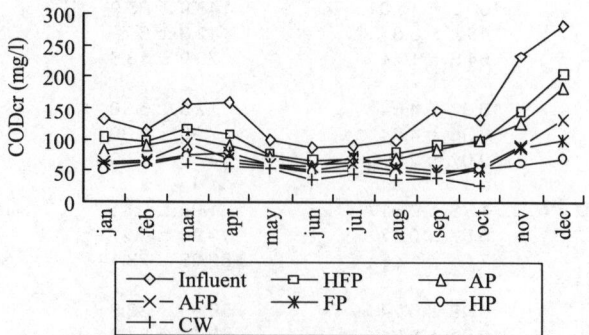

Figure 3 Monthly effluent COD_{Cr} in different stages

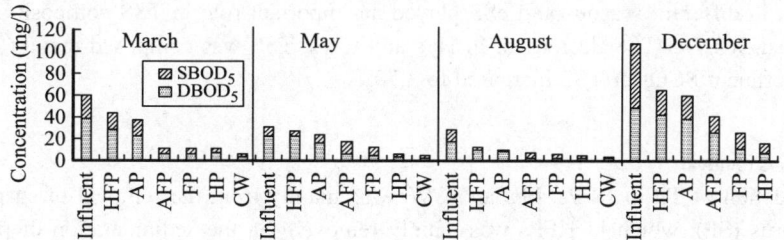

Figure 4 Comparison of BOD$_5$ and DBOD$_5$ in different seasons

SBOD$_5$ was related well to the composition of TSS in the system. In March, May, August and December 2002, the correlated coefficients between TSS and SBOD$_5$ were 0.918, 0.367, 0.121 and 0.690 respectively, which means that the degradation and sedimentation of iSS is the main removal mechanism of SBOD$_5$ in winter, while the effect of algae on SBOD$_5$ made the correlation unclear.

TSS removal

The TSS removal was efficient, being 84.8% in 2002 and 88.2% in 2003 as shown in Table 2. The final effluent of TSS was in the range 7.7–14.2 mg/l, and below 5.6 mg/l from August to October in 2002–2003.

In ponds system iSS usually decreased gradually as the water flowed through the system, mainly depending on hydraulic conditions, while eSS (external SS) varied significantly under different conditions, which often has a sharp increase due to the growth of plankton in a longer hydraulic retention time. The combined ponds–wetland system removed eSS effectively. In summer, most of the algae grown in ponds were eliminated in CWs through filtration by reed and degradation by bacteria. The effluent *chlorophyll* a and TSS concentrations were less than 4.3 mg/l and 18.3 mg/l, respectively. In winter, though the CWs were not in use, the concentration of eSS that was composed mainly of algae in ponds was low and contributed little to TSS, and the final effluent TSS concentrations ranged from 6.40 mg/l to 18.50 mg/l.

The iSS and eSS were characterized by iSS being removed at high efficiencies in ponds–wetland system with proper hydraulic conditions (Figure 5), with eSS showing different variations in different conditions. Therefore, the iSS and eSS varied in proportion of TSS with water flow along the combined system. In HFPs and APs, the TSS was composed mainly of iSS and the effect of eSS was ignored. In the two ponds, the sedimentation of inorganic iSS was the main removal mechanism of TSS and the ratio of SCODcr/TSS increased from 0.93 to 1.29. In AFPs and FPs, TSS has a remarkable

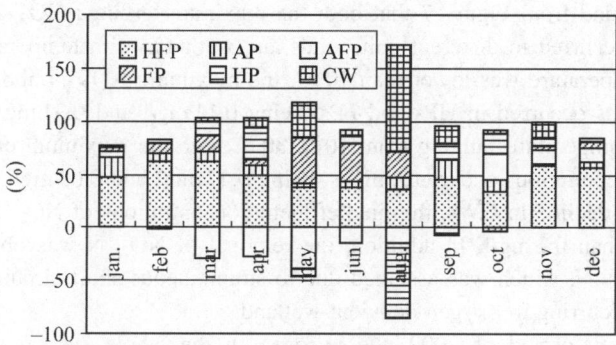

Figure 5 Monthly removal efficiencies of TSS in different stages

variation in different seasons and eSS played an important role in TSS composition, and the annual SCODcr/TSS decreased. In HPs and CWs, TSS was composed mainly of eSS and the effluent SCODcr/TSS increased to 3.78.

Phosphorus removal

The reduction of TP in HFPs, FPs and CWs was mainly from the removal of suspended phosphorus (SP), which in HFPs was mainly removed via the sedimentation of particulates. The average removal efficiencies for DP in HFPs were only 3.87%, and the TP removal via the sedimentation of SP accounted for 72.4%. The removal of TP in FPs and CWs was from flocculation, adsorption on the floccus and biofilm attached to the surface of reed stalk, filtration and uptake by aquatic plants, and DP in the final effluent accounted for 95.4% of TP. In APs, AFPs and HPs, the reduction of TP was mainly from removal of DP, because the contents of SP remain constant and the removal masses of TP and DP were similar.

In ponds–wetlands system, 85–93% of TP in sediments was inorganic phosphate. The order of P-fractions in terms of abundance was $FePO_4 > Ca_5(OH)(PO_4)_3 > (AlPO_4) > $ -organic-P $>$ residual-P. While in FPs and HPs, the $Ca_5(OH)(PO_4)_3$ was the dominant fraction in sediments, which was related to the higher pH (8.5–9.0) in the ponds.

The adsorption–desorption experiment of sediment illustrated that the balance of TP in sediments and bulk water was reached within two months. The potential of desorption or adsorption of phosphorus was affected by the characteristics of the sediments. The wastewater quality varied significantly between dry and wet seasons. The balance of adsorption–desorption in sediments fluctuated with influent TP variations. The minimum and the maximum removal efficiencies of TP occurred in May and November, respectively, when the influent TP concentrations varied significantly.

Even in the growing season, the vegetation did not show significant uptake capacity for phosphates as shown in Figure 6. Firstly, the phosphate taken up by vegetation roots system mostly from the sediments, and the phosphate balance between sediments and bulk water in ponds could alleviate the effect. Secondly, the growth of vegetation through photosynthesis usually results in an increase in pH and decrease in ORP of sediments, which could accelerate the release of phosphate from sediments (Horppila and Nurminen, 2003). Thirdly, with the increase of temperature, the phosphate contained in aquatic plants detritus could gradually release via leach-out and mineralization of organics.

Nitrogen removal

The increase of NO_3^--N and NO_2^--N in the treated wastewater was mainly from biochemical reactions in the system, especially NO_2^--N that was not contained in raw wastewater. It was found from Figure 7 that both the maximum effluent NO_3^--N and NO_2^--N concentrations occurred in different unit ponds and wetlands with temperature variation. When water temperature was lower than 10 °C, the maximum NO_2^--N and NO_3^--N effluent concentrations occurred in HPs and FPs, being 0.11 mg/l and 0.31 mg/l respectively, while at water temperature ranging from 10 to 30 °C both the maximum effluent NO_2^--N and NO_3^--N concentrations occurred in HPs, being 0.47 mg/l and 0.63 mg/l respectively.

With a sharp decline in CWs, the final effluent concentrations of NO_2^--N and NO_3^--N were both less than 0.2 mg/l. In addition, the removal of NO_2^--N was obviously higher than that of NO_3^--N, which was assumed due to simultaneous short-circuit nitrification–denitrification occurring in oxygen deficient wetland.

In the majority period of 2002, the nitrogen in the whole system was composed mainly of ammonia nitrogen (76–78%) and organic nitrogen (13.0–18.0%), which

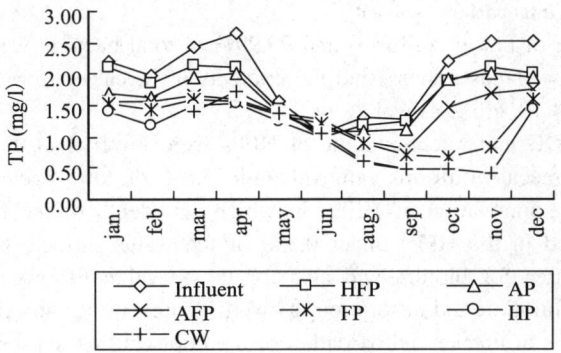

Figure 6 Monthly effluent TP in different stages

means that the TN concentrations depended on the variation of ammonia and organic nitrogen.

Water pH and temperature are the most important factors affecting the volatilization rate of NH_3-N. Although the water pH and temperature in May were similar to that in October, the removal efficiency of 71.6% in October was much higher than that in May (32.9%), which means that the volatilization was not a major removal mechanism for ammonia nitrogen. The concentrations of both NO_2^--N and NO_3^--N were less than 0.9 mg/l from April to November, which is ascribed to kinetic balance between nitrification and denitrification under the given conditions, thus resulting in a negligible accumulation of NO_x^--N.

In wetlands system, nitrification–denitrification was also the major process affecting removal efficiencies for NH_3-N and TN. In addition to re-aeration on the water surface of CWs, the reed roots also provided oxygen to the wetland and made the reed grow under aerobic conditions. The biofilm attached growth on the surface of reed stems and roots has a DO gradient from surface to bottom of the biofilm, i.e. from aerobic to anoxic and then anaerobic phase, which made nitrification and denitrification take place simultaneously in such numerous mini-bioreactors. In wetlands, the nitrification was the rate-limiting factor of nitrification–denitrification process.

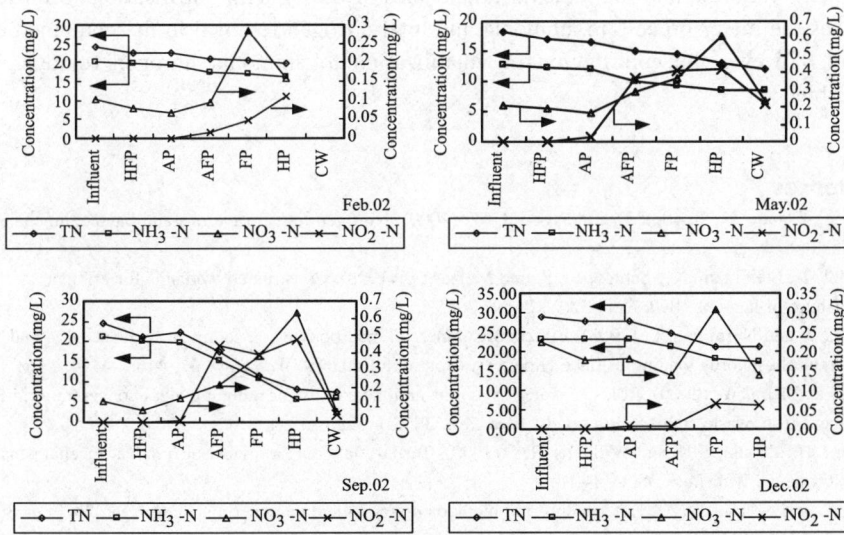

Figure 7 Variation of effluent nitrogens at different stages in different seasons

Faecal coliforms and total bacteria removal

More than 99.97% of faecal coliforms and 99.99% of total bacteria were removed in the combined ponds–wetlands system, and the contents of faecal coliforms were within the WHO limitation, $<10^4$ cfu/L as well.

Although pH, DO and retention time of HFPs were lower than that of conventional HFPs, 99.3% of faecal coliforms removal and 93.8% of total bacteria removal were reached due to the innovation of HFPs, in which 1 m depth of the ring-lace synthetic fibrous was packed in the HFPs under 0.3 m of the water surface, which provides an attached growth area for biofilm and also for the faecal coliforms and total bacteria removal through filtration and adsorption, as well as preventing the offensive odor from spreading by means of uptake and oxidation of the odor causing substances like H_2S and mercaptan by biofilm attached to the carriers. The correlation analysis was conducted between faecal coliforms and pH and HRT in the pond system (from APs to HPs), and indicated that no significant correlations between pH and faecal coliforms, while HRT had close correlation with the removal of faecal coliforms ($r^2 = 0.9955$).

Wastewater reclamation and reuse

100,000 m^3/d of wastewater is treated, reclaimed and reused in the combined ponds-wetland system, 20–30 t fish fries were harvested in the fish ponds each year, and the economic plants like duckweed and reed were harvested regularly, which complement the operation/maintenance costs to the extent that no operation fee is need from the local finances. The effluent quality of the system well meets the wastewater reuse standards for agricultural uses.

Conclusion

The performance of the combined eco-system is gradually improved with the maturation and the increase of species and biomass of biological community year by year. The iSS usually decreased gradually with the water flow through the system, while eSS varied significantly under different environmental conditions. The proportion and the concentrations of eSS were related to the removal of COD_{Cr}, BOD_5, $SBOD_5$, and TSS.

The adsorption–desorption is an effective process for TP removal, but was limited by quick saturation. The balance of TP between sediments and bulk water reduced the effect of influent variation and the removal efficiencies of the system. Nitrification–denitrification was the major process to ammonia and total nitrogen removal in the combined eco-system, however the contribution of volatilization to ammonia nitrogen removal was negligible.

References

Gross, A., Claude, E., Boyd, C.E. and Wood, C.W. (2000). Nitrogen transformations and balance in channel catfish ponds. *Aquacult. Eng.*, **24**, 1–14.

Gschlößl, T., Steinmann, C., Schleypen, P. and Melzer, A. (1998). Constructed wetlands for effluent polishing of lagoons. *Wat. Res.*, **32**(9), 2639–2645.

Horppila, J. and Nurminen, L. (2003). Effects of submerged macrophytes on sediment re-suspension and internal phosphorus loading in Lake Hidenvesi (Southern Finland). *Wat. Res.*, **37**, 4468–4474.

Kim, Y. and Kim, W. (2000). Roles of water hyacinths and their roots for reducing algal concentration in the effluent from waste stabilization ponds. *Wat. Res.*, **34**(13), 3285–3294.

Maynard, H.E., Ouki, S.K. and Williams, S.C. (1999). Tertiary lagoons: a review of removal mechanisms and performance. *Wat. Res.*, **33**(1), 1–13.

Odum, T. and Odum, B. (2003). Concepts and methods of ecological engineering. *Ecol. Eng.*, **20**, 339–361.

Pettersson, K. (2001). Phosphorus characteristics of settling and suspended particles in Lake Erken. *Sci. Total Environ.*, **266**, 79–86.

Rydin, E. (2000). Potential mobile phosphorus in lake Erken sediment. *Wat. Res.*, **34**(7), 2037–2042.

Salter, H.E., Boyle, L., Ouki, S.K. and Quarmby, J. (1999). The performance of tertiary lagoons in the United Kingdom: I. *Wat. Res.*, **33**(18), 3775–3781.

Selig, U., Hübener, T. and Michalik, M. (2002). Dissolved and particulate phosphorus forms in a eutrophic shallow lake. *Aquat. Sci.*, **64**, 97–105.

Standard Methods for the Examination of Water and Wastewater (1995) 19th edn, APHA/AWWA/WEF, Washington DC, USA.

Steinmann, C.R., Weinhart, S. and Melzer, A. (2003). A combined system of lagoon and constructed wetland for an effective wastewater treatment. *Wat. Res.*, **37**, 2035–2042.

Tanner, C.C. (1996). Plants for constructed wetland treatment systems-A comparison of the growth and nutrient uptake of eight emergent species. *Eco. Eng.*, **7**, 59–83.

Wang, B.Z., Wang, Lin and Yang, Luyu (1999). Case studies on pond ecosystems for wastewater treatment and utilization in China. *Global Wat. Wastewat. Tech.*, **8**, 64–71.

Wang, L., Wang, B., Yang, L., Qi, P. and Deng, W. (2001). Eco-pond systems for wastewater treatment and utilization. *Water 21*, **8**, 60–63.

Municipal wastewater treatment with pond–constructed wetland system: a case study

X. Wang*, X. Bai*, J. Qiu* and B. Wang**

*School of Environmental Science and Engineering, Shanghai Jiaotong University, 800 Dongchuan Rd, Shanghai, 200240, China (E-mail: *xinzewang@sjtu.edu.cn*; *xhbai@sjtu.edu.cn*; *jpq@sjtu.edu.cn*)
**School of Municipal and Environmental Engineering, Harbin Institute of Technology, 202 Haihe Rd, Harbin, 150090, China (E-mail: *baozhen@public.hr.hl.cn*)

Abstract The performance of a pond–constructed wetland system in the treatment of municipal wastewater in Kiaochow city was studied; and comparison with oxidation ponds system was conducted. In the post-constructed wetland, the removal of COD, TN and TP is 24%, 58.5% and 24.8% respectively. The treated effluent from the constructed wetland can meet the Chinese National Agricultural and Irrigation Standard. The comparison between pond–constructed wetland system and oxidation pond system shows that total nitrogen removal in a constructed wetland is better than that in an oxidation pond and the TP removal is inferior. A possible reason is the low dissolved oxygen concentration in the wetland. Constructed wetlands can restrain the growth of algae effectively, and can produce obvious ecological and economical benefits.
Keywords Ecological engineering; pond-constructed wetland system; wastewater reclamation

Introduction

It is important and necessary in China to find creative, cost-effective, and environmentally sound ways to control water pollution. According the statistical data of 2002, quantity of discharged wastewater in 2002 is $43.95 \times 10^9 \, m^3$ and the total capacity of wastewater treatment plants in China is just about 36.5% of discharge. In big cities such as Beijing and Shanghai, the sewage treatment rate is about 56% and 60%, and more wastewater treatment plants are under construction. But in the small towns and rural areas, the high costs (for construction, maintenance and operation) of the most conventional treatment processes have brought severe economic pressures to society (Wang *et al*, 2001). Different from big cities, small cities and rural areas can afford relative large land for wastewater treatment and reclamation. The pond–wetland system process can be one of the solutions in the treatment of wastewater from small towns and rural areas in China (Bai *et al*, 1998; Wang *et al*, 1999).

The main contents in this paper included the removal efficiency of organic matter, nitrogen and phosphorus by pond and wetland treatment systems, and the variation of DO in the oxidation ponds and wetland system. Ecological and economical effects of pond-constructed wetland systems were also evaluated.

Methods

The experimental pond-constructed wetland system is in Kiaochow city, which is located on the north west of Kiaochow Bay with $1313 \, km^2$ area and 740×10^3 citizens. About $20 \times 10^3 \, m^3/d$ wastewater is discharged to oxidation pond treatment systems (shown in Figure 1), in which 42.2% is from industrial discharge and 57.8% is from domestic sewage. The quality of raw water and the dimension of each pond are shown in Table 1 and Table 2 respectively.

In 1997, in order to improve the effluent quality of the pond system, a constructed wetland of $25 \, hm^2$ area, located at the east of the pond system, was employed to treat

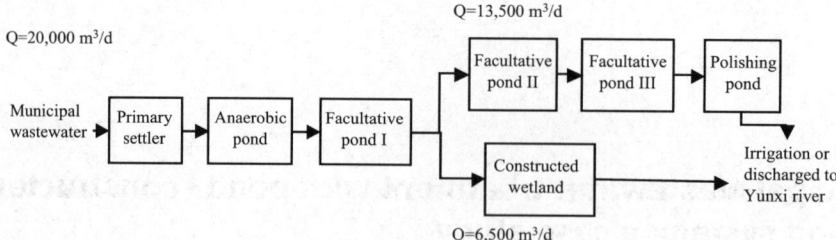

Figure 1 Treatment process

6000–7000 m³/d part of the effluent of facultative pond I, which had been pretreated by the primary settler, anaerobic pond and facultative pond I (shown in Figure 1). The average depth of the reed-bed wetland is 0.5 m.

Sampling and analyzing

The wastewater samples were analyzed once a week. The period of experiment was from 25th June to 28th October 1998.

Results and discussion
Organic matter removal

Organic removal by oxidation ponds and reed bed wetland is shown in Figure 2. During the actual operation, because of a large amount of algae grown on the surface of facultative ponds and polishing pond, the effluent COD_{Cr} of the oxidization ponds was sometimes higher than that of the constructed wetland. The COD_{Cr} removal of constructed wetland was 24%, higher and more stable than that of the oxidation ponds system. In constructed wetland systems, organic pollutants associated with settleable solids in a wastewater are removed by sedimentation, and the colloidal/soluble organic remaining in solution is removed as a result of metabolic activity of micro-organisms that may be: (1) suspended in the water column; (2) attached to the sediments; and (3) attached to the roots and stems of the aquatic plants. The most widely occurring bacteria in constructed wetland are *Pseudemonas, Alcaligenes and Flavobacterium* etc. All of them are rapid growing microorganisms.

Table 1 Characteristic of the raw wastewater (mg/L)

Index	Range	Mean	Index	Range	Mean
BOD_5	40.8–234	104.1	TP	1.127–13.3	6.6
COD_{Cr}	129–871	310	Total Cr	0.005–0.658	0.113
SS	41–156	101	As	0.023–0.456	0.21
NH_3-N	12–127	63.15	Volatile phenol	0.001–0.889	0.176
TN	27.4–189	72.36	pH	6.26–7.66	7.28

Table 2 Dimension of oxidation ponds system in Kiaochow city

Item	Sediment tank	Anaerobic pond	Facultative pond I	Facultative pond II	Facultative pond III	Polishing pond	Total
Length/m	56	533	482	310	310	305	
Width/m	7	360	355	290	290	187	
Depth/m	7	3.5	2.2	1.8	1.4	0.9	
Area/$10^4 m^2$	0.04	19.2	17.1	9.0	9.0	5.7	60
volume/$10^3 m^3$	3	672	376	162	126	51	1390

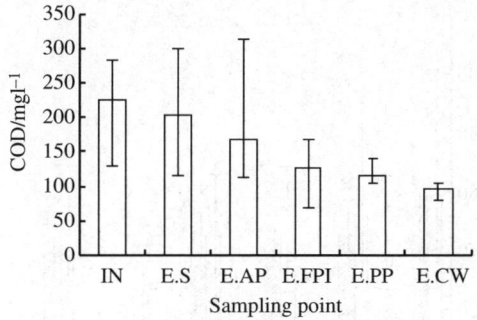

Figure 2 Organic removal by the pond and wetland: IN: Influent, E.S: effluent of settler, E.AP: effluent of anaerobic pond, E.FPI: effluent of facultative pond I, E.PP: effluent of polishing pond, E.CW: effluent of constructed wetland

Variation of dissolved oxygen

Variation of dissolved oxygen in the oxidation ponds and wetland system is shown in Figure 3. In the oxidation ponds system, dissolved oxygen in each pond increased gradually. The average DO of polishing pond effluent during four months was 8.27 mg/L. The maximum was 9.90 mg/l. But when the wastewater flowed into the constructed wetland, DO began to decrease. During four months, average DO concentration at the effluent outlet was 1.65 mg/l, and the minimum was 0.1 mg/l. The gradual increase in the oxidation pond system was because of the photosynthesis of the algae and the effect of air penetration. In the wetland, due to the growth of reed plant, reaeration by wind was limited. During the warm climate, rapid growing reed can cover the free water surface of the wetland rapidly, which affects the photosynthesis of algae in water. On the other hand, antagonistic interaction between aquatic plants and algae can suppress the growth of algae. Respiration of soil microbes and the growth of reed root and stem also consume a large amount of oxygen.

Nitrogen removal

NH_3-N, NO_3-N and TN removal by oxidation pond and reed bed wetland system is shown in Figure 4. It can be seen that NH_3-N can be removed efficiently by oxidation ponds and constructed wetlands. Average removal rate was 59.4% and 61.3%. But for the TN removal, the effect of reed bed wetland was superior to the oxidation pond system. Average TN removal rates were 58.5% and 37.6%, respectively. Nitrogen is removed

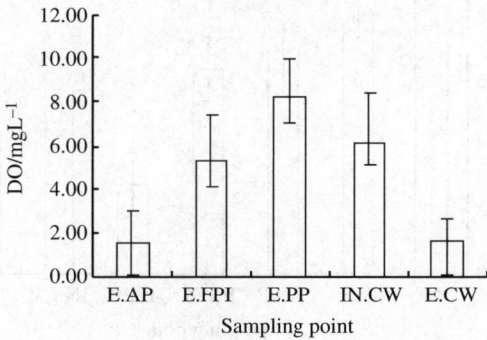

Figure 3 Variation of DO in the pond system and wetland: E.AP: effluent of anaerobic pond, E.FPI: effluent of facultative pond I, E.PP: effluent of polishing pond, IN.CW: influent of constructed wetland, E.CW: effluent of constructed wetland

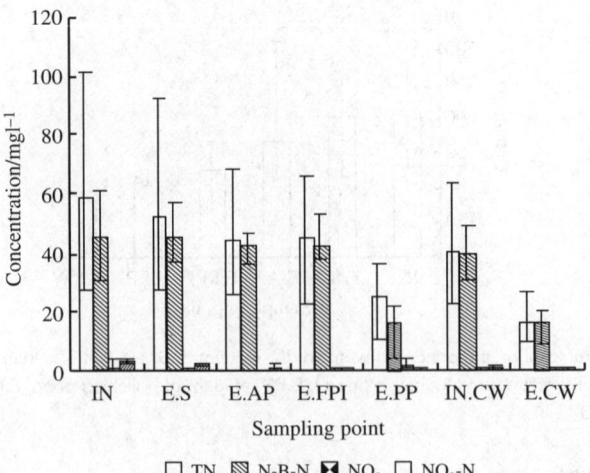

Figure 4 NH$_3$-N, NO$_2$-N, NO$_3$-N and TN variation in the oxidation pond system and constructed wetland: IN: Influent, E.S: effluent of settler, E.AP: effluent of anaerobic pond, E.FPI: effluent of facultative pond I, E.PP: effluent of polishing pond, IN.CW: influent of constructed wetland, E.CW: effluent of constructed wetland

from wastewater during aquatic treatment by a number of mechanisms: (1) uptake by plants and subsequent harvesting of plants, which is seasonal and inconsistent; (2) volatilization of ammonia; and (3) bacterial nitrification/denitrification. Of these, bacterial nitrification/denitrification appears to have the greatest nitrogen removal potential. Nitrifying bacterial grow readily on the submerged portions of aquatic plants. Denitrification occurs in the sediments of aquatic systems and there is a great deal of micro-environment consisting of aerobic and anoxic conditions around the plant rhizosphere (Ronald *et al*, 1997).

Phosphorus removal

Phosphorus removal by the oxidation ponds and constructed wetland is shown in Figure 5. All of the effluent TP from oxidation ponds and wetland systems was below 5 mg/l,

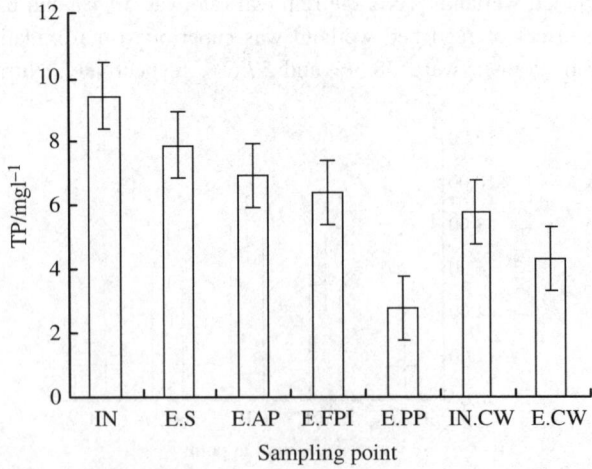

Figure 5 TP variation in the oxidation pond system and constructed wetland: IN: Influent, E.S: effluent of settler, E.AP: effluent of anaerobic pond, E.FPI: effluent of facultative pond I, E.PP: effluent of polishing pond, IN.CW: influent of constructed wetland, E.CW: effluent of constructed wetland